William Turner Coggeshall

Five Black Arts

A Popular Account of the History, Processes of Manufacture, etc.

William Turner Coggeshall

Five Black Arts

A Popular Account of the History, Processes of Manufacture, etc.

ISBN/EAN: 9783744724272

Printed in Europe, USA, Canada, Australia, Japan

Cover: Foto ©ninafisch / pixelio.de

More available books at **www.hansebooks.com**

FIVE BLACK ARTS.

A POPULAR ACCOUNT

OF THE

History, Processes of Manufacture, and Uses

OF

PRINTING,   GAS-LIGHT,
POTTERY,   GLASS,
IRON.

WITH NUMEROUS ILLUSTRATIONS.

CONDENSED FROM THE ENCYCLOPÆDIA BRITANNICA.

COLUMBUS:
FOLLETT, FOSTER AND COMPANY.
1861.

PREFACE.

The Encyclopædia Britannica* is standard authority on Science, and on Arts and Manufactures. It is an expensive work, and its circulation in America is confined to Libraries in prominent cities. It was, therefore, deemed advisable, by the Publishers of this Book, to put in form convenient for the humblest private Library, the information it furnishes on some of the Arts in which everybody is interested.

Printing, Pottery and Porcelain, Gas-Light, Glass, and Iron were selected by the Editor, because in their uses they are familiar to all the people, but in their history, and in the process of their manufacture, are mysteries to a large majority.

The articles, herewith published, are agreeable in style; they have been condensed only in parts pertaining especially to English processes of man-

---

* The Encyclopædia Britannica, or Dictionary of Arts, Sciences, and General Literature. Eighth Edition, with extensive improvements and additions, and numerous engravings. Adam and Charles Black, Edinburgh, Scotland. Little, Brown & Co., Boston, Massachusetts. 22 vols. quarto. 1861.

ufacture; and the Editor is confident that they will prove both instructive and interesting to all who have not made the Arts represented a particular study, and will be serviceable to many who practically pursue those Arts.

W. T. C.

COLUMBUS, OHIO, *January*, 1861.

CONTENTS.

	Page
HISTORY AND PROCESS OF PRINTING.	
HISTORY	4
PRACTICAL PRINTING	75
STEREOTYPING	96
POLYTYPAGE	106
PRINTING FOR THE BLIND	111
OTHER PROCESSES	114
NATURE PRINTING	115
PRINTING IN COLORS	118
BANK-NOTE PRINTING	124
PRINTING MACHINES	126
POTTERY AND PORCELAIN.	
HISTORICAL SKETCH	140
THE MATERIALS	166
THE MANUFACTURE	171
ORNAMENTATION	179
GLASS: ITS HISTORY AND MANUFACTURE.	
HISTORY	189
CROWN GLASS	202
SHEET	210
PLATE	214
STAINED OR PAINTED	219
GLAZING OF WINDOWS	228
THE CUTTING DIAMOND	231
FLINT GLASS OR CRYSTAL	234
BOTTLE	241

HISTORY AND PROCESSES OF MAKING GAS-LIGHT.

HISTORY	247
ARRANGEMENT OF APPARATUS	253
RETORTS FOR COAL	255
CONDENSING, MAIN AND DIP PIPES	264
TAR APPARATUS	268
APPARATUS FOR SEPARATING	266
GASOMETER	271
MAIN AND SERVICE PIPES	275
GOVERNOR OR REGULATOR	278
GAS METER	280
BURNERS	284
OIL, WATER, AND RESIN GAS	294
DETERMINING ILLUMINATING POWER	296
HINTS FOR IMPROVING COAL GAS	303
DETERIORATION OF " "	308
ECONOMY OF " "	309
SECONDARY PRODUCTS	311

IRON: HISTORY OF ITS MANUFACTURE, WITH AN ACCOUNT OF ITS PROPERTIES AND USES.

HISTORY OF IRON MANUFACTURE	317
THE ORES	323
THE FUEL	331
MANUFACTURE	336
CONVERSION OF CRUDE IRON INTO MALLEABLE	349
MACHINERY OF MANUFACTURE	363
THE FORGE	370
STRENGTH AND OTHER PROPERTIES OF CAST-IRON	374
MALLEABLE	381
STATISTICS OF THE IRON TRADE	391

ILLUSTRATIONS.

Index to Descriptive Pages.

	Page
PRINTING.	
Fig. 1	65
2	66
3, 4	68
5	69
6, 7	79
GLASS.	
Fig. 1	208
2, 3, 4, 5, 6	209
7, 8	210
GAS-LIGHT.	
Fig. 1	278
2, 3	280
4, 5	282
6	284
7, 8, 9, 10	287
11	292
12	296
13	298
IRON.	
Fig. 1	336
2, 3	337
4	343
5	363
6, 7, 8, 9, 10, 11, 12	364
13	365
14, 15, 16, 17	368
18, 19	370

HISTORY

AND

PROCESS OF PRINTING.

BY THOMAS C. HANSARD.

LETTER-PRESS PRINTING.

PRINTING is the art of taking one or more impressions from the same surface, whereby characters and signs, cast, engraven, drawn, or otherwise represented thereon, are caused to present their reverse images upon paper, vellum, parchment, linen, and other substances, in pigments of various hues, or by means of chemical combinations, of which the components are contained on or within the surface from which the impression is taken, or in the fabric of the thing impressed, or in both.

The most important branch of printing is what is called *letter-press* printing, or the method of taking impressions from letters and other characters *cast* in relief upon separate pieces of metal, and therefore capable of indefinite combination. The impressions are taken either by superficial or surface pressure, as in the common printing-press, or by lineal or cylindrical pressure, as in the printing machine and roller-press. The pigments or inks, of whatever color, are always upon the surface of the types; and the substances which may be impressed are various. Wood-cuts and other engravings *in relief* are also printed in this manner.

Copperplate printing is the reverse of the above, the characters being *engraven* in intaglio, and the pigments or inks contained within the lines of the engravings, and not upon the surface of the plate. The impressions are always taken by lineal or cylindrical pressure; the substances to be impressed, however, are more limited. All engravings in intaglio, on whatever material, are printed by this method.

Lithographic printing is from the surface of certain porous stones, upon which characters are *drawn* with peculiar pencils. The surface of the stone being wetted, the chemical coloring compound adheres to the drawing, and refuses the stone. The impression is taken by a scraper that rubs violently upon the back of the substances impressed, which are fewer still in number. Drawings upon zinc and other materials are printed by this process.

Cotton and *calico* printing is from surfaces engraven either in relief or intaglio. The chemical compounds are either on or within the characters, as pigments or chemical colors, or in the fabric to be printed, but mostly in both; the combination of chemical substances producing color when the fabric and the engraving are brought into contact. The impression is either superficial or lineal, but mostly lineal.

HISTORY.

The origin and history of an art which has exercised such an influence on civilization, and contributed in so essential a manner to the cultivation of the human intellect, have naturally become a matter of inquiry amongst the learned, and have almost as naturally been the source of earnest controversy; for there are few effects of human invention or industry that have been originated and brought to perfection at a particular epoch, without any previous train of thought or circumstance, so that the precise day or year could be noted in which the perfect Minerva started forth in full maturity. On the contrary, it is difficult to say at what period of time the germ of the art of printing did not exist. So obvious is the reproduction of similar appearances from an impression of the same surface, that the most early of mankind must have noted it; and even the impression of a foot or a hand must have suggested a simple and intelligible mode of conveying an idea, before the invention of any kind of writing. Accordingly, these and similar signs are found to compose the chief characters of the earliest writing.

Observing this general law of the gradual perfectibility of human arts, we must look back to the most remote ages for the first steps of that of printing. We shall accordingly find certain evidence, that more than two thousand years before our era, a method of multiplying impressions, rude and imperfect in the extreme, was certainly practiced.

The earliest practice which can with propriety be called *printing* was probably that of impressing seals upon a plastic material, the purpose being confined to the single effect of each single impression. The next step of which the diligence of inquirers has taken note, and which is a step thus much further in advance that its object was the multiplication of impressions for the purpose of diffusing information—the practice, namely, of impressing symbols or characters upon clay and other materials used in forming bricks, cylinders, and the walls of edifices—was an art confined, so far as our knowledge extends, to the ancient centers of civilization in Egypt and Asia. Some examples of this art found their way many years ago into the great public museums and chief private collections of Europe, where they were objects of curiosity and wonder. In the present day, the researches of Sir Gardner Wilkinson and others into the antiquities of Egypt, and of Sir Henry Rawlinson and Mr. Layard into the ruins of the buried cities of Asia, have produced a vast quantity of materials illustrative of the subject. The relative antiquity of the Egyptian and Asiatic remains belong to another inquiry. Among the Egyptian remains are numerous bricks of clay stamped with the *nomen* and *agnomen* of the king inclosed within a cartouche. The mode by which the impressions were made is manifest. The prints are very irregularly placed, without any reference to parallelism with the sides, and are always more or less awry, according to the manual skill and care of the workman: the surface of the bricks around the depression is forced up considerably, which is exactly the effect of pressing the hand or any substance into a plastic material; and the edges, both of the general depressions and of the figures, present the effect of the stamps having been drawn up whilst the clay was yet damp and adherent to it. It is therefore evident that the inscriptions were stamped in after the clay had been turned out of the mould, and were not produced by any part of it. To make the evidence complete, there have been found many stamps of wood, having on the face cartouches and inscriptions precisely resembling in kind those which must have been used for stamping the bricks. On some of these stamps and impressions there are slight traces of color. There have also been found in Egypt numerous figures of baked clay and porcelain on which hieroglyphic characters have apparently been impressed singly, side by side, by stamps; and on the walls of their ruder buildings hieroglyphic and pictorial figures of considerable size have been produced by the same means and afterward colored. Of articles of domestic use are certain instruments called *tesseræ*, having incised characters, the use of which has certainly been to stamp plastic materials; and there have also been found leather belts and ornaments on which figures have been impressed singly by tools.

The ruins of the cities of Asia supply us with numerous examples similar to those of Egypt, but carrying the art farther. The ruins contain countless bricks, on which are impressed inscriptions similar to those of Egypt, but much more elaborate. Mr. Layard says, that the characters on the Assyrian bricks were made separately: some letters may have been impressed singly with a stamp, but from the careless and irregular way in which they are formed and grouped together, it is more probable that they were all cut by an instrument and by hand; but that the inscriptions on the Babylonian bricks are generally inclosed in a small square, and are formed with considerable care and nicety; they appear to have been impressed with a stamp, on which the entire inscription, and not isolated letters, was cut in relief. From this circumstance, Mr. Layard ascribes greater antiquity to the Assyrian remains.

Mr. Layard's researches have further made evident that the ancient inhabitants of these cities practiced a more advanced and elegant usage of imprinting in their domestic and ornamental arts. He has discovered great quantities of tiles and tablets covered with *incised* or *incussed* characters, on which was impressed, while the clay was yet wet, a line of characters or symbols—apparently an authorization or verification—produced by the rolling of engraved cylinders; and other tiles, of which he says, "The most common mode of keeping records in Assyria and Babylon was on prepared bricks, tiles, or cylinders of clay, baked after the inscription was *impressed;*"—this impression must not be mistaken for the application of a stamp; it is effected by the use of an instrument in the hand, by which various combinations of the same form were indented into the moist clay, and therefore partakes more of the character of *impressed writing:* in many of the specimens thus impressed, the writing (or text) does not cover the entire tile or tablet, and the blank is filled up by repeated impressions of the same seal; and in some cases the entire text has been surrounded by an impression from a cylinder rolled round, forming an endless scroll, by which any addition to the text is rendered impossible. Great numbers of cylinders have been found. They are elaborately engraven on various stones; some are perfect cylinders, some barrel-shaped, others slightly curved inward. Others again are af baked clay, on which the characters have been *incussed* while the clay was yet moist. Many of them are perforated longitudinally, and revolve on a metal axis. In describing an engraved cylinder of great beauty found in the mounds opposite Mosul, Mr. Layard says, that on each side there were sixty lines written in such minute characters, that the aid of a magnifying glass was required to ascertain their forms. The habitual use of these elaborate articles is unknown,—by some they are supposed to be charms,—by others, records of family or personal transactions. The smaller examples were used to impress plastic materials as signets; but it is clear, from the shapes of the greater number, and from the circumstance that the characters they bear are invariably engraven or impressed in the order in which they are to be read, and not reversed, that they were not intended to multiply impressions on soft surfaces by way of diffusing information.

That a similar art was known to the inhabitants of the old world generally, may safely be assumed. It is therefore not a little remarkable that peoples so original and ingenious as the Greeks, and so imitative as the Romans, should have left almost no vestige of their having practiced any such means as this to multiply their beautiful creations of fancy, or to embellish the tasteful appliances of domestic life; especially when we consider the easy application of the art to pottery, and the beauty, taste, and ingenuity which they exhibited in that manufacture. For, excepting a few paltry designs *en creux* on some of the coarser specimens, and a few marks upon the Roman military vessels, evidently stamped, there is no appearance of either people having had any idea of this kind. They had, however, numerous instruments presenting a singular instance how very nearly we may approach to an important discovery, and yet pass on unheeding. These are stamps of various sizes, having on their faces inscriptions in raised characters reversed. The material is brass or bronze. The letters of the inscriptions are considerably raised, and the face of them is rough and rounded, as though they were rudely cast in a mould. To the back of most a handle has been fastened; some have a loop to allow the fingers to pass through; some a boss to rest in the palm of the hand; some a ring. One use of these stamps has probably been to press the inscription into a soft material; but the more common application, especially of the smaller specimens, has evidently been to *print* the inscription on surfaces by the aid of color. It has been suggested that their purpose was to imprint the coverings of bales of goods with the marks of their owners. Among relics of this kind is the signet of C. Cæcilius Hermias.* The face of this is two inches by four-fifths of an inch, and the inscription (reversed)

| CICAECILI |
| HERMIAE. SN. |

with a border, is in relief, the surrounding parts being cut away to a considerable depth. It should be especially noticed, that the surface of the background is very rough; and there is a ring at the back by which it could be handled or suspended. These circumstances render the use of it very clear. It would be very much easier to incise the required inscription, and to let the field stand (indeed the art of engraving *en creux* was well known and used), than to cut away the field and leave the letters in relief; and it would produce a much more beautiful effect if it were used to impress any soft substance; whereas, cut as it is, the impression sunk into wax or clay would not only be ugly, but illegible, and the rough surface of the background would present the most ungainly appearance upon the prominent parts of the wax, being the parts most presented to the eye. Its use therefore is evident. The relieved inscrip-

* In the British Museum.

tion, and no other part, being covered with ink or pigment, was impressed upon an even surface (papyrus, linen, parchment), and consequently left a perfect but reversed imprint of itself. This is the precise effect of printing with types. From the Greek *agnomen*, Cæcilius probably lived under the emperors, when literature had become one of the pursuits of the great, and when the difficulties and expense of procuring books by the slow process of copying were bitterly felt. It is singular, therefore, that the Romans should have overlooked so obvious an improvement upon their own signets as the engraving whole sentences and compositions upon blocks, and thence transferring them to paper—even if they had gone no farther than this.*

From this time a vast period elapses before any circumstance can safely be instanced as showing that the practice of transferring characters was known to any, even compara-

---

* The Chinese printing is not unlike this, and must by no means be supposed to have much similarity to the modern art. They assert that it was used by them several centuries before it was known in Europe; in fact, fifty years before the Christian era. They certainly may have used their method centuries before our art, for it differs in nothing but extent from that of the old Roman. The following is a description of their method at the present day, and it is probably the same in every respect as that in practice two thousand years ago in an empire where nothing is changed. As their written language consists of from eighty to one hundred thousand characters, it would be utterly impracticable to use movable types, and the use of block-printing would be the most easy and rapid. The sentences, therefore, desired to be multiplied, being drawn upon their thin paper, this is made to adhere with the face downward to a block of soft wood, so that the characters appear, though *reversed*. The plain wood is then cut away with most wonderful rapidity, and the drawing left in relief. Both sides of the block are similarly operated upon. The engraved wood is then properly arranged upon a frame, and the artist, with a large brush, covers the whole surface, the field as well as the relief, with a very thin ink; he then lays very lightly over it a sheet of paper, and passes a large soft brush over it, so slightly, yet so surely, that the paper is pressed upon the raised figures, and upon no other part. The rapidity with which this is performed is extraordinary; for Du Halde asserts that one man can print 10,000 sheets in one day, a number which would appear incredible, did not very good testimony exist at the present time that one man can print 700 sheets per hour. The method of putting the thin sheets together when printed is as different from ours as their printing and mode of reading. The sheets are printed on one side only; but instead of the blanks being pasted together to form one leaf, the sheet is so folded that no single edge of paper is presented to the reader, but only the double folded edge, the loose edges being all at the back of the book. The late emperor had punches or matrices cut, from which copper types were cast; but the number of characters required—about 60,000—is so great, that composition is almost impracticable.

tively civilized people. From the rough and imperfect attempts above indicated an early and obvious advance was engraving pictures upon wooden blocks. The first practice of this is involved in obscurity; but most writers on the fine arts agree that the art was invented toward the end of the thirteenth century, by a brother and sister of the illustrious family of Cunio, lords of Imola, in Italy. By some the whole narrative is considered as apocryphal, but it is nevertheless generally admitted. The engravings were discovered by a Frenchman of the name of Papillon, in the possession of a Swiss gentleman, M. de Grœder, who deciphered for him the manuscript annotations found on the leaves of the book in which they were bound. These purported that the book had been given to Jan. Jacq. Turine, a native of Berne, by the Count of Cunio, with whose family he, Turine, appears to have been intimately acquainted. Then follows a romantic history of the twins, and the cause of their invention. The book is entitled: "The Heroic Actions, represented in figures, of the great and magnanimous Macedonian king, the bold and valiant Alexander; dedicated, presented, and humbly offered to the most Holy Father Pope Honorius IV., the glory and support of the Church, and to our illustrious and generous father and mother, by us Alessandro Alberico Cunio, cavaliere, and Isabella Cunio, twin brother and sister; first reduced, imagined, and attempted to be executed in relief, with a small knife, on blocks of wood, made even and polished by this learned and dear sister; continued and finished by us together, at Ravenna, from the eight pictures of our invention, painted six times larger than here represented; engraved, explained by verses, and thus marked upon the paper, to perpetuate the number of them, and to enable us to present them to our relations and friends, in testimony of gratitude, friendship, and affection. All this was done and finished by us when only sixteen years of age." This title is here given in full length, because, if genuine, it presents us at once with the origin, execution, and design of these first attempts at block-printing. The book consists of nine engravings, including the title; the figures are tolerably well designed, and the draperies graceful, with here and there attempts at cross-hatching; under the principal personages are their names; above, are inscriptions indicating the subject, and below, four lines of poetical Latin explanatory of it; and in some parts of each print is an inscription indicating the share the twins respectively had in the execution. The color of the pigment is gray.

The first subject is Alexander on Bucephalus. Upon a stone, *Isabel. Cunio pinx. et scalp.*

The second subject, the passage of the Granicus. *Alex. Alb. Cunio Equ. pinx. Isabel. Cunio scalp.*

The third subject, Alexander cutting the Gordian Knot. *Alex. Albe. Cunio Equ. pinx. et scalp.*

The fourth subject, Alexander in the tent of Darius. *Isabel. Cunio pinx. et scalp.*

The fifth, Alexander giving Campaspe to Apelles. *Alex. Alb. Cunio Eques. pinx. et scalp.*

The sixth, the Battle of Arbela. *Alex. Alb. Equ. et Isabel. Cunio pictor. et scalp.*

The seventh, Porus brought to Alexander. *Isabel. Cunio pinx. et scalp.*

The eighth, the Triumph of Alexander upon his Entry into Babylon. *Alex. Alb. Equ. et Isabel. Cunio pictor. et scalp.**

From the dedication of this book to Pope Honorius IV., it is deduced that these engravings must have been executed between 1284 and 1285, inasmuch as this pope only enjoyed the pontificate two years; and it is suggested that a copy of it might be found in the library of the Vatican. The narrative appears to be confirmed by many incidental circumstances, which could not be the invention either of Papillon or his informer. The name of Alberico seems to have been a favorite with the family of Cunio, and a Count of that name actually figures in history in the very years of the presumed invention; a relative of the twins, of course, not the male artist himself.

The interval between the time of the twin Cunio and the next mention of any similar usage is very perplexing; but upon examination it will appear that that long period was not altogether a blank in the art. The next earliest evidence

---

* It is not unlikely that the twins may have been directed in the choice of their subject by the identity of the name of the great conqueror with that of the brother; at least such coincidences are not without parallel in the history of literature.

is a document of the government of Venice, discovered amongst the archives of the Company of Printers in that city. It bears the date of 1441, and as it throws some degree of light upon the controversy relative to the invention of printing, it is here given from Ottley's *History of Engraving*.

"MCCCXLI. October the 11th. Whereas the art and mystery of making cards and printed figures, which is used at Venice, has fallen into total decay; and this in consequence of the great quantity of playing-cards, and colored figures printed, which are made out of Venice; to which evil it is necessary to apply some remedy; in order that the said artists, who are a great many in family, may find encouragement rather than foreigners. Let it be ordered and established, according to that which the said masters have supplicated, that from this time in future, no work of the said art that is printed or painted on cloth or on paper, that is to say, altar-pieces (or images), and playing-cards, and whatever other work of the said art is done with a brush or printed, shall be allowed to be brought or imported into this city, under the pain of forfeiting the works so imported, and xxx livres and xii soldi, of which fine one-third shall go to the state, one-third to the Signori Giustizieri Vecchi, to whom the affair is committed, and one-third to the accuser. With this condition, however, that the artists who make the said works in this city may not expose the said works to sale in any other place but their own shops, under the pain aforesaid, except on the day of Wednesday at St. Paolo, and on Saturday at St. Marco, under the pain aforesaid."

From this it seems manifest that the art of printing from wood-blocks was not lost, but, on the contrary, had been so long practiced as to become an extensive and profitable business in Venice, and had spread over the Continent to such a degree as to destroy the trade of the Venetian artists. The establishment of an important manufacture, and its decay, necessarily infer a long period. From the constant conjunction of the two arts of painting and printing in this document, we may infer (what the existence of prints and cards of later date prove) the method in which these figures and cards were manufactured, namely, that the outline was first printed, and that the colors and shading were filled in by the painter and illuminator. The history of playing-cards now becomes of some importance to the narrative. When cards first came into use is uncertain; but mention is made of them in the year 1254, when they were interdicted by St. Louis on his return from the Crusade: they were also forbidden by the Council of Cologne in 1281. In 1299 they are expressly mentioned under the name *carte;* and in *Das Gulden Spiegel*, printed by Gunther Zainer in the year 1472, it is said that cards first came into Germany in 1300. An old French poet, who wrote " En l'an mil iij cent xxviij," has the line, " Jouent aux dex, aux cartes, aux tables." There is no evidence earlier than the Venetian decree to connect the art of printing from wood-blocks with the art of making cards; but as it is evident from that document that such connection did exist, it is a fair presumption that it originated not very long after the introduction of the game; and as the sum paid by Charles VI. for " troix jeux de cartes " was so small as fifty-six Parisian sols, it has been conjectured that they must have been illuminated prints. The Venetian decree against the importation of painted and printed figures from abroad now brings us to the country from which the chief export was made. It appears, therefore, that in the Low Countries the manufacture was carried on to a great extent; and we shall also find that in Holland and Germany, and probably over most of Europe, religion had called this art to her aid; that whilst the noble and wealthy recreated the mind and delighted the eye with the exquisite productions of the scribe and illuminator, the more humble were equally gratified with rude and simple illustrations of interesting portions of Scripture, or pictures of favorite saints. It is probable that the poorer classes hung up these drawings in their dwellings, where they excited as true and heart-felt devotion as the masterpieces of the painter's art in the oratories of the great. There is no evidence how early the art was practiced, nor whether the outlining the figures of saints and sacred subjects preceded the printing of cards, or was suggested by the latter; but it is certain that at the end of the fourteenth and the commencement of the fifteenth century the practice was very common. The impressions were taken by means of a burnisher, the gloss caused by the friction being distinctly visible on the backs both of cards and prints preserved to this time. As facility in practice increased, a distich or quotation illustrative of the print became a natural improvement; and to this was frequently added a coat of arms, the name of the saint, or the title of the subject, all in the field, or over the head of the figure; and, lastly, sometimes a date. The earliest print of which the date is given within the print itself,* is a woodcut of St. Christopher carrying the infant Jesus across the sea. It is of folio size, and colored in the manner of our playing-cards. At the bottom is the inscription—

𝕮ristofori faciem die quacunque tueris   Millesimo cccc°
   Illa nempe die morte mala non morieris.   rr° terno.

It was found in the Monastery of Buxheim, near Meiningen, and is now in the possession of Earl Spencer.

The next advance was obvious. Instead of a single block, a series of blocks were employed, with additional literary illustrations; and thus were the first printed books formed. The earliest and most memorable of these are the *Historia Sancti Johannis Evangelistæ*, the *Ars Memorandi*, the *Ars Moriendi*, the *Biblia Pauperum*, the *Historia Virginis Mariæ*, and the *Speculum Humanæ Salvationis*. The most important of these works is the *Historiæ Veteris et Novi Testamenti seu Biblia Pauperum*—truly the Poor Man's Bible. It consists of forty leaves printed upon one side of the paper only, by friction, from as many blocks; the color is brown; the prints are placed opposite to each other, and the blank backs are pasted together into one strong leaf. The cuts are about 10 inches in height, and 7½ in width. Each print contains three sacred subjects in compartments, and four half-length figures of prophets in smaller divisions, two above

---

* There is said to be a print at Lyons with the date 1384, but its existence is doubtful. There has lately been discovered a print with the date of 1418, but its authenticity is yet under discussion. It was found by an inhabitant of Malines, who, in breaking up an old coffer which had been used to contain the archives of the former *Grand Conseil* of Malines, observed an ancient-looking print pasted inside the lid. The subject is the Virgin and Child, with Saints Catherine, Dorothy, Barbara, and Margaret, within a palisaded inclosure. On the top-bar of the gate is the date m : cccc °rviii. distinct and unmistakable. The design and execution are very superior to those of the St. Christopher and the block-books. The *London Athenæum* of 1844 contains a full description of the print, and the volume of 1845 a *fac-simile*. The earliest dated print taken from an engraved metal plate is by Maso Finiguerra, 1460.

and two beneath the principal subjects. Latin inscriptions
are on either side of the upper figures, rhythmical verses on
either side of the lower, and additional inscriptions are on
labels at the bottom of the whole. The central subjects are
from the New Testament, the others from the Old, and in some
manner allusive to the former. There are many copies of
this work, evidently from different blocks, and of different
dates. Indeed it appears to have been a most popular book,
and was printed repeatedly long after the introduction of le-
gitimate printing; there are several editions in which the inscriptions are actually printed with movable types. The
exact date of these curious works is not ascertained; but Dr.
Horne possessed a copy contained in one volume with the *Ars
Moriendi* and the *Apocalypse*, all works of the same style,
the binding of which bore the date of 142-. The original
composition and design of this work is attributed, and not
without some show of reason, to Ansgarius, who was Bishop
of Hamburg and Bremen in the ninth century.

A similar book is the *Canticles*, a small folio volume of
thirty-two subjects, two being painted on each leaf, and on
only one side of the paper, and the leaves also pasted back
to back. It differs from the *Biblia Pauperum* in that the inscriptions are engraven on scrolls fantastically dispersed
amongst the figures. This is generally allowed to be of
somewhat later date than the preceding, and to hold an intermediate space between it and the *Speculum Humanæ
Salvationis*, to which a larger space must be devoted, on account of its importance in the controversy relative to the invention of printing.

This is not, strictly speaking, a block-book; for whilst the
form of the design and the portion of Scripture represented
are engraven on wood, the inscription is in some cases engraven on wood also, but in others is printed in *movable type*.
The Latin edition, *perhaps* the first, consists of sixty-three
leaves, divided into five unequal gatherings. The subjects
are chiefly from the Old and New Testament; but sometimes such stories have been selected from ancient history as
might seem in some way appropriate to the events recorded
in sacred writ. Each subject has a short Latin inscription
underneath it, and the text occupies the remainder of the
page. Its size is folio; the impressions are taken with a

burnisher, on one side of the paper; the color of the ink is brown, and the backs are pasted together, as in the books previously described. The work is certainly of nearly the same date, though probably a little later, than the *Biblia Pauperum;* and it may even have been in part executed by the same artist, for in the earlier portions there is so much general resemblance, both in design and execution, as to make it probable that the same graver was employed in both. The latter part, however, is the work of another artist; the lines are not so bold, and there is an attempt at fineness of execution, of shading, and of distance, which the earlier master did not attempt; the design, though in better drawing, is not so spirited; the drapery is more correct, though not so graceful; and in fact the engraver was a better workman, but not so great an artist. It must be understood, that there are numerous editions of this work, many differing in essential particulars, but some so nearly similar as to require a microscopic eye to detect the variations. Of four of these, two are in Latin, two in Dutch; and between these four lies the contest for antiquity. Mr. Ottley (whose beautiful *History of Engraving* contains a well-drawn-up account of his inquiry, illustrated by most convincing examples) has, from a minute and laborious examination, decided that the two Latin and two Dutch are printed from the self-same blocks, and by comparing them, and finding evidences of fractures in the one which do not exist in the other, he has very satisfactorily awarded the palm of antiquity. First, although the Latin inscriptions in the earlier part of the *first* Latin edition (so called by commentators) are engraven on blocks of wood, these blocks are not of the same piece as the figures, the work having been divided between two artists, the one more skilled in engraving figures, and the other in engraving letters. *Secondly*, parts of the engraving broken in the *first* Dutch are perfect in the *first* Latin; parts imperfect in the *first* Latin are unbroken in the *second* Dutch, whilst the *second* Latin is the most perfect of all; from which the conclusion is drawn that the *second* Latin is the most ancient, then the *second* Dutch, next the *first* Latin, and lastly the *first* Dutch. This order of succession is of considerable importance, because the *first* Latin is printed with movable—some commentators say fusil—

types. The printing of this work is claimed for Laurence Koster.

But by whomsoever these curious works were printed, they bring us to the very threshold of the invention of printing, in the proper sense of the word. Bibliographers agree that the pictorial parts of the *Biblia Pauperum*, the *Canticles*, and the *Speculum* were engraven by the same engraver, but from the designs of different artists; and that while of the *first* Latin edition (placed *third* by Ottley) the plates numbering 1, 2, 4, 5, 6, 7, 8, 9, 10, 11, 13, 14, 16, 17, 21, 22, 26, 27, 46, are printed entirely from wooden blocks, the five leaves of which the preface consists, and the text of the remaining leaves—(there are 63 in all)—are printed from *movable type*. Therefore, between the printing of the first edition of these three works, and the third of the *Speculum*, the art of printing with movable type had become known to the printer.

We have now come fairly to the practice of printing in the real sense of the word; and we have also arrived at the long-pending, long-controverted question, of who invented it, and where? The honor is disputed by as many cities as contended for the birth of Homer. Only three of these can show the slightest argument for their pretensions: Harlem, Strasburg, and Mentz. Harlem claims it for her citizen Laurence Koster, or Laurent Janszoon Koster (or Custos). The claim rests principally upon the narrative in the *Batavia* of Hadrianus Junius, a native of West Friesland, who dwelt at Harlem. The work was written in 1575, but not published until 1588. The following is a close translation of the narrative:

"There lived, a hundred and twenty-eight years ago, at Harlem, in houses sufficiently splendid (as a workshop, which remains to this day entire, can serve as proof), overlooking the forum from the neighborhood of the royal palace, Laurentius Joannes, by surname *Ædituus*, or Custos* (which at

---

* In the original, Koster is simply said to have been surnamed *Ædituus*, seu *Custos*, but no mention is made of the Cathedral. The statement, therefore, that he was curtos of the cathedral is a gratuitous insertion of after narrators. The word Custos has been Dutchified into Coster or Koster; but there is no apparent reason why we may not suppose that *Custos* was a barbarous Latin word for keeper, or constable, or any other translation the word will bear.

that time lucrative and honorable office an illustrious family of that name held by hereditary right), the person who now seeks back by just avouchments and oaths the lapsing glory of the invention of printing, nefariously possessed and seized upon by others [the man], with the greatest right to be presented with the greater laurel of all honors. He by chance, walking in a suburban grove (as was the fashion of citizens in easy means to do after dinner in those days), began first to fashion beech-bark into letters, which being impressed upon paper, reversed in the manner of a seal, produced one verse, then another, as his fancy pleased, to be for copies to the children of his son-in-law; which when he had happily accomplished, he began (for he was of great and acute genius) to agitate higher things in his mind, and first of all devised with his son-in-law, Thomas Peter, who left four children, all of whom obtained the consular dignity (a thing which I mention that all may understand the art arose in an honorable and talented, not a servile family), a more glutinous and tenacious species of writing ink, which he had commonly used to draw letters; thence he expressed entire figured pictures with characters added; in which sort I have myself seen *Adversaria* printed by him, the traces of the works being only on opposite pages, not printed on both sides. That book was in the vernacular tongue by an anonymous author, bearing for title *Speculum Nostræ Salutis;* in which it is to be observed among the first beginnings of the art (for never any is found and perfected at once), that the reverse pages being smeared with glue, were stuck together, lest they, being blank, should present a deformity. Afterward he changed beech-blocks for lead; afterward he made them of tin, because it was a material more solid and less flexible, and more durable: from the relics which remained of which types very ancient wine-flasks being made, they are to this day to be seen in those houses of Laurentius which I have mentioned looking upon the forum, inhabited afterward by his grandson Gerard Thomas, whom I name for honor's sake, a noble citizen, who departed this life a few years ago. The studies of men favoring, as it happened, the new art, since a new merchandise never before seen, brought buyers from every side, with most eager quest, at once the love of the art increased, the establishment increased, workmen in the art being added to the family, the first touch of evil; among whom was a certain Joannes, either (as the suspicion is) that Faustus of ominous name, faithless and unlucky to his master, or some other of the same name, I do not greatly care which, because I am unwilling to disquiet the shades of the silenced, touched with the plague of conscience while they lived. He being sworn by oath to the processes of printing, after he had (as he thought) learned thoroughly the art of putting the characters together, the knowledge of fusil types, and whatever else may relate to the matter, taking an opportunity, than which he could not have found one more fit, on the very eve which is sacred to the birth of Christ, on which all in common are accustomed to labor at the sacred ceremonies, stole the whole materials,* tied up a package of the instruments of his master used in that art; thence with a servant hurried from the house, went in the beginning to Amsterdam, thence to Cologne, until he arrived at Mayence, as to the altar of an asylum, where he might live safe beyond the reach of arrows (as the saying is), and having opened an office, enjoyed the rich fruit of his robberies. Indeed, from it, in the space of the (or a turning) year, in the year 1442 from the birth of Christ, with the same types which Laurentius had used at Harlem, it is certain that he produced to light the *Doctrinale* of Alexander Gallus, which grammar was then in most famous use, with the *Tractates* of Peter Hispanus, his first productions. These are, for the most part, things which I have formerly heard from aged men worthy of belief, who have received them as things delivered from hand to hand, as a torch in a race, and have found others relating and attesting the same things. I remember that Nicholaus Galius, the instructor of my youth, a man with iron memory, and venerable for his long years, related to me, that when a boy he had heard, not once only, a certain Cornelius, a bookbinder, and rendered serious by age, nor less than eighty years old (who had lived as an underworkman in that office), relating with much mental anger, and with fervor, the course of the proceeding, the manner of the invention, (as he had received it from his master), the improvement

---

* Or whatever else *choragium* may mean; literally it signifies the *properties* of a theater.

and increase of the art, and other things of the kind; and that the tears would burst from him against his will at the shame of the affair, as often as he talked of the robbery. Which things do not differ from the words of Quirinus Talesius Con., who confessed to me that he had formerly the same from the mouth of the same bookbinder."

Beyond this narrative of Hadrian Junius there is little, or rather no testimony to the truth of Koster's claim, all subsequent argument being either drawn from or referred to this statement. Many very learned bibliographers have given full credence to Hadrian; while others not less acute absolutely deny Koster any pretense whatever—Santander calling in question his very existence; and there is a third party who, being unable to decide between the opposing arguments, and willing to take refuge in a middle course, allow to Koster the credit of having invented printing from blocks, but assign to his rivals that of printing from movable types.

The whole argument may, however, be reduced into a reasonable compass. The probability of Hadrian's narrative will naturally be the subject of inquiry. *First*, the roundabout way in which this hearsay evidence reached Hadrian, is in itself an unsatisfactory circumstance. Little belief can be accorded to an uncertain bookbinder, even had any circumstances been adduced besides the name Cornelius, by which this bookbinder could be identified. *Secondly*, Talesius was many years secretary to Erasmus, who, although a Dutchman and resident in Holland, repeatedly and unhesitatingly ascribes the invention to John Gutenberg of Strasburg at Mentz.* It is not at all probable that, had Erasmus ever heard of this story, or given the slightest credence to it if he had, he would have omitted some mention of a circumstance so gratifying to his national vanity; or that he should have remained in ignorance of a story well known to his secretary, and commonly bruited about, and therefore known to some of the learned men amongst whom Erasmus lived. *Thirdly*, the story of the engraving on beech-bark

---

* Anno Christi 1440. Magnum quoddam ac pene divinum beneficium collatum est universo terrarum orbi, a Johanne Gutenberg Argentinensi, novo scribendi genere reperto. Is cum primus artem impressoriam, quem Latini vocant *excusoriam*, in urbe Argentinensi invenit; inde Moguntiam veniens eandem feliciter complevit. (*Epit. Rerum Script.* 1502, cap. 95.)

accidentally, when it is quite certain that the art of taking impressions from wood-blocks of the figures of cards and of saints and sacred subjects, with religious and legendary inscriptions, had been known and extensively practiced, not only in Italy and Germany, but in Holland itself, for more than a century, is absurd. *Fourthly*, every author who has written upon the matter has given up all claim on Koster's behalf for the invention of cast type, the evidence in favor of others being too strong to be got over. *Fifthly*, the tale of the conversion of the relics of these types into drinking-cups, which were yet to be seen (1575), is discredited by the circumstance that no one has since seen or heard of them, although a controversy for the honor of a discovery in which they would have been evidence, was even then and has ever since raged furiously. *Sixthly*, the story of John Fust having stolen all his printing materials on the eve of Christmas, and decamped, first to Amsterdam, then to Cologne, and lastly to Mentz, and his publishing there within the same year, is self-contradictory; for type is not a very portable commodity; nor would he easily have escaped pursuit at Amsterdam, a town under the same government. Again, John Fust was originally no printer, but a wealthy goldsmith of Mentz, and certainly never worked as any printer's journeyman. Indeed this is such a palpable misstatement, that commentators upon Hadrian have boldly supposed that the thief was John Gutenberg—not he of Mentz, but a brother, also named John. Unfortunately Gutenberg's brother was not named John, but Friele; there was a cousin John; but the only evidence by which we become aware of the existence of these persons excludes the supposition that either practiced the art; nor is it at all likely that members of a noble family, and wealthy men, should have worked in the service of any man. If it should be asserted that it was *the* John Gutenberg, his time is so well accounted for that it is impossible, since he was then resident at Strasburg, and never was at Amsterdam or Cologne. Thus, then, the narrative of Hadrian Junius appears upon examination to be utterly incredible, being at once at variance with itself and with all probability.

Arguments for or against the claim of Harlem may be urged not derived from this narrative. Although these circumstances are not to be believed, the main facts *may* nevertheless be correct. Koster *may* have printed the *Speculum* and other block-books attributed to him. Ottley says that they were certainly printed in Holland, for that the types are not those used in Germany, but closely resembled such as were afterward cut or cast in Holland; and that they are of greater antiquity than any books printed by those who afterward used the art in the Low Countries. He also attempts to show, by the water-marks in the paper, that the works in question were produced in these parts. Water-marks, however, and some bearing a general resemblance to these, were common in the papers used by printers of Cologne, Louvain, and elsewhere; and the argument is worth little or nothing, for no evidence can be given even of the dates of these works, and much less of the printer. The *Speculum* was printed again and again after the invention of letter-press printing; nor is there the slightest evidence, supposing these assertions to be correct, to connect them with the name of Koster. It is a conclusive argument *against* him, that those other works ascribed to him and his descendants are executed with the self-same types used at Utrecht in 1473 by Ketelaer and De Leempt. Van Mander, who lived at Harlem in 1580, in his *History of the Lives of Dutch Painters and Engravers*, treats the claim of Harlem with contempt; for, speaking of printing, he describes it as an art " of which Harlem, with much presumption, arrogates to herself the honor of the invention;" nor does he make the slightest mention of his famous fellow-citizen. There is not the least evidence that his *three* grandsons (not *four*, as Hadrian says) ever carried on his business; for where are their works? and in their time printers had become so proud of their art as not only to put their names to every work, but even to add a long history of their undertaking and progress. Where are the books ascribed to them? what mention is made of them by their cotemporaries? In a subsequent part of this article it will be seen that Caxton, the first English printer, is asserted to have been sent to Harlem to learn the art, and if possible to carry off one of the workmen. These things being also matter of controversy, cannot be used in argument; nevertheless it is of some value that Caxton, who, supposing it to be true, would be an excellent witness in favor of Harlem, upon all occasions refers the invention to Gutenberg, and makes no mention whatever of Harlem or Koster.

Santander labors to disprove the very existence of any such person. But there is no necessity to go so far as Santander: we may allow Koster's identity; we may even allow that he practiced the art of taking impressions from wood-blocks; but this is very different from acknowledging his claim to the invention of the art of printing. The most strenuous champion of Koster is Meerman, an eminent French bibliographer of the last century, who, in his *Origines Typographicæ*, published at the Hague in 1765, strongly maintains this narrative of Hadrian; which is not a little singular, seeing that the Newcastle Typographical Society published a letter from him to Wagenaar, of eight years' prior date, in which he expresses a precisely contrary opinion. He calls Seitz's (Hadrian's) story a mere supposition, and the chronology a romantic invention; gives to the *Speculum* the date of 1470 as the earliest possible; attributes the honor to Gutenberg, and incidentally mentions his intention of publishing a pamphlet on the subject. Notwithstanding this, in his work, without any new fact whatever, he accredits Hadrian's story, finds consistency in the dates, believes the *Speculum*, and denies John Gutenberg—completely reversing his previous conclusion, though his premises remain the same.

The statement of Ulric Zell, given in the *Cologne Chronicle*, though always referred to by bibliographers, has not received the attention it seems to deserve. Ulric Zell is supposed to have been one of the workmen employed in the office of Fust and Schœffer at Mentz, when that city was taken by the Count of Nassau in 1462. On this event Zell betook himself to Cologne, where he established a press, from which in 1467 he issued his first work. He continued to carry on the art in this city for many years. The *Cologne Chronicle* was printed by Koelhoff in 1499. Under the head of "Invention of Printing," it contains an account of its discovery, communicated by Ulric Zell, which, considering the place where it was published, the nearness of the time, and the intimate connection of the narrator with the first movements of the art, carries great weight.

"*Item*, this most worthy art aforesaid [was] first of all invented in Germany, at Mayence on the Rhine; and that is a great honor to the German nation, that such ingenious people are to be found there; and that happened in the year of our Lord 1440.

"*Item*, although the art was invented at Mayence as aforesaid, in the manner it is now commonly used, yet the first idea originated in Holland from the Donatuses, which were printed there even before that time; and from out of them has been taken the beginning of the aforesaid art, and has been invented much more masterly and cunningly than it was according to that same method, and is become more and more ingenious."

Now we know that the *Donatuses* were block-books of a rude form, in no way resembling the art used by Zell and his cotemporaries; and such as they are, there is no evidence that Koster printed any one of them.

All evidence, then, and the general consent of the learned, in failure of Koster, unhesitatingly ascribe this invention to

JOHN GUTENBERG, surnamed Genzfleisch, Gensfleisch, or Gensefleisch, von Solgenloch or Sorgenloch. He was a native of Mentz, and of a noble family, possessed of considerable property in various places in the neighborhood. Fortunately the life of Gutenberg does not rest merely upon hearsay evidence, or the doubtful guesses of bibliographers from dateless wood-cuts; legal documents supply most important information. It appears that, for some reasons unknown, he resided for many years at Strasburg, and had even acquired rights of citizenship. The first document presents him in no amiable light. It is a lawsuit instituted to compel him to perform his marriage-contract with Anne von Isernen Thür; and it would appear that he was compelled to make good his promise, the name of Anne Gutenberg being found in the same register of the nobility liable to the wine-duty in the city of Strasburg, in which Gutenberg's name also appears. The next document is so curious that an ample abstract of it cannot but be interesting.

It appears that he had contracted an engagement with Andrew Dritzehen, John Riffe, and Andrew Heilmann, to instruct them in the secrets of certain arts, and had entered into partnership with them for their better advantage. Andrew Dritzehen and Andrew Heilmann having called upon him one day, perceived that he was engaged in a wonderful

and unknown art, the secret of which he was desirous of
keeping to himself; that, moved by their importunities, he
consented to enter into partnership with them for the term of
five years, on two conditions—first, that they should pay him
the sum of 250 florins, 100 immediately, and the remainder
at a certain fixed period; second, that if any one of the
partners should die during the term of the copartnership, the
survivors should pay to his heirs the sum of 100 florins, in
consideration of which the effects should become the property
of the surviving partners. Andrew Dritzehen died before
the expiration of the period agreed on, being still indebted
to Gutenberg in the sum of 85 florins. George and Nicholas, brothers of the deceased, demanded to be admitted to
the partnership, and on refusal, brought an action against
Gutenberg as principal partner. The magistrates gave judgment on the 12th of December, 1439, relieving Gutenberg
from the demand of the sum of 15 florins, being the difference
of the sum of 100 florins, stipulated to be paid to the heirs
of a deceasing partner, and the sum of 85 florins due to
Gutenberg by Andrew on the original contract. The following evidence was produced on the trial:

"Anna, the wife of John Schultheiss (*holzman, marchand
de bois*), deposed, that on one occasion Nicholas Beildeck
came to her house to Nicholas Dreizehen, her relation, and
said to him, 'My Nicholas Dreizehen, Andrew Dreizehen, of
happy memory, has placed four *stucke* (pages?) in a press,
which Gutenberg has desired that you will take away and
them from one another put off, that no man may know what
it may be, for he is not willing that any one should see.'

"Also John Schultheiss says, that Laurence Beildeck
sometime came to his house to Nicholas Dreizehen, when
Andrew Dreizehen his brother was dead, and that the said
Laurence Beildeck thus spoke to said Nicholas Dreizehen:
'Andrew Dreizehen, your brother, now happy, had four
stucke lying underneath in a press. Therefore John Gutenberg desires you that you will take them therefrom and
upon the presses take from one another so that no man can
see what that is.'

"Also Conrad Sahspach deposed, that sometime Andrew
Heilmann came to him upon the Street of Merchants and
said, 'Dear Conrad, as Andrew Dreizehen is departed, as

you made the presses, and know about the matter, do you go thither, and take the *stucke* from the presses, and thoroughly separate them from one another, so that no man may know what it is.'

"Laurence Beildeck says that he was sent by John Gutenberg to Nicholas Dreizehen, after the death of Andrew his brother, to say to him, 'That he the presses which he under his care has to no man should show; which also this witness did. And he further conversed with me, and said he should take so much trouble as to go to the presses, and with the two screws upon or from them so separate the *stucke* from one another, and these *stucke* he should then in the presses [*or*, on the presses] separate, so that thereafter no man can see nor understand.'

"The same witness also said that he knew well that Gutenberg, a little before the feast of the Nativity, had sent his servant to both Andrews to take away all *stucke*, which were broken up in his sight, that none of them might be found perfect. Moreover, after the death of Andrew, this witness was not ignorant that many were desirous of seeing the presses, and that Gutenberg had commanded that some one should be sent who might hinder any one from seeing the presses, and that his servant was sent to break them up.

"Also John Dunne, goldsmith, said, that three years or thereabouts previous he had received from Gutenberg about 300 florins for materials relating to printing." *

From this curious document may be learnt, that separate types were used; for if they were blocks arranged so as to print four pages, how could they be so pulled to pieces that no one should know what they were, or how could the abstraction of two screws cause them to fall to pieces? It appears that some sort of presses were used, and the transfers no longer taken by a burnisher or roller; and, lastly, that the art was still a great secret at the time when Koster was at the point of death. Hence it is manifest that the ingenuity of Gutenberg had made a vast advance from the rude methods of the time, and had in fact invented a new and hitherto unknown art.

---

* The original German text of these documents is given in M. Leon de Laborde's interesting tracts on the origin of printing.

These documents would be decisive in favor of Strasburg as the place in which printing was invented, had it appeared that any effects were produced by this establishment. This, however, does not seem to have been the case, as Gutenberg and his successors make no mention of the fact, but, on the contrary, claim for themselves the production of the first book at Mentz. Indeed the partnership appears to have expired without any attempt at entering into fresh engagements; for, about the year 1450, Gutenberg returned to his native city with all his materials, without any opposition from his partner. In this place he entered into partnership with John Fust, a wealthy goldsmith and citizen, who engaged, upon being taught the secrets of the art (a fact that completely overthrows the fable of his having been one of Koster's workmen, and of his having stolen his types), and being admitted into a participation of the profits, to advance the necessary funds; and he did accordingly advance the considerable sum of 2020 florins. The new partnership immediately commenced operations, and hired a house called Zum Jungen, and took into their employ Peter Schœffer and others. Their subsequent operations we again find curiously chronicled in the records of another lawsuit,[*] in which Gutenberg was soon engaged with his new ally; for Fust, dissatisfied with their proceedings, sought to recover from Gutenberg money advanced, with interest, including 800 florins of the sum advanced in virtue of the deed of partnership. Gutenberg in defense alleged, that the 800 florins had not been paid at once, as stipulated; and that they had been expended in preparation for the work (apparently meaning thereby that this sum of money should have been paid down for his own use, in consideration of his communicating the secrets of his art, and that instead of so applying it to his private purposes, he had expended it for the joint benefit); whilst, as to the other sums, he offered to give an account of their appropriation, but denied that he was liable for the interest. The judges awarded that Gutenberg should pay the interest, as well as the part which his accounts showed he had applied to his individual use. This decision took place

---

[*] Wolfii *Monumenta Typographica.* Fournier, *Origine de l'Imprimerie.*

on the 6th of November, 1455. Upon this, Fust obtained from the public notary the following document:

"To the Glory of God, Amen. Be it known unto all those who shall see or hear read this instrument, that in the year of Our Lord 1455, third indiction, on Thursday the sixth day of November, the first year of the Pontificate of our very Holy Father the Pope Calixtus III., appeared here at Mayence, in the great parlor of the Barefooted Friars, between eleven o'clock and midday, before me, the Notary, and the undersigned witnesses, the honorable and discrete person, James Fust, citizen of Mayence, who, in the name of his brother, John Fust, also present, has said and declared clearly, that on this same day, and at the present hour, and in the same parlor of the Barefooted Friars, John Gutenberg should see and hear taken by John Fust an oath, conformable to the sentence pronounced between them. And this sentence read in the presence of the honorable Henry Gunter, Curé of St. Christopher of Mayence, of Henry Keffer, and De Bechtoff de Hanaw, servant and valet of the said Gutenberg; John Fust, placing his hand upon the Holy Evangelists, has sworn between the hands of me, the Notary Public, conformable to the sentence pronounced, and to a letter which he has sent to me, and has taken the following oath, word for word: I, John Fust, have borrowed 1550 florins which I have transmitted to John Gutenberg, which have been employed for our common labor, and of which I have paid the rent and annual interest, of which I still owe a part. Reckoning, therefore, for each hundred florins borrowed, as above is recited, six florins per annum, I demand of him the repayment and the interest, conformably to the sentence pronounced; which I will prove in equity to be legal, in consequence of my claim upon the said John Gutenberg. In presence of the honorable Henry Gunter, of Henry Keffer, and of Bechtoff de Hanaw aforesaid, John Fust has demanded of me an authentic instrument, to serve him as much and as often as he hath need, in the faith of which I have signed this instrument, and have set thereto my seal."

From this it would appear (indeed the mortgage of his printing materials to Fust, mentioned in this document, proves) that Gutenberg had expended the whole of his considerable private fortune in his experiments, and had fallen into the power of his more wealthy associate; for in consequence of this judgment, and owing probably to his being unable to repay the sums demanded, the whole of his materials, constructed with so much perseverance, fell into Fust's hands; for the initial letters used by Gutenberg and his partners, in works known and supposed to have been executed between 1450 and 1455, are likewise used by Fust and Schœffer in the Psalter of 1457 and 1459. After such a mortifying result of so many years' labor, it would have been no matter for wonder had Gutenberg abandoned the unprofitable pursuit. On the contrary, he appears to have immediately started anew with fresh vigor, and this time with success. Another legal document gives curious information:

"We, Henne (John) Genszfleisch de Sulgeloch, named Gudinburg, and Friele Genszfleisch, brothers, do affirm and publicly declare by these presents, and make known to all, that, with the advice and consent of our dear cousins, John, and Friele, and Pedirmann Genszfleisch, brothers, of Mentz, we have renounced and do renounce, by these presents, for us and for our heirs, simply, totally, and at once, without fraud or deceit, all the property which has passed by means of our sister Hebele, to the convent of St. Claire of Mentz, in which she has become a nun, whether the said property has come to it on the part of our father Henne Genszfleisch, who gave it himself, or in whatsoever manner the property may have come to it, whether in grain, ready money, furniture, jewels, or whatever it may be, that the respectable nuns, the abbess, and sisters of the said convent, have received in common or individually, or other persons of the convent (have received), from the said Hebele, be it little or much; and we have promised and do promise, by these presents, in good faith, for us and for our heirs, that neither we, nor any person on our part, nor yet our said cousins, nor any of their heirs, nor any person on their part, shall either demand, gain, nor claim of the said convent, nor of the abbess, nor of the convent in general, nor of the persons who may be found therein individually, the said property, of whatever kind it may be, either wholly or in part, and that we will never demand it again, either through an ecclesiastical or civil court, or without the aid of the law; and that neither we nor our heirs will ever molest the said convent, either by words or deeds, either secretly or publicly, in any manner. And as to the books which I, the said Henne, have given to the library of the convent, they are to remain there always and forever; and I, the said Henne, propose also to give in future, without disguise, to the library of the said convent, for the use of the present and future nuns, for their religious worship, either for reading or chanting, or in whatever manner they may wish to make use of them according to the rules of their order, *all the books which I, the said Henne, have printed up to this hour, or which I shall hereafter print, in such quantities as they may wish to make use of;* and for this the said abbess, the successors and nuns of the said convent of St. Claire, have declared and promised to acquit me and my heirs of the claim which my sister Hebele had to the sixty florins, which I and my said brother Friele had promised to pay and deliver to the said Hebele, as her portion and share arising from the house which Henne our father, assigned to him for his share, in virtue of the writings which were drawn up thereupon, without fraud or deceit. And in order that this may be observed by us and by our heirs, steadfastly and to its full extent, we have given the said nuns and their convent and order these present writings, sealed with our seals. Signed and delivered the year of the birth of J. C. 1459, on the day of St. Margaret."

From this it will appear, that his new establishment had actually produced the long wished-for effect. He appears to have carried on the business ten years; for in 1465 he entered into the service of Elector Adolphus of Nassau, as one of his band of gentlemen pensioners, with a handsome salary, as appears from the letters-patent, dated the 17th January, 1465, and finally abandoned the pursuit of an art which, though it caused him infinite trouble and vexation, has been more effectual in preserving his name and the memory of his acts, than all the warlike deeds and great achievements of his renowned master and all his house. Gutenberg died on the 24th of February, 1468. His printing-office and materials had passed into the hands of Conrad Humery, syndic of Mentz, who had probably assisted him with money, and who appears to have been in some degree his partner.

He afterward sold them to Nicholas Bechtermunze of Elfield, whose works are greatly sought after by the curious, as they afford much proof, by collation, of the genuineness of the works attributed to his great predecessor.

There does not appear to be any record of the early life of JOHN FUST or PETER SCHŒFFER before their partnership with Gutenberg, save that the former was a wealthy goldsmith and an ingenious man, and that Schœffer, surnamed de Gernsheim, was a scribe. It is very likely that the combination of character and qualifications of these three men may afford a good clue to the wonderful taste and beauty which distinguish the works issued from their press, and consequently to the great general improvement of the art during their life. The ingenuity of Gutenberg would readily suggest a new and expeditious method of manufacturing types; the practical skill of Fust as a worker in metals (and the working in gold and silver had at that time attained a most extraordinary nicety and beauty), and his large pecuniary resources, would readily provide the necessary appliances, while the taste of Schœffer would give all possible grace and beauty to the new forms. For Schœffer, it must be recollected, was a scribe, one of the ancient and honorable craft whose occupation was destined to fall before the new art; a transcriber, perhaps an illuminator, of the manuscript works in use before printed books; and those who have had the happiness of viewing those exquisite specimens of skill which beguiled our ancestors into study and devotion (when will modern typography produce such feasts for mind, and eye, and imagination?) will readily conceive that Schœffer's eye was already schooled for the conception, and his hand for the execution, of all the beauty the trammels of a new art and limited skill would allow. Aided by his own taste and his partners' invention and wealth, Schœffer proceeded to a new enterprise, namely the *casting* of type. The entire conception and execution of this invention has been generally attributed and allowed to Schœffer. It seems most probable, however, that where three ingenious men are bound together by art and interest, no one of them can lay exclusive claim to any invention or undertaking executed in the workshops and for the mutual benefit of all. Allowing, therefore, to Schœffer, the honor of having suggested some such plan, the other two may fairly put in a claim for their portion of the credit on the score of their suggestion and assistance; especially since Fust, as a worker in metals, would have been the party to engage workmen to elaborate the conceptions of his partners' brains. Accordingly the only evidence upon the subject appears to show that the partners had for some time practiced a method of taking casts of types in moulds of plaster; for it must be remembered that the types of Gutenberg's earlier efforts, both at Strasburg and at Mentz, were cut out of single pieces of wood or metal with infinite labor and imperfection. This method of casting, however, although a great improvement, was at best but a slow and tedious process. Almost every type cast would require a new mould; no skill or care could enable the workman to impress so small a thing as a type is at the face, yet so elongated in the shank, fully, freely, and steadily, into a soft material; and it would be necessary afterward, under the most favorable circumstances, that the squareness and sharpness so indispensable in type should be given by another slow process; so that at best this advance was but an imperfect and tedious operation. Schœffer has therefore an undoubted claim to be considered as one of the three inventors of printing; for he it was who first suggested the cutting of punches, whereby not only might the most beautiful form of type the taste and skill of the artist could suggest be fairly stamped upon the matrix, but a degree of sharpness and finish quite unattainable in type cut in metal or wood could be given to the face; whilst to the shank, by the very same process by which the face was cast, the mould would give perfect sharpness and precision of angle. Add to this, that the punch being once approved of, could be kept ready to stamp a new matrix in precisely the same condition and form as the first, should that be worn out or mislaid, or make a duplicate should the demands of business require it. It is nevertheless rather singular, that the mould represented on the right side of the press of Ascensius, shortly after the time of Schœffer, should be precisely the same in form and manner of use as that of the present day. This was evidently an immense stride toward perfection; let Schœffer therefore take a place on the right hand of the inventor.

Whatever may have been the several shares of the masters in perfecting their art, their joint labors were effectual. The first productions of their press—passing over an *Alphabet*, the *Doctrinale* of Alexander Gallus, and a *Donatus*, which are of doubtful authenticity, and are merely block-books—were three editions of Donatus, the first *books* known to have been printed entirely with movable types. In 1455 they printed the celebrated *Litteræ Indulgentiæ Nicolai V. Pont. Max.*, which is the first work—it is only a single page—printed with movable types which is dated. In 1455, or thereabouts, for it has no date, they printed the famous *Biblia Latina Vulgata*, generally known as "the Mazarine Bible." It has no colophon or *Explicit*. And it should be noted, that there is no book known which bears the conjoint names of Gutenberg, Fust, and Schœffer, nor any which has the imprint of Gutenberg alone.

Within eighteen months of their separation from Gutenberg, Fust and Schœffer produced the celebrated Psalter. This was printed with large cut type. As it is impossible that a new font could have been prepared, and so splendid a work printed within that short space, it must be evident that the partners did great injustice to Gutenberg in suppressing his name from the colophon. This book was produced in the month of August, 1457, and is the first book which bears the name of the place where it was printed, those of the printers, and the date of the year in which it was printed. This Psalter was reprinted in 1459, 1490, and 1502, and always in the same type, which, it is remarkable, was never used for any other work, probably because its great size made it unfit for any other works than those not intended for popular reading, but to lay on desks like our church Bibles. On the 16th of October, 1459, Fust and Schœffer published the *Durandi Rationale Divinorum Officiorum*, with an entirely new font of type; in 1460 the *Constitutiones Clementis V.;* and in 1462 the celebrated Latin Bible. In 1465 they printed *Cicero de Officiis*, in which occur the first printed Greek types. Fust enjoyed this successful and glorious practice of his art but ten short years; yet in this period what an immense advance from the misshapen and irregular lumps of their first efforts, ugly in themselves, and more ugly in their utter want of relative proportion and alignment, to the well-proportioned, evenly-stand-

ing type of the Bible! The plague carried him off in Paris
about the year 1466, full of years, and perchance full of
honors. Schœffer survived many years, and, in conjunction
with Conrad Henlif, produced a great number of works.
His name is found in the colophon of the fourth edition of
the Bible of 1402, about which time he is supposed to have
deceased. There are ten books which are known to have
been printed by Fust and Schœffer conjointly. Schœffer
continued to print during a period of thirty-five or thirty-six
years after the death of Fust, and his productions are very
numerous.

Were we to take tradition for our guide as regards the
character of Fust, we should regard him as a conjuror and
an adept in the black art. The popular story (and many
"grave and discreet old men" have given credit to the tale)
runs, that having kept these proceedings profoundly secret,
as soon as their Bible was finished, Fust transferred himself
to Paris with many copies of the new work, and palmed
them upon the learned as manuscripts—to which, as they
were printed on vellum, in a type bearing much resemblance
to the written books of the period, and the vignettes and
initial letters were splendidly illuminated, they were not very
dissimilar; that some eager scholar or devotee became the
possessor of the first copy, supposing it to be a rare chance,
at the moderate price of four or five hundred crowns; that
as he brought the work into the market, the price fell rapidly
to sixty, and then to thirty crowns, by which time the extraordinary glut produced suspicion, and Fust was accused of
multiplying Holy Writ by the aid of the Devil, and was accordingly persecuted by the priesthood, whilst the laity, looking to their temporal interests, prosecuted him for his inroad
into their pockets; and that from these things Fust was
obliged to quit Paris precipitately.

Having thus given a sketch of the origin and history of
the art of printing, a brief account of the works issued by
the illustrious triumvirate will not only be proper here, but
will give the general reader a better idea of the astonishing
perfection to which the art rose under the taste and genius
of its inventors. As before remarked, there is not a single
work of Gutenburg which bears his name; yet there are
several which bear such internal evidences that the literati

of all parties and opinions are unanimous in attributing them to his press.

Of these works, Dr. Dibdin, the well-known bibliographer, gives the following account:

"First, as to the *character of the type* used by the early Mentz printers. This appears to have been uniformly what is called *Gothic;* and if we except the varieties of the larger type (from three-eighths to two-eighths or to a quarter of an inch), which appear in the Psalters of 1457, 1459, and 1490 (the type common to the most works executed about the same period), we shall observe three distinct sets or forms of letters used in the printing-office of Faust and Schoiffher. Of these three typographical characters, two only (if we except the one with which the Bible of 1455 was executed) are visible in the publications which appear to have been printed in the lifetime of Faust; that is to say, the larger Gothic used in the Bible of 1462, and the smaller Gothic in the *Offices* of Cicero, of the dates of 1465 and 1466. These appeared united, the former, for the first time, in the *Constitutions of Pope Clement V.*, of the date of 1460. Schoiffher introduced a type of an intermediate size, which may be seen, among other works, in the *Rudiments of Grammar* of 1468, and in the *Decretals of Pope Gregory the Ninth,* of the date of 1479. This intermediate type is of a narrower form, and prints very closely. Of the three types here mentioned, the largest is undoubtedly of the handsomest dimensions; but they all partake of the *Secretary Gothic,* and may be said to be the model of that peculiar character which was adopted by the early Leipsic printers, Thanner and Boëttiger, and was more especially used by John Schoiffher and the other German printers for nearly the whole of the sixteenth century. Shew me, Lisardo, one book, nay, one leaf only, printed in the *Roman type,* in the colophon of which the name of Faust or of Peter Schoiffher appears, and you shall immediately have the amount of the balance in my favor, at my banker's, be it great or small, be it 200*l.* or 20*l.*, for such a precious and unheard-of curiosity.

"We shall now, in the second place, say a few words as to the *character of the printing,* or of the *mechanical skill,* of the early Mentz press. There can be but one opinion upon this point. Every thing is perfect of the kind, the paper, the ink, and the register, or regularity of setting up the page. The Bible of the supposed date of 1455 is quite a miracle in this way;* but the Psalters are not less miraculous, nor is less praise due to the *Constitutions of Pope Clement V.*, of the date of 1460, and the *Bible* of 1462; while the *Durandus*, of the earlier date of 1459, exhibiting the first specimen of the smallest letter, strikes one as among the most marvelous monuments extant of the perfection of early typography. Almost all the known works before the year 1462 are printed *upon vellum*, doubtless because they ventured upon limited impressions; and even of the Bible of 1462 more copies have been described upon vellum than upon paper. Upon the whole, the vellum used by Faust and Schoiffher, although inferior to the Venetian, is exceedingly good, being generally both white and substantial.

"In the third place, let us notice the nature or character of the works which have issued from the press of Faust and Schoiffher. Whatever may be our partiality toward that establishment from which the public were first gratified with the sight of a printed book, candor obliges us to confess that the fathers of printing were not fortunate, upon the whole, in the choice of books which issued from their press.

"In the fourth place (for I told you I should be somewhat tautologous), consider what is the typographical appearance of these books which Gutenberg is really supposed to have executed. It is quite unique. A little barbarous, and certainly wholly dissimilar from any thing we observe in other cotemporaneous productions of the Mentz press. You will please to understand that I think very doubtfully of the *Donatuses*, which are considered to have been printed by him; as well as of the *Speculum Sacerdotum*, and *Celebratio Missarum;* concluding the *Catholicon* of 1460, and the *Vocabularies* of 1467 and 1469, to be the more genuine produc-

---

* This is even sober praise. The mechanism of the press-work, and appearance of the ink, beautiful, regular, and glossy as the whole appears, does not strike one with more astonishment than the manufacture of the paper. "Charta," says Tungendres, "ejusdem est crassitudinis, qualem illo tempore libris imprimendis consumere mos fuit." And again, "Charta ob ejus densitatem atque spissitudinem haud ingratam ubique se maxime commendat." (*Disq. de Not. Charact. Libror.* p. 27, p. 46.) And see Meerman's testimony in favor of the paper of the Soubiaco press, *Orig. Typog.* vol. i. p. 9, note.

tions of his press, or of the types used by him. Is it not surprising, I ask, that these works are executed in types quite different from any thing in the Mentz productions? and this from a man who is considered as the parent of printing in that city. No wonder, if they *be* the actual productions of Gutenberg, that Faust and Schoiffher thought so meanly of his talents, and that on a dissolution of partnership they adopted a different and a very superior character."

In confirmation of these remarks of the learned bibliographer, we shall here insert a specimen of Gutenberg's *Balbus de Janua*, which will also be a curious illustration of ancient art. Notwithstanding the appearance of these types, the reader is assured that the original is really printed from separate pieces of metal.*

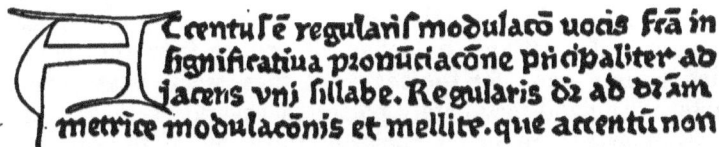

Dr. Horne, in the appendix to his *Introduction to Bibliography*, says of the Psalter, " This precious work, as Santander justly calls it, is one of the most known among early printed books, from the various and correct descriptions of it which have been given by different bibliographers. Until the discovery of Pope Nicholas's *Literæ Indulgentiarum*, this was supposed to be the very first article ever printed with a date affixed; the book is executed on vellum, and of such extreme rarity that not more than six or seven copies are known to be in existence; all of which, however, differ from each other in some respect or other. The most perfect copy known is that in the imperial library at Vienna; it comprises 175 leaves, of which the Psalter occupies the 135 first and the recto of the 136th. The remainder is appropriated to litany, prayers, responses, vigils, etc. The psalms are executed in larger characters than the hymns, similar to those used for missals prior to the invention of printing; but

* The initial A is illuminated in a very brilliant blue. The reader who is desirous of obtaining the full effect of this specimen can fill up the printed outline in water-color.

all are distinguished for their uncommon blackness. The capital letters, 288 in number, are cut on wood with a degree of delicacy and boldness which are truly surprising; the largest of these, the initial letters of the psalms, which are black, red, and blue, must (as Lichtenberger has remarked) have passed three times through the press. Copies are now in the Queen's library at Windsor, and in that of Earl Spencer at Spencer House."

The extraordinaay praise awarded by these eminent bibliomaniacs to the first productions of the Mentz press may perchance excite in the minds of the more sober public a suspicion that these writers have been led away by their enthusiasm beyond the limits of matter-of-fact truth, and have seen merit in defects, beauty in deformity, and luster in antiquity. Assuredly, nevertheless, such is by no means the case; and the happy individual who gains access to the *chef-d'œuvres* of Fust and Schœffer will return from the inspection a wiser man; for the beauty of these works is inconceivable. England fortunately possesses several of these treasures of art, there being copies of the Bible of the supposed date of 1450-55 in the Royal Library, in the Bodleian, and in those of Earl Spencer and Henry Perkins, Esq.; whilst of the six known copies of the Psalter of 1457, two are in England, namely, one at Windsor, and one in the possession of Lord Spencer. Of the Latin Bible of Fust and Schœffer, 1462 (the first bearing date), there are copies *on vellum* at Blenheim, in the libraries of Lord Spencer, the Earl of Jersey, one formerly belonging to Sir M. Sykes, in the British Museum, and in the Bodleian (imperfect). Copies *on paper* are rarer still, there being but three in England, viz., those in the Royal Library and the British Museum, and one lately in the possession of Mr. Willett.

Apparently, in retaliation for the injustice done to Gutenberg by his partners in depriving him of any share of the honor of producing the Psalter of 1457, which, as before stated, must be the joint production of all three, although it was not finished until after the secession of Gutenberg, bibliographers have generally agreed in attributing the printing of the Bible of 1450-55 to Gutenberg alone, when it is equally manifest that Fust and Schœffer had as much claim to the honor as their coadjutor. It is an exceedingly beautiful book, in two very large folio volumes, in two columns, containing from forty-one to forty-three lines each, in very large well-cut types. It consists of six hundred and forty-one leaves; it has no title, paging, signatures, or catchwords; the initial letters are not printed, but painted in by illuminators, and the initial letters of each verse of the psalms are painted alternately red and black, by way of guide to the priests in their alternate reading. From the luster and blackness of the ink, its evenness of color, and beautiful execution, it is a very superb book; but it is nevertheless surpassed by the Fust and Schœffer edition of 1462, when they had attained greater experience in the practice of the art. By far the choicest, however, of these *editiones principes*, is the Mentz Psalter or *Codex Psalmorum* before mentioned. Dr. Horne says that the six known copies of this edition differ from each other in some respects, and proceeds to give some particulars in which variations are found; but by collating the copies in the Royal Library, that at Windsor, and that at the British Museum, it will be found that, although bearing the same date, they are in fact three distinct editions. It would have excited no surprise had it been found that the printed ornaments differed, as nothing would be more easy than to change the colors with which the different blocks were worked; and in fact in the Museum copy the initial B is printed in a bright blue, and the scroll-work is red; but the text varies in such a manner that there can be no doubt of their perfect distinctness.

It must also be noted that in the Windsor copy each line is "justified out," which is not the case in Earl Spencer's copy; and that in the Museum copy the page commences with rubrical matter, which is continued down the two first lines of text, which are shortened. The difference is effected by variations in the contractions of many of the words. The book* is a very large folio, on vellum, consisting of about a hundred and thirty leaves, printed on both sides. There are generally twenty-three lines in a page, in Gothic type. Every psalm begins with a splendid initial letter, *printed* in two colors in almost every case. Occasionally,

---

* The copy described is that at Windsor; the illuminations, no doubt vary in every copy.

however, this appears to have been neglected, and then the letter is painted in by the illuminator, but *not* in imitation of the printed letters. The initials consist of a bold character, of Gothic cut, surrounded by a scroll, which is sometimes of great length, but that of the B extending from the top to the bottom of the page. The same wooden block is used as often as the letter occurs, but it is not always in the same colors. Moreover, every *verse* commences with a smaller initial printed in a red color. Nor is this work destitute of the embellishments of the illuminator; for at the commencement of every psalm is a rubric, painted in a most brilliant red, in a smaller letter, of precisely the same character as the text, and also the music of the chant, with the words underneath it painted in black. The initial letters of both are spendidly illuminated in various colors. The paint is used in such profusion that the letters are absolutely in relief, often to the extent of one-sixteenth of an inch; and besides these, the letter following the grand initial has a broad bar painted down it, and very frequently the first letter after the pauses indicated in our authorized version by a colon is illuminated in a similar manner. One page is particularly splendid; it consists of short verses, in which the first words are constantly repeated. It commences with a grand initial, and there are twenty-two smaller initials to the verses; the second letter of the first verse, and the first letter after every pause (twenty-three in number), having the broad illuminated bar. Wherever the psalm commences too near the bottom to allow of the full exuberance of the scroll, a piece of paper appears to have been laid over a portion of the cut, to prevent the impression from appearing; and in one psalm where the chant is of unusual length, the lower part of the initial O, and a corresponding portion of the scroll, are thus suppressed; the music being illuminated in its place, and the scroll continued below it. Sometimes the illuminator has omitted to add his initial letter; and in this copy the double device is omitted. The accuracy with which the colored blocks are printed within the text and within each other is perfectly astonishing. From this description it may be conceived how very superb is the *first* book ever printed, the date, and place, and artist, of which can be accurately ascertained. Dr. Dibdin in the *Bibliotheca Spenceriana*,

Mr. Savage in his work on *Decorative Printing*, Dr. Horne, whose wood-block is not colored, and several other writers, have given fac-similes of the same copy (Lord Spencer's), which, however, all differ from one another.

The capture of the city of Mentz by Count Adolphus of Nassau in the year 1462, had the effect of interrupting the labors of Fust and Schœffer; and moreover the distracted state of the city enabled, perhaps compelled, the workmen initiated in the mysteries of the art to flee into the neighboring states, and thus spread its practice over the whole civilized globe. Such, indeed, was the fame it had already acquired, and such the idea entertained of its importance, that every community with the slightest pretensions to literature appears to have sought a knowledge of it with the greatest avidity. Thus, within six years of the publication of the Psalter, it had spread to several cities having some connection with Mentz, and within fifteen years to almost every town of consideration in Christian Europe. A chronological list of the cities which first seized upon the invention would be greatly too long for this article; it may be interesting, however, to extract a few of the principal, with a notice of such printers as are remarkable either for the beauty or the scarcity of their works. The reader is not to suppose that all, or indeed any great number of these, learned the practice of the art under the tuition of the first masters. A few are known to have been pupils of the inventors, and it is probable that many others of them were so; but the majority, in all likelihood, were men of learning, enterprise, or capital, who derived their typographical knowledge from such facts as had transpired, or from inferior workmen of Fust and Schœffer or Gutenberg supplying deficiencies by their own ingenuity.

Strasburg. Mentelin. Some writers have claimed for Mentelin the invention of printing, representing that Gutenberg was his servant, without, however, showing the slightest ground for their assertions; but others, more reasonable, say that he was acquainted with Gutenberg, and instructed by him, and that on the latter's quitting Strasburg he established a printing-office, and carried on the business successfully. Mentelin most probably printed about the year 1458. His type is rude and inelegant. The only book bearing his name is Beauvais's *Speculum Historiale*, of date 1473. Schæpflin says, that he, as well as Fust and Schœffer at Mentz, printed 300 sheets per day.

In 1461. *Bamberg.* Albert Pfiister. He printed a collection of *Fables*, of date 1461. This book is excessively rare; it is printed with cast metal type, and is illustrated with 101 wood-cuts, in much the same style as the old *Biblia Pauperum*. All his other works are printed in the same type.

1465. *Subiaco* and *Rome.* Schweynheym and Pannartz. Their known works are, a *Donatus*, without date; Lactantius, 1465; St. Augustin on *the City of God*, 1467; *Cicero de Oratore*, without date; and the *Commentary of De Lyra on the Bible*, 1471, all in folio. These works were printed in a new letter, very closely resembling the type now in use called *Roman*, and of which they were the introducers. In *De Lyra* are the earliest specimens of Greek types worthy of the name; some few letters appear in the *Cicero de Officiis* printed at Mentz, but so wretchedly imperfect that they are unworthy of mention. It is curious that the Greek font of Schweynheym and Pannartz at Subiaco was evidently very small; but upon their removal to Rome they cast a much larger font. The cut and appearance of this Greek is more than respectable. There is a very curious petition from them to the pope, praying for assistance on the ground that they had entirely ruined themselves by printing *De Lyra*, for which there was no sale, and representing that they had on their hands no less than eleven hundred folio volumes of that work. Subiaco is the first place in Italy in which printing was practiced. At Rome Ulric Han and Lignamine were cotemporaries. Their works, particularly those of Han, are excessively rare.

1467. *Elfield.* Henry and Nicholas Bechtermunze. They purchased from Conrad Humery the types and materials of Gutenberg. Their works are not at all remarkable for beauty, but are very rare, and much sought for as affording evidence of Gutenberg's works.

1467. *Cologne.* Ulric Zell. His type is Gothic, and of no beauty; but his works are rare.

1468. *Augsburg.* Ginther Zainer printed the first book in Germany with Roman type.

1469. *Venice.* John de Spira, whose works are of the utmost beauty. His edition of Pliny is splendid, and enormous sums have been given for those printed in vellum. He did not use Greek characters; but Greek passages are composed in Roman types. In the same city, at the same time, printed Nicholas Jenson, whose works are equal, if not superior, to those of Spira; they are not so rare, but are almost equally sought after. A copy of his folio Latin Bible of 1479, printed in Gothic type, was sold at Mr. Edwards's sale for 115*l*. 10*s*. Venice was also the residence of Christopher Valdarfar, whose works gave rise to a most extraordinary event connected with bibliography, viz., the sale of the first edition of *Il Decamerone di Boccaccio*, printed by him in 1471. For many years it had been known that a single copy of this work was in existence, and the most devoted bibliomaniacs had used their utmost endeavors to discover it, but in vain. At length, about 1470, an ancestor of the Duke of Roxburghe obtained possession of it for the sum of one hundred guineas. In lapse of time it became the property of John duke of Roxburghe, the accomplished, indefatigable, and undaunted bibliomaniac, after whose death his gorgeous library was dispersed by the auctioneer in the year 1811. The interest excited amongst the learned by this sale was intense. It was known that the collection contained the most superb specimens of every kind of ancient lore; that the illuminated manuscripts were the most brilliant, the ballads the most obscure, the *editiones principes* the most complete that the world could produce; that the rarest Caxtons, the finest Pynsons, and grandest specimens of the foreign printers, were here to be found; above all, it was rumored that a mysterious edition of Boccaccio's Decameron would become a bone of contention amongst the noblest of the literati. The public, learned and unlearned, were infected with the mania, and the daily papers teemed with notices of the sale. At length the important day arrived, the 17th of June, 1811. St. James's Square was the place. Mr. Evans presided. The room was crowded; Earl Spencer, the Marquis of Blandford, the Duke of Devonshire, and an agent of Napoleon, were amongst the most prominent. The book was a small folio, in faded yellow morocco binding, black-letter. " Silence followed his (Mr. Evans's) address," says Dibdin.

"On his right hand, standing against the wall, stood Earl
Spencer; a little lower down, and standing at right angles
with his lordship, appeared the Marquis of Blandford. The
duke, I believe, was not then present; but my Lord Althorpe
stood a little backward, to the right of his father, Earl Spen
cer. Such was 'the ground taken up' by the adverse hosts.
The honor of firing the first shot was due to a gentleman of
Shropshire, unused to this species of warfare, and who seemed to recoil from the reverberation of the report himself had
made. 'One hundred guineas,' he exclaimed. Again a
pause ensued; but anon the biddings rose rapidly to five
hundred guineas. Hitherto, however, it was manifest that
the firing was but masked and desultory. At length all random shots ceased, and the champions before named stood
gallantly up to each other, resolving not to flinch from a trial
of their respective strengths. 'A thousand guineas' were
bid by Earl Spencer; to which the marquis added 'ten.'
You might have heard a pin drop. All eyes were turned;
all breathing well nigh stopped. Every sword was put home
within its scabbard, and not a piece of steel was seen to move
or to glitter save that which each of these champions brand-
ished in his valorous hand. See, see; they parry, they lunge,
they hit; yet their strength is undiminished, and no thought
of yielding is entertained by either. 'Two thousand pounds'
are offered by the marquis. Then it was that Earl Spencer,
as a prudent general, began to think of an useless effusion of
blood and expenditure of ammunition, seeing that his adver-
sary was as resolute and fresh as at the onset. For a quarter of a minute he paused, when my Lord Althorpe advanced
one step forward, as if to supply his father with another spear
for the purpose of renewing the contest. His countenance
was marked with a fixed determination to gain the prize, if
prudence in its most commanding form, and with a frown of
unusual intensity of expression, had not bade him desist.
The father and son for a few seconds converse apart; and
the biddings are resumed. 'Two thousand two hundred and
fifty pounds,' said Lord Spencer. The spectators are now absolutely electrified. The marquis quietly adds his usual 'ten,'
and there is an end of the contest. Mr. Evans, ere his hammer fell, made a due pause, and, indeed, as if by something
preternatural, the ebony instrument seemed itself to be

charmed or suspended in 'in mid-air.' However, at length down dropped the hammer, and, as Lisardo has not merely poetically expressed himself, 'the echo' of the sound of that fallen hammer 'was heard in the libraries of Rome, of Milan, and Saint Mark.' Not the least surprising incident of this extraordinary sale is, that the marquis already possessed a copy of the work, which wanted a few leaves at the end; he therefore paid this enormous sum for the honor of possessing a few pages. The prize of this contest is now in the possession of Earl Spencer."

1469. *Milan.* Lavagna. In 1476 Dionysius Palavasinus printed the Greek Grammar of Constantine Lascaris, in quarto, which is the first book printed entirely in Greek. The first printing in Hebrew characters was performed at Soncino, in the duchy of Milan, in 1482.

1470. *Paris.* Ulricus Gering, M. Crantz, and M. Friburger.

1471. *Florence.* Bernard Cennini. In 1488 Demetrius of Crete printed the first edition of Homer's works, in most beautiful Greek.

1474. *Basle.* Bernardus Richel.

1474. *Valencia.* Alonzo Fernandes de Cordova.

1474. •*Louvain.* Joannes de Westphalia.

1474. *Westminster.* William Caxton, the Game of Chess.

1475. *Lubeck.* Lucas Brandis.

1476. *Antwerp.* Thierry Martins of Alost.

1476. *Pilsen* in Bohemia. *Statuta Synondalia Pragensia;* printer's name not known.

1476. *Delft.* Maurice Yemantz.

1478. *Geneva.* Adam Steinschawer.

1478. *Oxford.* Theodericus Rood.

1480. *St. Albans. Laurentii Guillielmi de Saona Rhetorica Nova;* printer's name not known.

1482. *Vienna.* John Winterburg.

1483. *Stockholm.* Johannes Snell.

1483. *Harlem. Formulæ Novitiorum,* by Johannes Andriesson. This is the earliest book printed at Harlem *with a date.* In giving this as the first work *known* to be printed at Harlem, the claims of Koster, his grandsons and successors, must, of course, be reserved.

1493. *Copenhagen.* Gothofridus de Ghemen.
1500. *Cracow.* Joannes Haller.
1500. *Munich.* Joannes Schobzer.
1500. *Amsterdam.* D. Pietersoen.
1507. *Edinburgh.* A Latin Breviary; no printer's name. From a patent of James IV. it appears that the first printing-press was established at Edinburgh in 1507. From the style and types, it is probable that they were imported from France.
1551. *Dublin.* Ireland was apparently the last country in Europe into which printing was introduced. The first book printed is a black-letter edition of the Book of Common Prayer, printed by Humphrey Powell.
1569. *Mexico.* Antonio Spinoza, *Vocabulario en Lengua Castellana y Mexicana.*
1639. *United States*, at the town of Cambridge, in the State of Massachusetts. Printer, Stephen Daye.*

---

*The first printing-press "worked" in the American Colonies was "set up" at Cambridge, Massachusetts, in 1639. Rev. Jesse Glover procured this press by "contributions of friends of learning and religion" in Amsterdam and in England, but died on his passage to the new world. Stephen Daye was the first printer. In honor of his pioneer position, Government gave him a grant of three hundred acres of land.

Pennsylvania was the second colony to encourage printing. William Bradford came to Pennsylvania with William Penn, in 1686, and established a printing-press in Philadelphia. In 1692, Mr. Bradford was induced to establish a printing-press in New York. He received 40*l.* per annum and the privilege of printing on his own account. Previous to this time, there had been no printing done in the Province of New York. His first issue in New York was a proclamation bearing the date of 1692.

It was nearly a century after a printing-press had been set up in New England before one would be tolerated in Virginia.

The southern colonists had no printing done among them till 1727. There was a printing-press

    At New London in Connecticut, in - - 1709.
    " Annapolis in Maryland, - - - - 1726.
    " Williamsburg in Virginia, - - - 1729.
    " Charleston in South Carolina, - - - 1730.
    " Newport in Rhode Island, - - - 1732.
    " Woodbridge in New Jersey, - - - 1752.
    " Newbern in North Carolina, - - - 1755.
    " Portsmouth in New Hampshire, - - - 1756.
    " Savannah in Georgia, - - - - 1662.

The first printing-press established in the North-West Territory was worked by William Maxwell, at Cincinnati, in 1793. The first printing

It was the custom of the early printers to distinguish their books by the most fantastic devices; and by these their works may be readily recognized. Many of them were of exceeding beauty, and all the skill and appliances of their art were employed to render them striking; they are really an ornament to their works. The invention of these has been ascribed to Aldus; but the very first printers, Fust and Schœffer, used each for himself, yet conjoined, devices of rare excellence.

Our chronological arrangement has prevented us from mentioning some of the most skillful typographers. Their works, however, are so numerous, and their efforts so well known, as to render it unnecessary to do more than mention their names. Such men as the Aldi,* Frobenius, Plantinus, Operinus, the Stephani, the Elzeviri, the Gryphii, the Giunti, the Moreti, and hosts of peers, have universal fame. The printing-office of Plantinus, in the Place Vendredi, at Antwerp, exists in its full integrity, and in the possession and use of his descendants the Moreti; the same presses, the same types, with the addition of every improvement modern skill has effected, are still in use, and an inspection of these singular relics of olden art will well repay the investigation of the curious.

THE FIRST PRESSES.

Of the mechanical means by which these beautiful impressions of the old printers were produced there is little or no record; but it is quite evident that they must have been effected by some more skillful process than mere manipulation, that is, than the appliance of a burnisher, as is evident

---

executed west of the Mississippi was done at St. Louis, in 1808, by Jacob Hinkle.

There had been a printing-press in Kentucky in 1786, and there was one in Tennessee in 1793—in Michigan in 1809—in Mississippi in 1810. Louisiana had a press immediately after her possession by the United States.

Printing was done in Canada before the separation of the American Colonies from the Mother Country. Halifax had a press in 1751, and Quebec boasted of a printing-office in 1764.—*Sketch of the Origin and Progress of Printing, by William T. Coggeshall*—NEWSPAPER RECORD. *Lay & Brother, Philadelphia,* 1857.

*It should be mentioned that Aldus Manutius invented the beautiful character of type called *Italic* at the end of the fifteenth century. The first book printed with it is a Virgil, 1501.

in the first wood-cuts, or of a roller, or superficial pressure applied immediately by hand. It is very probable that one of the difficulties which Gutenberg found insuperable at Strasburg, was the construction of a machine of sufficient power to take impressions of the types or blocks then employed; nor is it at all wonderful that the many years he resided at that city were insufficient to produce the requisite means; for, with cutting type, forming his *screws*, inventing and making ink, and the means of applying his ink when made, his time must have been amply occupied. Moreover, the construction of a press would require a versatile genius, and excellent mechanical skill, not to be looked for in one man. But upon his junction with Fust and Schœffer, the gold of the former, and the invention of all the three, would soon supply the defect; and, for aught that appears to the contrary, the press used in their office differed in no essential point from those in use until the improvements of Blaew in 1600–20. Fortunately, amongst the singular devices with which it pleased the earlier printers to distinguish their works, Badius Ascensius of Lyons (1496–1535) chose the press; and there are cuts of various sizes on the title-pages of his works. It appears from these, that, like that of Gutenberg, they could print only four pages at a time, and that at two pulls; the table and tympan ran in, and the platen was brought down by a powerful screw, by means of a lever inserted into the spindle.

The color which the earliest typographers used was probably made according to the style of work in hand. The earliest copies of the *Speculum* and *Biblia Pauperum* were printed in a brown color, of which raw umber is the principal ingredient. It appears to have been well ground and thin. It was, most likely, of the same tint as the old drawings of the same subjects, and would be better adapted for the filling up in various colors, as appears to have been the practice, than a black and harsh outline of ink. Fust and Schœffer, however, introduced, and their followers adopted, black ink, and were so skillful in compounding it that their works present a depth and richness of color which excites the envy of the moderns; nor has it turned brown, or rendered the surrounding paper in the slightest degree dingy. The method of applying it to the types was by means of balls of skin stuffed with wool, in every respect the same as those used fifty years ago. The ink was laid in some thickness on a corner of a stone slab, and taken thence in small quantities and ground by a muller, and thence again taken by the balls and applied to the types. The types appear to have been disposed in cases very much the same as ours. The composing-stick differs somewhat, but cannot now be very clearly made out. The different operations of casting the type, composing, reading, and working, are mostly represented in the same apartment; but, it is probable, more for the sake of pictorial unity, than because such was really the custom. There must have been many workmen engaged in most of the old establishments; and they well knew the value of cleanliness, which is unattainable where all the operations are carried on together.

The general and original belief is that William Caxton, who for thirty years resided in the Low Countries, under the reign of Charles the Bold, and who had taken every opportunity of learning the new art, and had availed himself of the capture of Mentz to secure one of the fugitive workmen of Fust and Schœffer, established a printing-office at Cologne, where he printed the French original and his own translation of the *Recuyell of the Histôryes of Troye;* that whilst at Cologne he became acquainted with Wynkyn de Worde, Theoderick Rood, both foreigners, and Thomas Hunte his countryman, who all subsequently became printers in England; that he afterward transferred his materials to England; that Wynkyn de Worde came over with him, and probably was the superintendent of his printing establishment; that his first press was established at Westminster, perhaps in one of the chapels attached to the abbey, and certainly under the protection of the abbot; and that he there produced the first book printed in England, the *Game of Chess,* which was completed on the last day of March, 1474.

The correctness of these facts is not matter of dispute, all writers agreeing that Caxton did so set up his press at Westminster, and print his *Game of Chess* in 1474; but it has been asserted that Caxton was not the first printer, nor his book the first book printed, in this country. Neither does the controversy rest upon the contradictory statements of many writers, for all authors of the same and succeeding period agree in ascribing the honor to Caxton; and when, in 1642, a dispute arose between the Stationers' Company and certain persons who printed by virtue of a patent from the crown, concerning the validity of this patent, a committee was appointed, who heard evidence for and against the petitioners, and throughout the proceedings Caxton was acknowledged as incontestibly the first printer in England. Thus Caxton seemed to be established as the first English typographer, when, soon after the Restoration, a quarto volume of forty-one leaves was discovered in the library at Cambridge, bearing the title of *Exposicio Sancti Jeronymi in Symbolum Apostolorum ad Papam Laurentium*, and at the end, " Explicit Exposicio Sancti Jeronymi in Simbolo Apostolorum ad papam Laurentium, Oxonie Et finita, Anno Domini M.CCCC.LXVIII. *xvii.* die decembris." Upon the production of this book the claim for priority of printing was set up for Oxford. In the year 1644 Richard Atkyns, who then enjoyed a patent from the crown, and whose claims consequently brought him into collision with the Stationers' Company, and who was desirous of establishing the prerogative of the sovereign, published a thin quarto work, entitled *The Original and Growth of Printing, collected out of the History and the Records of the Kingdome; wherein is also demonstrated that Printing appertaineth to the Prerogative Royal, and is a Flower of the Crown of England.* The book was published "*by order and appointment of the Right Hon. Mr. Secretary Morrice.*" In support of this proposition Atkyns asserted that he had received from an anonymous friend a copy of a manuscript discovered at Lambeth Palace, amongst the archiepiscopal archives. The following is an abstract of this document: "Thomas Bouchier, archbishop of Canterbury, earnestly moved the king, Henry VI., to use all possible means to procure a *printing* mould, to which the king willingly assented, and appropriated to the undertaking the sum of 1500 merks, of which sum Bouchier contributed 300. Mr. Turnour, the king's master of the robes, was the person selected to manage the business; and he, taking with him Mr. William Caxton, proceeded to Harlem in Holland, where John Guthenberg had recently invented the art, and was himself personally at work; their design being to give a considerable sum to any person who should draw away one of Guthenberg's workmen. With some difficulty they succeeded in purloining one of the under-workmen, Frederick Corsellis; and it not being prudent to set him to work in London, he was sent under a guard to Oxford, and there closely watched until he had made good his promise of teaching the secrets of the art. Printing was therefore practiced in England before France, Italy, or Germany, which claims priority of Harlem itself, though it is known to be otherwise, that city gaining the art from the brother of one of the workmen, who had learned it at home of his brother, and afterward set up for himself at Mentz." The *Exposicio* is asserted by inference to be the work of Corsellis. That this document is a forgery may be safely assumed; because of the more than unsatisfactory manner in which it is said to have been obtained; because no one ever saw this copy; because no one, except the unknown, ever saw the original, for it is not amongst the archives nor in the library of Lambeth Palace, nor was it when the Earl of Pembroke made diligent search for it in 17—, nor was it found when the manuscripts, books, and muniments were moved into a new building; because Caxton himself, who took so important a share in the alleged abduction of the workman, states that twelve years afterward he was diligently engaged in learning the art at Strasburg, and repeatedly ascribes the invention to Gutenberg, "at Mogunce in Almayne;" because, when three years afterward the Stationers' Company instituted legal proceedings against the University of Cambridge, to restrain them from printing, this document was rejected, as resting only on Atkyns's authority; because Archbishop Parker, in his account of Bourchier, mentions the invention of printing at Mentz, but makes no claim for his having introduced it into England; and Godwin, *de Præsulibus Angeliæ*, says that Bourchier, during his primacy of thirty-two years, did nothing remarkable, save giving 120*l*. for poor scholars, and some books to the university, and that he minutely examined two registers of his proceedings during this term, without making any mention of his having found therein any record of so remarkable a transaction; because, since these transactions must have taken place before 1459, Henry VI. was at that time struggling fearfully for his throne and life, Edward IV. being crowned in that year; from internal evidence of the document itself, for, not to mention the weak evidence for the city of Harlem, it is quite certain that Gutenberg never printed there, and by Junius the theft is ascribed to John Fust, who certainly was a rich goldsmith of Mentz; whereupon Meerman, finding these statements at variance with possibility, boldly invents another theory, making the sufferers Koster's grandsons, who never printed, as far as is known, and the robber Corsellis himself; and, lastly, because six years elapsed between this asserted introduction and the publication of his *Exposicio*, and eleven years between this and any other publication from any Oxford press. Although these facts entirely confute the pretensions of Corsellis, there nevertheless remains the book itself, and unless some evidence can be produced, Oxford will still maintain the distinction of having printed the earliest book in England. Some of the most learned bibliographers entirely refuse their assent to the genuineness of the book. Middleton asserts that there must be an error of an x in the imprint, and produces many remarkable instances of similar typographical errors. This, however, is mere assertion; and, as in the Lambeth record, the best evidence is to be sought in the production itself; accordingly the work is printed with *cast* metal types, which are not proved to have been used by Koster at all, that art being invented by Gutenberg, Fust, and Schœffer at Mayence. The letter is of very elegant cut, the pages regular, and the whole work has the appearance of having been executed at a considerably advanced era of the art. Another and a good argument is, that the work has signatures, or marks for the binder, at the foot of the page, which were not used on the Continent before 1472, by John Koelhoff at Cologne. The evidence in favor of Caxton is direct and strong; the date of the Oxford book is contradicted by internal evidence, and discredited by the story set up in its support; there seems, therefore, no sufficient ground for withdrawing from Caxton the fame of being the introducer of printing into England.

WILLIAM CAXTON was born about the year 1412, in the Weald of Kent. His father was a wealthy merchant, trading in wool. He was brought up to the business of a mer-cer, and conducted himself so much to his master's satisfaction, that on his death he bequeathed him the then considerable sum of twenty marks. Caxton then proceeded, probably as the agent of the Mercers' Company, into the Low Countries. He must have been a man of some wealth and consideration, for in 1464 he and Richard Wethenhall were appointed by Edward IV. "embassadors and special deputies" to continue and confirm a treaty of commerce between him and Philip, duke of Burgundy; and, upon the marriage of Edward's sister Margaret with Charles duke of Burgundy, he was appointed to the household retinue of the princess, by whom he appears to have been treated with much familiarity and confidence; for at her instigation he first commenced his literary labors, and he mentions her as repeatedly commanding him to amend his English. His first work was a translation of the *Recuyell of the Historyes of Troye*, which he afterward printed at Strasburg, when his leisure had allowed him to turn his attention to the study of printing. The first production of his press is allowed to be the French *Recuyell* above mentioned, his second the *Oracion of John Russell on Charles Duke of Burgundy being created a Knight of the Garter*, which took place in 1469. Of his transactions between 1471 and 1474 there is no record; probably he was engaged in the diligent pursuit of the art, and preparing to transfer his materials to England, which he accomplished some time before 1477, when we find him printing in or near the Abbey of Westminster, of which Thomas Milling, bishop of Hereford, was at that time abbot. The first production of his English press was the *Game of Chess*, bearing date 1474, which work, however, some assert to have been printed by him at Cologne. His next production was the *Boke of the hoole lyf of Jason;* but his first book bearing date and place in the colophon is the *Dictes and Sayings of Philosophres*, a translation from the French by the gallant Earl Rivers, "at Westmestre, the yere of our lord M. CCCC. lxxvij." From this time he continued both to print and translate with great spirit. His "capital work" was a *Book of the noble Historyes of Kyng Arthur*, in 1485, the most beautiful production of his press.

There is but one copy of any of Caxton's works printed upon vellum; it is the *Doctrinal of Sapyence*. "Translated out of Frensshe in to Englysshe by wyllyam Caxton at Westmestre. Fynyshed the vij day of May the yere of our lord M.CCCC.lxxix. Caxton me fieri fecit." This unique copy is in the library at Windsor, and it is in beautiful preservation. It is moreover doubly unique, for it contains an additional chapter, to be found in no other copy whatever, and which is entitled "Of the negligencies happening in the Masse and of the Remedies. Cap. lxiiij." It is a curious treatise of minute omissions and commissions likely to occur in the service of mass, with directions how to remedy such evils. Of their importance here are two specimens, "If by any negligence fyl (fall) any of the blood of the Sacrament on the *corporus*, or upon any of the vestments, then ought to cut off the piece on which it is fallen, and ought well to be washen, and that piece to be kept with the other relics." "And if the body of Jesu Christ, or any piece, fall upon the palé of the altar, or upon any of the vestments that ben blessed, the piece ought not to be cut off on which it is fallen, but it ought right well to be washen, and the washing to be given to the ministers for to drink, or else drink it himself." This singular treatise finishes with this grave confession, "This chapitre to fore I durst not sett in the booke, by cause it is not convenient ne appertaining that every lay man should know it et cetera."

The Royal Library possesses another work of Caxton, which, as a *perfect* copy, is also unique. This is the "Subtyl Historyes and Fables of Esope. Translated out of Frenshe in to Englyshe by Wyllyam Caxton at Westmynstre In the yere of our lord M CCCC lxxxiij Emprynted by the same the xxvj daye of Marche the yere of our lorde M CCCC lxxxiiij And the fyrste yere of the regne of kyng Rycharde the thyrde." It consists of 142 leaves. Each fable is illustrated by a rude wood-cut, all of which are said to have been executed abroad, where similar editions of Æsop were frequently printed. They are, however, most probably copied; for there is nothing either in their design or execution that a most moderate artist might not perform; and this will equally apply to other wood-cuts interspersed in Caxton's works.

It has been said that the works of Caxton have been eagerly sought for by English bibliomaniacs. The most remarkable instances of this are the enormous prices given for some of them at the sale of the Duke of Roxburghe's library before mentioned. The *Chastysing of God's Children* was knocked down to Earl Spencer for 146*l*. The *Sessions Papers* were bought for the Society of Lincoln's Inn for 378*l*. The Duke of Devonshire gave 351*l*. 15*s*. for *The Mirrour of the World*, and 180*l*. for the *Kalendayr of the Shyppers*. Gower's *Confessio Amantis* produced 366*l*.; *The Boke of Chyvalry*, 336*l*. The *Recuyell of the Historyes of Troye* gave rise to a startling contest. It was the identical copy presented by Caxton to Elizabeth Grey, queen of Edward IV. and sister of his patroness. " Sir Mark Sykes vigorously pushed on his courser till five hundred guineas were bidden; he then reined in the animal, and turned him gently on one side 'toward the green sward.' More hundreds are offered for the beautiful Elizabeth Grey's own copy. The hammer vibrates at nine hundred guineas. The sword of the marquess is in motion, and he makes another thrust—' One thousand pounds.' ' Let them be guineas,' said Mr. Ridgway, and guineas they were. The marquess now recedes. He is determined upon a retreat; another such victory as the one he has just gained (the Valdarfar Boccaccio) must be destruction; and Mr. Ridgway bears aloft the beauteous prize in question." (Dibdin.) At Mr. Willett's sale *Tullius of Old Age* produced 210*l*., and became the property of the Duke of Devonshire.

Caxton must have been a man of wonderful perseverance and erudition, cultivated and enlarged by an extensive knowledge of books and the world. Of his industry and devotedness some idea may be formed, when Wynkyn de Worde, his successor, states, in his colophon to the *Vitæ Patrum*, that Caxton finished his translation of that work from French into English *on the last day of his life*. He died in 1491, being about fourscore years of age. His epitaph has been thus written by some friend unknown: " Of your charite pray for the soul of Mayster Willyam Caxton, that in hys tyme was a man of moche ornate and moche renommed wysdome and connynge, and decesed full crystenly the yere of our Lord MCCCCLXXXXI.

    Moder of Merci shyld him from thorribul fynd,
    And bryng hym to lyff eternal that neuer hath ynd."

The type used by Caxton is in design very inferior to that used upon the Continent even earlier than his period; but in the latter part of his life he very materially improved his fonts, and some of his later productions are very elegantly cut. The design is peculiar to him, and is said to be in imitation of his own handwriting; it bears, however, some resemblance to the types of Ulric Zell, from whom Caxton derived most of his instruction, and is something between *Secretary* and *Gothic*. He appears to have had two fonts of *English*, three fonts of *Great Primer*, one *Double Pica*, and one *Long Primer*.* He used very few ornamented initial letters, and those he did employ are very inferior in elegance to those of foreign printers. He preferred inserting a small capital letter within a large space, and leaving the interval to be filled up according to the taste of the illuminator, owing to which many excellent performances are destitute of these beautiful ornaments. Caxton's ink was not remarkable for depth of color or richness; his paper was excellent; and he probably used presses of the same construction as the continental printers. His works are not very rare, but are highly prized by English collectors. Copies of one or more of his works are to be found in most collections of any pretension, and are well worthy of inspection. The number of his productions is sixty-two. Although Caxton was the first English printer, he was not the only one of his day, Wynkyn de Worde, Lettou and Machlinia, Hunte, Pynson, the Oxford printer whoever he may have been, and he of St. Alban's, being his cotemporaries.

WYNKYN DE WORDE came, as we have already seen, from Germany with Caxton, and remained with him in the superintendence of his office until the day of his death, when he succeeded to the business. He was a native of Lorraine, and evidently a man of considerable information and taste, and of great spirit in the conduct of his affairs. After his succession to Caxton's business, he carried on in the same premises for about six years, when he removed to the "Sygn of the Sonne in flete strete, against the condyth." De Worde appears to have immediately commenced a complete renovation of the art, cutting many new fonts of all sizes,

---

* These are terms by which modern printers distinguish the sizes of their type.

with vast improvement of the design and proportion; he moreover provided his cotemporaries, then becoming very numerous, with type; and it is even said that some of the letter used by the English printers less than a century ago are from his matrices, nay, that the punches are still in existence. He was the first (or Pynson) to introduce Roman letters into England, which he made use of amongst his Gothic to distinguish any thing remarkable, in the same manner as Italic is used in the present day. His works amount to the extraordinary number of four hundred and eight. "His books are, in general, distinguished by neatness and elegance, and are always free from professed immorality. The printer has liberally availed himself of such aid as could be procured from the sister art of engraving; although it must be confessed that by far the greater, if not the whole, number of wood-engravings at this period are of foreign execution; nor is it without a smile that the typographical antiquary discovers the same cut introduced into works of a directly opposite nature."

In his *Instruction for Pilgrims to the Holy Land*, printed in 1523, the text of which is in Roman, and the marginal notes in Italics, he makes the first use in England of Greek, which is in movable type, of Arabic and Hebrew, which are cut in wood; and the author complains that he is obliged to omit a third part, because the printer had no Hebrew types. Appended to the work are three Latin epistles, in which he makes use of Arabic.

His works are, of course, not so rare as those of his predecessor, but are nevertheless much sought after; and, when sold by the side of the Caxtons at the Duke of Roxburghe's sale, produced large prices. *Bartholomæus de Proprietatibus Rerum*, the first book printed on paper made in England, was bought by the Duke of Devonshire for 70*l*. 7*s*. Chaucer's *Troylus and Cresseide*, 43*l*.; Hawys's *Exemple of Vertu*, 60*l*.; *Passetyme of Pleasure*, 81*l*.; *Castell of Pleasure*, 61*l*.; *The Moste Pyteful Hystorye of the Noble Appolyon, Kynge of Thyre*, 110*l*.

De Worde died about the year 1534. In his will, still in the Prerogative Office, dated 5th June, 1534, he bequeaths many legacies of books to his friends and servants, with minute directions for payment of small creditors and forgiveness of debtors, betokening a conscientious and kindly disposition. His device is generally that of Caxton, with his own name added to the bottom; but he also used a much more complicated one, consisting of fleurs-de-lis, lions passant, portcullis, harts, roses, and other emblazonments of the later Plantagenets and the Tudors.

JOHN LETTOU and WILLIAM MACHLINIA printed separately and jointly before the death of Caxton, but were very inferior to him in every respect; their type being most especially barbarous. Their works are not very numerous, and are principally upon legal subjects; they printed the first edition of Lyttleton's Tenures.

RICHARD PYNSON was a Norman by birth, and studied the art of printing under his "worshipful master William Caxton." It would seem that he was an earlier printer than Wynkyn de Worde, having established an office before the death of Caxton. His first work is of date 1493, and was printed "*at the Temple-bar of London.*" He enjoyed high patronage, and was appointed by Henry VII. to be his printer before 1503. He is perhaps inferior to De Worde as a typographer, his first types being extremely rude. He afterward used a font of De Worde's, and another peculiar to himself in this country, probably imported from France. Some of his larger works, Fabian's Chronicle, Lord Berner's translation of Froissart (which are the first editions of these important additions to English literature), and some of his law-works, are very fine specimens of the art. His device was a curious compound of R and P, on a shield which is sometimes supported by two naked figures.

RICHARD GRAFTON claims especial notice. He was by trade a grocer, although of good family. Of his education nothing appears; but he was one of the most voluminous authors of his time, having, by his own account, written a considerable portion of Hall's Chronicles, an Abridgment of the Chronicles of England, and a Manual of the same, a Chronicle at Large, and other books of historical character, under what circumstances is not known. In 1537 Grafton published Thomas Mathew's translation of the Bible, which was printed abroad, but where is not satisfactorily ascertained; and in 1588 the Testament translated by Miles Coverdale, which was printed at Paris by Francis Regnault. At this time it would not appear that English printers were in high estimation; for Lord Cromwell, desirous of having the Bible in the English language, thought it necessary to procure from Henry VIII. letters to the king of France for license to print it at Paris, and urged Bonner to tender his earnest assistance. Bonner entered upon the undertaking with such zeal, that in recompense he was soon afterward appointed to the bishopric of Hereford. Miles Coverdale had charge of the correctness (see his letter, Gent.'s Mag. 1791), and Richard Grafton and Edward Whitchurch were the *proprietors;* but under what arrangement does not appear. When the work was on the point of completion, the Inquisitors of the Faith interfered, seized the sheets, and Grafton, Whitchurch, and Coverdale were compelled to make precipitate flight. The avarice of the lieutenant-criminal induced him to sell the sheets for waste paper instead of destroying them, and they were in part repurchased. Under the protection of Cromwell they next, after many difficulties, obtained their types and other materials from Paris, and the Bible was completed at London in 1539. "Thus they became printers themselves, which before this affair they never intended." The edition consisted of 2500 copies. Cromwell next procured for them a privilege (not an exclusive one, however) for printing the Scriptures for five years. Very shortly after the death of Lord Cromwell, Grafton was imprisoned for printing Mathew's Bible and the Great Bible, his former friend Bonner much exaggerating the case against him. The prosecution, however, was not followed up; but in a short time he was, with Whitchurch, appointed printer to Prince Edward, with special patents for printing all church-service books and primers. The document is curious. It recites that such "bookes had been prynted by strangiers in other and strange countreys, partly to the great losse and hynderance of our subjects, who both have the sufficient arte, feate and treade of prynting, and partly to the setting forthe the bysshopp of Rome's usurped auctoritie, and keping the same in contynuall memorye;" and that, therefore, of his "grace especiall, he had granted and geven the privilege to our wel-biloved subjects Richard Grafton and Edward Whitchurch, citezeins of London," exclusive liberty to print all such books for seven years, upon pain of forfeiture of all such books printed elsewhere.

One Richard Grafton, supposed to be the above, was member of parliament for the city of London in 1553–54, and also in 1556–57, and in 1562 was member for Coventry. He is supposed to have died about 1572, and not in very affluent circumstances. He used a punning, or, as the heralds would call it, a canting device, of a young tree or *graft* growing out of a *tun*. His works are distinguished for their beauty, and are very numerous and costly. He was one of the most careful and meritorious of English printers.

These are the titles of a few of his early Bibles, etc.

The Byble, 1537, folio. "The Byble, which is all the holy Scripture: In whych are contayned the Olde and Newe Testament truly and purely translated into Englysh by Thomas Mathew. Esaye 1 ☞ Hearcken to ye heauens, and thou earth geaue eare: For the Lorde speaketh. M.D.XXXVII." The title of the New Testament is, "The newe Testament of our sauyor Jesu Christ, newly and dylygently translated into Englyshe, with Annotacions in the Mergent to help the Reader to the vnderstandyng of the Texte." This was printed in France.

The New Testament, Latin and English. 1538. Octavo. "The new testament both in Latin and English after the vulgare texte; which is red in the churche. Translated and corrected by Myles Couerdale: and prynted in Paris, by Fraunces Regnault. M. ccccc. xxxviii in Nouembre. Prynted for Richard Grafton and Edward Whitchurch, cytezens of London. Cum gratia & priuilegio regis."

The Byble in Englysshe. 1539. Folio. "The Byble in Englyshe, that is to saye the content of all the holy Scrypture, bothe of yᵉ olde, and newe testament, truly translated after the veryte of the Hebrue and Greke textes, by yᵉ dylygent studye of dyuerse excellent learned men, expert in the forsayde tongues. *Prynted by Rychard Grafton, and Edward Whitchurche. Cum priuilegio—solum.* 1,539." This is a very superb book, and is the one which was commenced at Paris and finished at London under the circumstances before related.

NEWE TESTAMENT IN ENGLYSSHE. 1540. Quarto.

"Translated after the texte of Master Erasmus of Roterodame."

THE PRYMER. English and Latin. 1540. Octavo.

THE BYBLE IN ENGLYSHE. 1540. Folio. A noble volume, called, from the preface, Cranmer's Byble.

THE BYBLE IN ENGLYSHE. 1541. Folio. "The Byble in Englyshe of the largest and greatest volume, auctorised and appoynted by the commaundement of oure moost redoubted prynce and soueraygne Lorde, Kynge Henrye the VIII, supreme head of this his churche and realme of Englande: to be frequented and vsed in euery Churche within this his sayd realme, accordyng to the tenoure of hys former Jniunctions geuen in that behalfe. Ouersene and perused at the comaundement of the kynges hyghnes, by the ryght reuerend fathers in God Cuthbert byshop of Duresme, and Nicholas, bisshop of Rochester." The lines of the title are printed alternately red and black.

Such, with many other manuals, primers, etc., were the productions of this most eminent British typographer.

The first complete edition of Shakspeare's plays was printed by ISAAC JAGGARD and EDWARD BLOUNT, in folio, in 1623. Of his single plays, the earliest is "The first part of the Contention betwixt the two famous Houses of Yorke and Lancaster," which was printed by "THOMAS CREED for Thomas Millington, and are to be sold at his shop, under Saint Peter's Church, Cornwall" (Cornhill), in 1594. These plays were printed by various typographers, amongst whom appear the names of George Eld, Valentine Simmes, R. Young, John Robson, and others who only give their initials.

The first edition of Milton's Paradise Lost was printed in quarto by PETER PARKER in the year 1667; the Paradise Regained in 1671.

During the troublesome times that preceded the great rebellion, the Puritans, jealously watched and persecuted, introduced the anomaly of ambulatory presses, which were constantly removed from town to town to escape the vigilance of the Star Chamber. At these presses many of Milton's controversial pamphlets were printed; and it is even said that the identical press at which the *Areopagitica* was printed is still in existence, and was lately in the possession of Mr. Valpy, the well-known printer of the Variorum Classics.

It is a very pleasing reflection, that the earlier practitioners of the art did, by their uniform good character and religious turn, tend much to render their profession productive of a highly moral class of literature, and to raise it in the estimation of all men. Had they been less respectable, had they turned their attention to the many ribald and tasteless writings of those times, the effect of the new art would have been to degrade literature and lower morals, to delay the spread of knowledge, and to give a depression to the character of the art and its practitioners, from which possibly they might never have recovered. These excellent and learned men appear to have received their temporal reward, in public estimation, sufficient wealth, and a length of years beyond the ordinary term of mortality.

Setting aside the claim of Corsellis, printing was first practiced at OXFORD by Theoderic Rood and Thomas Hunte from 1480 to 1485. In Rymer, vol. xv. is a grant by Queen Elizabeth to Thomas Cooper, clerk of Oxford, for the exclusive printing of his Latin Dictionary. In 1585 a printing-press was established at the expense of the Earl of Leicester, chancellor of the university. Joseph Barnes was appointed printer to the university in 1585.

At CAMBRIDGE John Siberch printed in 1521, when Erasmus resided there, and probably executed some of his books. Thomas Thomas, M.A., was the printer to the university in 1584.

At ST. ALBAN'S printing was very early practiced, certainly in the year 1480. It would appear that the printer was a schoolmaster. It has been asserted, but without shadow of argument, that printing was introduced here many years before Caxton.

Printing was not introduced into Scotland till thirty years after Caxton had set up his press at Westminster. Under the patronage of James IV., who was a zealous encourager of learning and the useful arts, WALTER CHEPMAN and ANDRO MYLLAR established the first printing-press at Edinburgh, as appears by a royal privilege granted to them in 1507.*

---

* "James, &c. To al and sindrj our officiaris legis and subdittis quham it efferis, quhais knawlage thir oun lettres salcum greting; Wit ye that

The only publications known to have issued from the press of Myllar and Chepman are a collection of pamphlets, chiefly metrical Romances and ballads, in 1508, of which an imperfect copy is preserved in the Advocates' Library ;* and the Scottish Service Book, including the Legends of the Scottish Saints, commonly called the Breviary of Aberdeen, in 1509.†

It is difficult to account for the discontinuance of printing in Scotland for about twenty years after this time; probably the disastrous events at the close of the reign of James IV. may have contributed to render it an unprofitable trade; but in its revival by DAVIDSON there was no deterioration, either in the magnitude and importance of the works attempted, or in the mode in which the mechanical part was executed. It was probably about the year 1536 that he printed, in a black-letter folio, " The History and Croniklis of Scotland, compilet and newly correkit be the Reuerend and Noble Clerke Maister Hector Boece. Translatit laitly be Maister Johne Bellenden. Imprentit in Edinburgh be Thomas Davidson, dwelling fornent the Frere Wynd ;" and in 1540 he printed the whole works of Sir David Lindsay.

Davidson was succeeded by Lekprevik, Vautrollier, and others; but none were distinguished as printers till the time of Ruddiman.

forsamekill as our lovittis servitouris Walter Chepman and Andro Millar, burgessis of our burgh of Edinburgh, has at our instance and request, for our plesour, the honour and proffit of uur Realme and lieges, takin on thame to furnis and bring hame ane prent, with all stuff belangand tharto, and expert men to use the samyne, for imprenting within our Realme of the bukis of our Lawis, actis of parliament, croniclis, mess bukis, and portuus efter the use of our Realme, with addicions and legendis of Scottis sanctis, now gaderit to be ekit tharto, and al utheris bukis that salbe sene necessar, and to sel the sammyn for competent pricis, be our avis and discrecioun thair labouris and expens being considerit," etc.

" Geven under our priv Sel at Edinburgh the xv day of September, and of our Regne the xx^d yer."

* These pamphlets were reprinted in a handsome quarto volume, edited by Mr. David Laing. The preface contains much accurate information regarding early printing in Scotland.

† Of this Service Book, which forms two volumes octavo, handsomely printed with red and black letter, in the years 1509 and 1510, a beautiful copy is preserved in the University Library of Edinburgh. As the name and device of Walter Chepman occur in the work, without any mention being made of his partner, we are led to the conclusion that Andro Myllar, if then alive, had relinquished his share in the concern.

A mere catalogue of printers would afford little amusement and less instruction; especially since the productions of the English press, save in the works of the printers above named, not only exhibited no advance, but even much deterioration, in most requisites of good printing. Indeed, to so low a point had the art fallen, and so little spirit was exhibited by English typographers, that the regeneration was left to an alien, whose perception of the inferiority and capacity of improvement at once raised the art to the level of the finest productions of Bodoni and Barbou.

This was JOHN BASKERVILLE, a japanner of Birmingham, who, having realized a considerable fortune, turned his attention to cutting punches for type, and succeeded in producing a series of fonts of remarkable beauty, so excellently proportioned, and standing so well, that the best of modern type-founders (and this seems the Augustan age of type-founding) have done no more than vary the proportions and refine the more delicate lines and strokes. Added to this, his press-work is of most excellent quality; his paper the choicest that could be procured; and his ink has a richness of tone, the mode of producing which has died with him. The works of Baskerville are amongst the choicest that can adorn a library. He died in 1775. His types and punches were purchased to print the splendid edition of Voltaire's works at Paris. He was worthily succeeded by BULMER, whose magnificent Shakspeare and Milton are amongst the most superb books ever issued from the press, and, with Macklin's Bible and Ritchie's, Bensly's Hume, and other works, may be fearlessly produced to win for this country the palm of fine printing; whilst in Scotland, THOMAS RUDDIMAN and the two FOULIS may challenge the prize of classical typography from Aldus and the Stephani. Indeed, the larger Greek types of the Foulis are without parallel for grandeur, their press-work is beautiful, and their correctness beyond all praise.

Modern printers, with all their faults, are not degenerate successors of these worthies. The works from present offices that make pretensions to fine printing need not be ashamed of comparison with these *chefs-d'œuvres;* whilst, from the vast improvements in the mechanism of the art in all its branches, paper, presses, ink, type, and other adjuncts, the average of

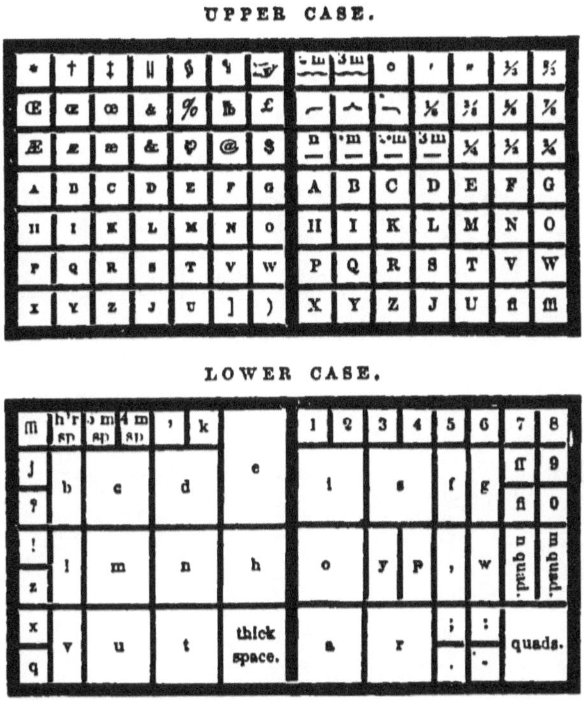

[PRINTING.] [PLATE 1.]

UPPER CASE.

LOWER CASE.

Fig 1.

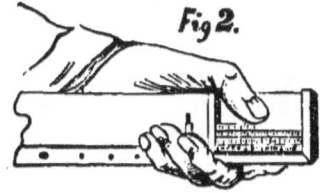

Fig 2.

the printing of the present day is infinitely superior to that of the last century. But in what relates to practical skill, correctness, taste, and diligence, we cannot hope to excel, though we may perhaps equal, these departed masters.

PRACTICAL PRINTING.

The first operation when the new font* has entered the doors of the printing-office, is to *lay* it in the *cases* (fig. 1). These are always in pairs; the *upper case* being divided into equal spaces or *boxes;* the part on the left of the broader division being appropriated to CAPITAL letters, figures, diæresis vowels, particular *sorts*, etc.; that on the right to SMALL CAPITALS, accented letters, and references. The letters and figures are arranged in alphabetical and numerical order, from left to right. The *lower case* is divided into unequal portions, according to the average occurrence of the particular letters; for the compositor (the workman whose duty it is to lay the font, and afterward to place together or *compose* the separate types into words) never looks at the face of the letter he picks up, but unhesitatingly plunges his fingers into any box, being sure that the letter he picks out thence is the one to which that box is appropriated, and consequently the one he requires. As there is no external mark or guide attached to the different boxes to denote the letters they contain, a stranger is not a little surprised and puzzled at the eccentric movements of the workman's hand. Accordingly, it will be observed, upon looking at fig. 1, that the letter e has a box one-half larger than c, d, m, n, h, u, t, i, s, o, a, r; and these are twice the size of b, l, v, k, f, g, y, p, w, or the comma; and four times the size of z, x, j, q, or the [] crotchets, full points, etc. These boxes are not arranged in alphabetical order, but those of most frequent occurrence are placed about the middle of the case to diminish the distance the hands of the compositor have to travel in picking up and receiving the types. There are also other pairs of cases similarly arranged for the *italic* letters. The

---

* A font is any weight of type of the same body and face, consisting of every letter, stop, figure, etc., in certain proportions, as stated on page 66, together with spaces and quadrats.

following are the proportions of some of the letters in a font of pica* of 800 lbs. weight:

Capitals, from 400 to 600 of each, but of J 80, and Q, X and Z, 180.
Small capitals, from 150 to 300 of each, excepting J, Q, X and Z, which, as in the capitals, are reduced in number.

a	8,500	b	1,600	j	400	z	200
e	12,000	c	3,000	k	800	&	200
i	8,000	d	4,000	m	3,000	,	4,500
o	8,000	f	2,500	n	8,000	.	2,000
u	3,400	h	6,400	q	500		

In a whole font there are about 150,000 letters, spaces, and figures.

The compositor, having placed his copy upon a part of the upper case little used, and having received the necessary directions, takes up an instrument called a *composing-stick* (fig. 2), (which, as well as the way of holding it and its use, will be better understood by reference to the drawing than by description), and sliding the inner movable portion wider or closer according to the desired width of the page, he fastens it with a screw; he then cuts a piece of brass rule to fit in easily between the end of the stick and slide, and which is called the *setting-rule*. This rule causes the letters to slip down without any obstruction from the screw-holes of the stick, or the nicks which serve to distinguish one font from another and enable the compositor, by turning them out-outward, to place the letters in their proper position. He then reads the first few words of his copy, takes first a capital letter from the upper case, the succeeding letters from the lower case, and at the conclusion of the word a *space*, which is merely the shank of a letter without any face, and not so high as a letter by about one-fourth part; and therefore, not receiving the ink, forms the blank space between words; but sometimes, through carelessness, it is allowed to stand up, in which case it is a fearful blotch upon a fair page, and must have been observed by most readers. He then proceeds with his next word, which will probably consist of lower case letters only; and so on until he has arrived at the end of his line. It is most likely, however, that the words he has occasion to compose, with the necessary spaces, will not fill up

* This is pica.

the exact width of the line, and that there will be sometimes too much, sometimes too little room, for getting in the whole or part of the next word. In this case he has to consider whether it will be better to crowd the line and get in the word or syllable, or make the line more open and take it over to the next line; his care being that his matter, when composed, shall not look too open or too close. Having decided, he takes out the spaces he has inserted, and puts in their stead others of greater or less width, as the case may require, in such a manner that on the face of the line being touched, it shall not feel loose, or require any particular pressure to force down the last letter into its proper place. This being accomplished in an artist-like manner, he takes out his setting-rule and places it in front of his line, and with a gentle pressure of his thumb forces both back into the composing-stick; he then proceeds in a similar manner with other lines until his stick is full, when, placing it upon a *frame* on which the cases rest, his setting-rule being in front, he lifts his lines out of the stick and places them upon a proper instrument called a *galley*. If, however, the matter is to be *leaded*, that is, if the lines of types are to be more apart than usual, the process is a little different. The compositor then has before him a quantity of pieces of metal called *leads*, of the exact width of the page, only one-fourth, one-sixth, or one-eighth of the body of the type, and not higher than spaces. After composing a line, before moving his setting-rule, he takes one or more of these and places it before the line; he then takes out the setting-rule, and proceeds as above described. Having thus gone on until a considerable quantity of matter is composed, the compositor next makes it up into pages, and then into sheets. First, taking by portions as many lines of his matter as are to be contained in a page, he adds thereto at the bottom a line of quadrats, which are of the same height as spaces but much larger, varying in length from one to four m's, and places at the top the folio of the page and the *running head* or line which indicates the title of the work or the subject of the page or chapter, and then adds such leads or other things as may be necessary; taking care that in the first page he places the *signature* (a letter of the alphabet intended for a guide to the binder, because by keeping this always outside, and the second signature on the next leaf, he cannot fold the sheet wrong). He next ties it tightly round with page-cord, and places it upon a piece of coarse paper. Having made up as many pages as the sheet consists of, viz., four if folio, eight if 4to, sixteen if 8vo, he next lays them down upon the *imposing-table** (a large plate of iron screwed on to a frame) in the necessary order. This is, to a stranger, a very curious arrangement; they appear to him to be placed at random, without any design or fixed rule, and as they are necessarily laid down in two divisions, one for each side of the sheet, one is of consequence the very reverse of the other. He may easily instruct himself, however: for if he take a sheet of paper and fold it into any required size, marking the folios with a pencil, and then open it without cutting, he will find they fall in curious irregularity. The pages are laid down on the table reverse of the order they have on the paper; for it must be remembered that every type and every page is like a seal, the reverse of the impression it leaves; consequently, were the pages laid down as on the marked paper, viz., the first page on the right hand, it would, in type, be at the extreme left, and so on. The schemes (figs. 3 and 4) of the *laying down* or *imposition of forms*, will give some idea of the apparent confusion of this process.

The pages being correctly laid down upon the imposing-table, the compositor removes the papers from under them, and next takes in both hands a *chase* (a frame of iron divided by cross-bars into four compartments, the inner angles of which are made rectangular with much care) and places it over them; and then having ascertained the size of the paper to be used, adjusts pieces of wood or metal, called *furniture*, between them. Within the chase, but next to the pages, he places other pieces of wood or iron called *side* and *foot sticks*, which are rather wider at one end than the other, and between these and the chase small pieces of wood, which decrease in width in the same proportion as the side-stick, and which are called *quoins*. He now takes off the cords from the pages, and, as he removes each cord, he tightens the adjacent

---

* A large slab of marble or stone is used for this purpose; but it is liable to split, and to have its smooth surface indented. A plate of iron turned into a lathe is now very generally substituted in England, but the marble is commonly in use in America.

PRINTING.] [PLATE 2.

PROOF PRESS.

quoins that the letters at the sides of the pages may not slip down. When all the pages are untied, and the quoins pushed up with his finger and thumb, he planes down the pages gently with a *planer* (a piece of beech perfectly plane and smooth on the face, about 9 inches long, $4\frac{1}{4}$ inches wide, and 2 inches thick), to prevent any of the letters from standing up. With a *shooting-stick* (which formidably-named weapon is merely a piece of hard wood,* a foot in length, an inch and a half in width, and half an inch in thickness) and a mallet he forces the quoins toward the thicker ends of the side and foot sticks, which consequently act as gradual and most powerful wedges, forcing the separate pieces of type to become a compact and almost united body, so that, the pages being securely *locked up* and again planed down, the whole mass, consisting of many thousand letters, may be lifted entire from the table. This united mass is called a *form;* that one which contains the first page being called the *outer form*, the other the *inner* (fig. 5).

The compositor is paid by the number of thousands of letters he composes, which is thus ascertained: The letter m, being on a shank which is supposed to have its four sides parallel and equal, is taken as the standard; he ascertains how many m's the page is in length, including the running head and the white line at the bottom; that is, in fact, how many lines of the particular type used there would be in a page of the given size, supposing it were all solid type; next, how many m's (laid on their side) it is in width, that is, how many times the letter m would be repeated in a line of the given length were it to consist of nothing but m's so laid. This latter sum is then doubled, because experience shows that the average width of the letters is one-half of the depth, or one-half of that of the letter m. The length of the page is then multiplied by the product of this doubled width, then by the number of pages in the sheet, and the result will give the average number of letters in the sheet. This will be much better understood by the following *casting-up* of a sheet of 8vo in pica:

* Iron or gun-metal is now generally substituted, as being more durable.

```
              Number of m's long........................    47
                  "    m's wide, 24 × 2................    48
                                                         ─────
                                                          376
                                                          188
                                                         ─────
                                                         2256
              Number of pages in a sheet of 8vo..........   16
                                                         ─────
                                                        13536
                                                         2256
                                                         ─────
                                                        36096
```

The compositor therefore is paid for composing 36,000 letters; for the odd figures are dropped, unless they amount to or exceed 500, when they are paid for as if they completed another 1000. If the sheet be of solid type, of the ordinary size, the price paid in London is sixpence per 1000 letters; if in the small type called minion, sixpence farthing; in nonpareil,* sevenpence; in pearl, eightpence. If the work be composed from print copy, the price is three farthings per 1000 less than it would be paid if the copy were manuscript. If, however, the type be leaded, the price is a farthing per 1000 less for fonts above pearl. If the work is to be stereotyped, and high spaces are used, it is subject to an additional charge of a farthing per 1000; if low spaces, of a half-penny per 1000. Works in foreign languages, in type of the ordinary size and character, are paid one half-penny per 1000 more, and three farthings per 1000 more in the smaller. Greek, with leads and without accents, eightpence three farthings; with accents, tenpence farthing. is eightpence half-penny per 1000; without leads or accents, Hebrew, Arabic, Syriac, etc., are paid double.† The com-

---

* This is nonpareil.

† In 1804, after a protracted litigation before the Court of Session, the journeymen compositors of Edinburgh succeeded in obtaining the sanction of the Court for an advance of one penny per thousand letters, or, upon an average, about one-fourth on the prices of their work. The grounds upon which the Court rested this decision were, that the wages were much too low; that they had remained for forty years unaltered, whilst the price of the necessaries of life had very much increased; and although it was proper to avoid a rise of wages which might lead to idleness, yet it was equally necessary to place the workmen upon a respectable footing, so as to enable them to do their work properly, and also to encourage them in cultivating and acquiring that degree of literature by which the public must infallibly be benefited; and that the fair criterion was, to make the wages of Edinburgh bear the same proportion to those of London which they did in the

positor, it appears, must therefore pick up 72,000 letters before he can receive an ordinary week's wages, must make up his matter into pages and impose them, and, moreover, correct all the blunders mischance or carelessness may have occasioned, with great expenditure of time also in many other particulars; but, as is hereafter described, he must have previously placed every one of these 72,000 into the appropriate boxes whence he has withdrawn them in composition. Now it is usually reckoned that this latter operation, called *distributing*, occupies one-fourth of a compositor's time, and the other operations another fourth; he has therefore only one-half of his time for composition; consequently he must pick up letters at the rate of 144,000 per week, 24,000 per day, or 2000 per hour. His rapidity of motion is therefore wonderful, and the exertion is so long continued, that the business, although apparently a light one, is in fact extremely laborious.

The number of thousands of letters in a sheet necessarily varies with the size of the type, width and length of the page, and the number of the pages. The example above given is the casting-up of an octavo sheet of pica solid, the page being of moderate size; a similar sheet of brevier* would contain 81,000 letters, and the cost of composing it would be 2*l.* 0*s.* 6*d.* Single tables, forming one uninterrupted mass of type, will sometimes contain 250,000 letters; and the labor of the compositor being very great in getting them up, he is paid double. Consequently the cost of composing such a table in pearl or diamond § would be not less than 16*l.* 13*s.* 6*d.*, without extra charges. Yet this large number of types, by the

---

year 1785, before the London prices were raised. That a court of law, whose province it is not to legislate, but to apply and enforce existing statutes, should have entertained a question regarding the price of labor, for the regulation of which there not only existed no law, but which had never been deemed a fit subject for legislative interference, appears to be a very singular incident in the history of judicial procedure. The prices thus fixed, however (namely, 4½*d.* per 1000 for book-work, with an additional halfpenny if nonpareil, and a penny if pearl, and 5½*d.* for law-papers and jobs), being regarded as not unreasonable, have ever since been adhered to by every respectable establishment in Edinburgh. The price for composition in New York and other American cities averages 25 cents per thousand. Compositors at night and on rule and figure work are paid extra.

* This is brevier.
§ This is diamond.

power of the wedge-formed side and foot sticks and quoins, is compressed into so solid a mass that it can be moved without much danger of disruption.

The sheet being now imposed, an impression is taken called a *proof*, which is carried down to the reader, who having folded the proof in the necessary manner, first looks over the signatures, next ascertains whether the sheet commences with the right signature and folio, and then sees that the folios follow in order. He now looks over the running heads, inspects the proof to see that it has been imposed in the proper furniture, that the chapters are numbered rightly, and that the directions given have been correctly attended to, marking whatever he finds wrong. Having carefully done this, he places the proof before him, with the copy at his left hand, and proceeds to read the proof over with the greatest care, referring occasionally to the copy when necessary, correcting the capitals or italics, or any other peculiarities, noting continually whether every portion of the composition has been executed in a workmanlike manner; and having fully satisfied himself upon these and all technical points, he calls his reading boy, who, taking the copy, reads in a clear voice, but with great rapidity and often without the least attention to sound, sense, pauses, or cadences, the precise words of the most crabbed or intricate copy, inserting, without pause or embarrassment, every interlineation, note, or side-note. The gabble of these boys in the reading-room, where there are three or four reading, is most amusing, a stranger hearing the utmost confusion of tongues, unconnected sentences, and most monotonous tones. The readers plodding at their several tasks with the most iron composure, are not in the least disturbed by the Babel around them, but follow carefully every word, marking every error, or pausing to assist in deciphering every unknown or foreign word. This first reading is strictly confined to making the proof an exact copy of the manuscript, and ascertaining the accuracy of the composition; consequently first readers are generally intelligent and well-educated compositors, whose practical knowledge enables them to detect the most trivial technical errors. Having thus a second time perused the proof, and carefully marked upon the copy the commencement, signature, and folio of the succeeding sheet, he sends it by his

[PRINTING.] [PLATE 4.

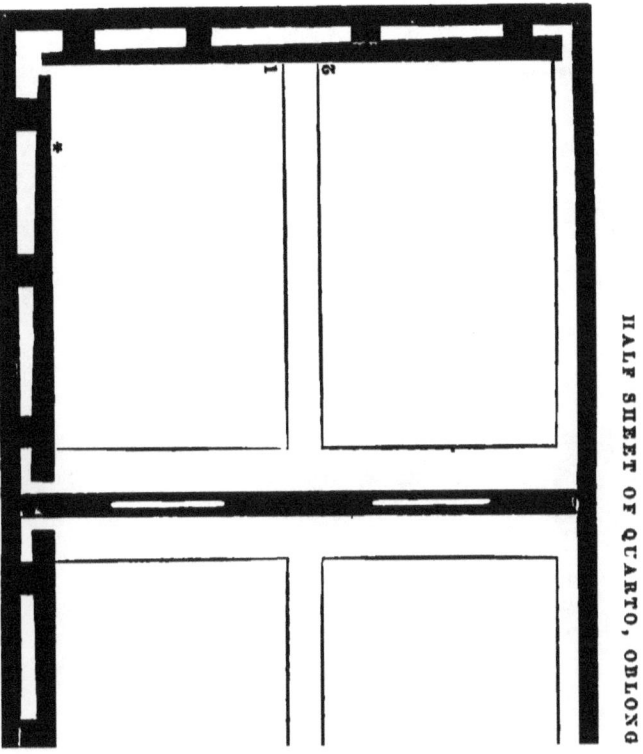

HALF SHEET OF QUARTO, OBLONG

reading-boy to the composing-room to be corrected by the workmen who have taken share in the composition. These immediately divide the proof amongst them, and each, taking that portion of it which contains the matter he had composed, and going to his cases, gathers the letters marked as corrections in the margin, together with a quantity of spaces of all sizes, and returns to the forms, which in the meanwhile one of them has *laid up* on the imposing-table and unlocked. He then with a bodkin lifts up each line in which a correction is required, draws out the wrong letter and inserts the right one, adjusting the spaces in such a way as to compensate for the increased or diminished size of the letter substituted, overrunning carefully several lines should any word have been added or struck out, so that the spacing may be uniform, and the corrected matter exhibit no indication of any alteration having been made. This is an operation requiring much practice and skill; and here is shown the value of attention in the preliminary operations. Should the types have been carelessly laid or inaccurately distributed, should the workman have been negligent in composition, capitaling, or spacing, he will consume as much time in amending his errors as in composing his matter, to the great detriment of his work, the injury and inconvenience of his employer and his companions, and great delay in every department of the printing-office. When every compositor has corrected his matter, that one whose matter is last in the sheet locks it up, and another proof is pulled, which, with the original proof, is taken to the same first reader, who compares the one with the other, and ascertains that his marks have been carefully attended to, in default of which, he again sends it up to be corrected; but should he find his revision satisfactory, he sends the second proof with the copy to the second reader, by whom it undergoes the same careful inspection; but this time, most technical inaccuracies having been rectified, the reader observes whether the author's language be good and intelligible; if not, he makes such queries on the margin as his experience may suggest; he sends it up to the compositor, where it again undergoes correction, and a proof being very carefully pulled, it is sent down to the same reader, who revises his marks and transfers the queries. The proof is then sent, generally with the copy, to the author for his perusal, who, having made such alterations as he thinks necessary, sends it back to the printing-office for correction. With the proper attention to these marks, the printer's responsibility as to correctness ceases, and the sheet is now ready for press. Such at least is the process of proof-reading which ought to be adopted; but now, from the speed with which works are hurried through the press, the proofs are frequently sent out with only one reading, the careful press-reading being reserved until the author's revise is returned.

It need scarcely be remarked that "correctness of the press" is a very material feature in every work, and more especially in those of a scientific nature. When the attention and the mind are devoted to the train of some close argument or passage of surprising beauty, it is surprising how easily an error of the press, even although it may not injure the sense, and may be as evident "as the sun at noon," will destroy the charm, and break the "thread of the discourse;" and even in works of ordinary reading they are exceedingly offensive. Many curious anecdotes are related of the methods which the earlier printers adopted to attain correctness. It was the glory of the early literati to take charge of the accuracy of new works; and, in return, the value and sale of each edition varied with the skill and reputation of the corrector. Of these, Erasmus is an illustrious example. Many of the first printers were led to the practice of the art by their love of learning, and their anxiety to promote it by the production of classic authors. Hence several are better known in the world of learning than in the circle of bibliographers; as the editors and correctors of valuable works, than as the careful or beautiful printers of them. Aldus, it is true, has so admirably succeeded in both characters, that he has fully established his double fame; but whether he was most valued himself upon his learning or his skill may be doubted. It would appear from his letters that he considered it as his chiefest duty to correct every sheet that passed through his press. In all his bustle in preparing every material in use in his art, in all his occupations public and private, this important duty was never neglected. He tells us "that he has hardly time to inspect, much less to correct, the sheets which are executed in his office; that his

[PLATE 5.

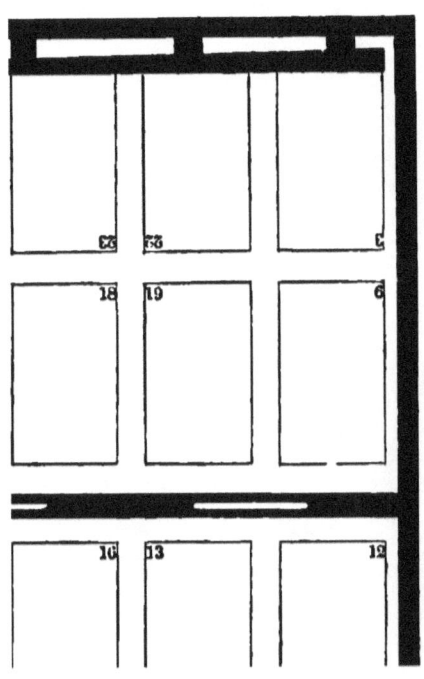

SHEET OF TWELVES: INNER FORM.

PRINTING.] [PLATE 6.

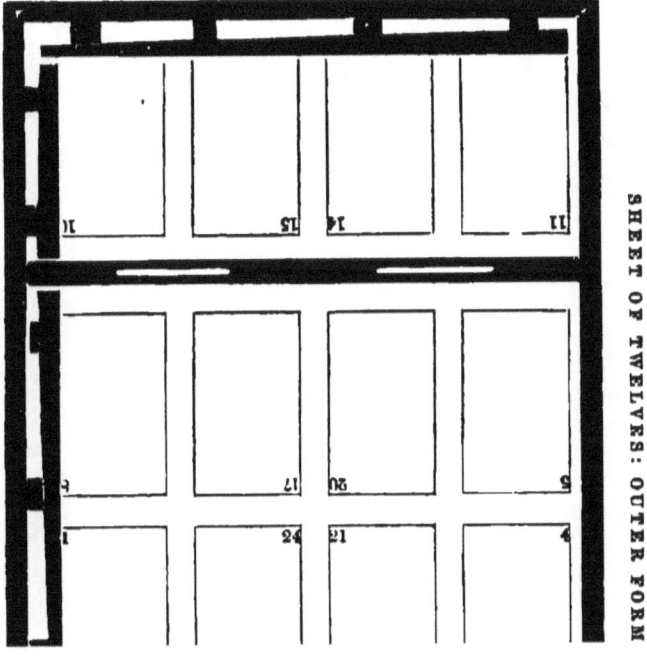

SHEET OF TWELVES: OUTER FORM

days and his nights are devoted to the preparation of fit materials; and that he can scarcely take food or strengthen his stomach, owing to the multiplicity and pressure of business; meanwhile," adds he, "with both hands occupied, and surrounded by pressmen who are clamorous for work, there is scarely time even to blow one's nose:" nor did his son or grandson depart from his ways, but did themselves insure the correctness of their works, even when the latter had risen to wealth and eminence, and enjoyed the laborious dignity of a professor's chair. The beautiful Greek works of the Stephani are especially valued for their correctness. Stephens corrected his own press with intense labor and minuteness, and is reported to have adopted a singular plan for obtaining perfect similarity to the copy, by employing females who had not the slightest knowledge of the Greek characters or language to compare every letter of the proof with the manuscript; a labor so intense as to be almost incredible. He is moreover said to have hung up proofs on the doors of his printing-office, and to have amply rewarded any who could detect inaccuracies therein. Coverdale, it will be recollected, corrected the first English Bible and Testament, and received a bishopric as his reward. Foulis, the celebrated printer at Glasgow, adopted the same plan to insure the accuracy of his edition of Horace, which is styled immaculate; in which, however, one error escaped detection, the ode commencing SCRIBERIS Vario, being printed, as originally issued, SCRIBFRIS Vario.

The experience of every printer will furnish a host of laughable errors; and indeed these defects have been deemed of such importance as to deserve preservation. (D'Israeli's *Curiosities of Literature*.) The omission of the word *not* from the seventh commandment, in an edition of the Bible printed by the Stationers' Company, is well known; and the company richly deserve the severe fine they incurred for spreading the immoral command, "Thou shalt commit adultery." The Bible so misprinted has received the name of the "Adultery Bible;" and a copy is preserved in the British Museum, the edition having been carefully suppressed. There is another Bible known as the "Vinegar Bible," from a misprint in the 20th chapter of St. Luke, where "Parable of the Vinegar" is printed for "Parable of the Vineyard;"

this proceeded from the Clarendon press. In the reign of Charles I. a very curious traffic in Bibles, etc., arose; they were printed by any one who chose, and imported in vast numbers from abroad. It will readily be imagined that these were made for *sale*, not for *use*, and that they abounded with egregious errors: but, what is worse than this, they were full of mistranslations and interpolations, and the omissions were fearful. All these were done as much by design as by accident, the Romanists and sectaries taking the opportunity of advancing their own tenets by interpolating and altering texts to suit their views. These monstrous anomalies produced, however, some good; they occasioned the necessity of the authorized version now in use, and printed under such authority as insures perfect fidelity, whilst there is sufficient competition to make it impossible that the Word of God can ever become a sealed book to the humblest and poorest Christian. Some of the blunders in these editions are sufficiently absurd to overcome the repugnance which must naturally be felt at such license. Thus, in Luke xxi. 28, *condemnation* has been misprinted for *redemption*. In Field's Bible of 1653, called the Pearl Bible, Rom. vi. 13, we find " Neither yield ye your members as instruments of *righteousness* unto sin," instead of *unrighteousness;* and 1 Cor. vi. 9, " Know ye not that the unrighteous *shall inherit* the kingdom of God ?" for *shall not inherit*. It is said that these corruptions are in a great measure owing to Field's cupidity, and that he received a bribe of 1500*l.* from the Independents to alter the text in Acts vi. 3, to sanction the right of the people to appoint their own pastors, " Wherefore, brethren, look ye out among you seven men of honest report, full of the Holy Ghost and wisdom, whom *ye* may appoint over this business," instead of *we*. This Bible is notorious, and, strange to say, valued, for its gross incorrectness. It is asserted that no less than six thousand errors of greater or less magnitude have been noted in it. But the most extraordinary example of carelessness is presented by the Vulgate, the printing of which was sedulously superintended by no less an authority than Sextus V., a curious example of the infallibility of the Pope. To the astonishment of the world, it swarmed with errors ; and a whimsical attempt was made to remedy the defects by pasting printed slips of paper over

PRINTING.]

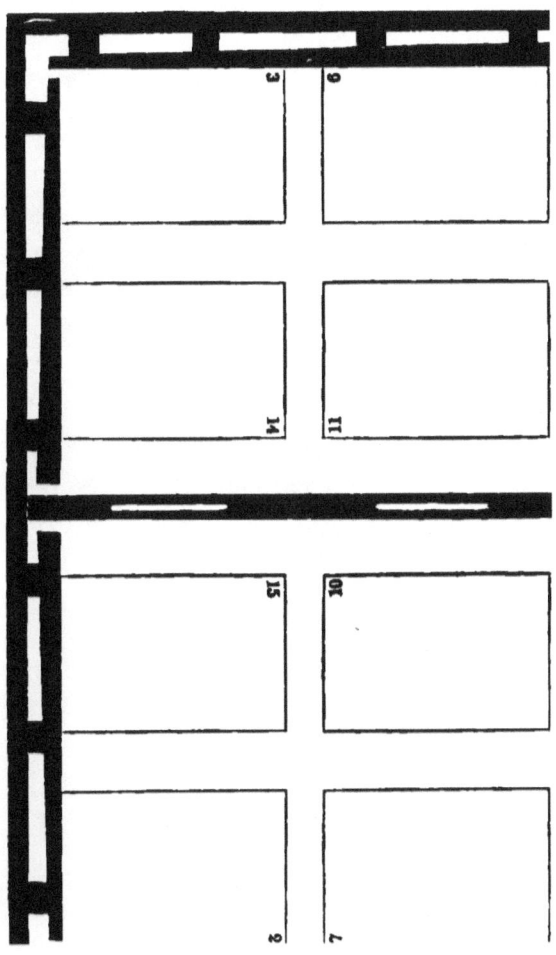

PRINTING.]

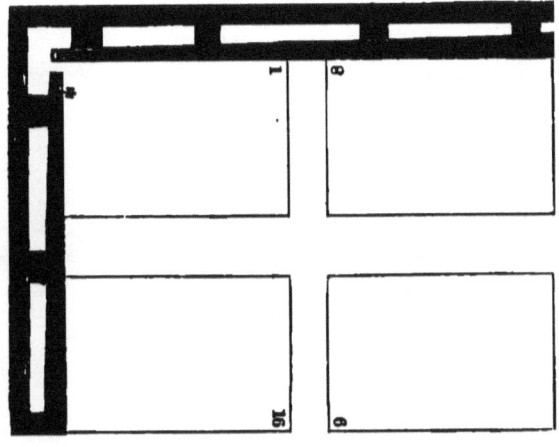

the erroneous passages. As this, however, was exceedingly laughable, the papal authority was exerted to the utmost to call in the edition, and with such effect that it soon became very scarce, and a copy of it has produced the sum of sixty guineas. To add to the absurdity, the volume contains a bull from the Pope anathematizing and excommunicating all printers who, in printing it, should make any alteration in the text. The monkish editor of *The Anatomy of the Mass*, printed in 1561, a work consisting of 172 pages of text and fifteen pages of errata, very amusingly accounts for these mistakes by attributing them to the artifice of satan, who caused the printers to commit such numerous blunders; but he does not inform us whether it was really the archangel fallen, or only his minor satellite, the printer's devil. The editor of an Ethiopic version of St. Paul's Epistles innocently confesses, in palliation of his errors, " that they who printed the work could not read, and we could not print: they helped us and we helped them, as the blind helps the blind."

The sheet being printed off in the way hereafter to be described, and the forms returned by the pressmen to the composing-room, and very carefully washed with lye, and rinsed with water, the compositor lays them up on a letter-board in the sink, and there unlocks them; he then passes one hand backward and forward over the pages so as effectually to loosen the type, and at the same time with the other pours on water, till, the lye and ink being washed away, it runs off clear. The forms are then allowed to drain, and carried to the bulks at the end of the frames. Each compositor employed on the work then takes a share of the letter, and, wetting the face of it plentifully with a sponge, which causes the types to adhere sufficiently to prevent accidents, yet not so much as to retard the workman, takes up a portion on his setting-rule, with the nick upward, and the face turned toward him; he then takes between his fingers and thumb a few letters, gives a rapid glance at the face to see what letters they are, and then, passing his hand rapidly over the cases, drops each into its appropriate box. In this operation the greatest attention is necessary, for it must be remembered that every letter dropped into a wrong box in distributing is sure to cause an error in composing; for the workman, as before stated, never looks at the letter he takes up, relying upon the correctness of the distribution. Compositors, therefore, should be especially careful, when learning their business, not to sacrifice accuracy to swiftness; for in this instance most especially is it found that too much haste is little speed. If the rapidity of motion in composition strikes the stranger with wonder, what must that of distribution occasion? Most compositors distribute four times as rapidly as they compose; if, therefore, he pick up two thousand letters in an hour, he would distribute eight or ten thousand, or about three per second. His letter being properly distributed, he again proceeds to compose in the manner before described, until the work is finished. The number of times the types are returned to their cases must depend upon the size of the font. A thousand pounds' weight of types would get up five or six sheets; and therefore, in an ordinary octavo volume, the types would be returned five or six times.

Many attempts have been made to substitute machinery for the manual labor of the compositors. The machines of Messrs. Young and Delcambre (1842), and of Major Rosenborg, deserve mention for their great ingenuity; and Major Beniowski has attempted a process by which, by the use of a new description of type, logotypes, cases, and machinery, a great saving of time and money may be effected. But there are requirements in the process of composing which are independent of mechanism, and which have hitherto rendered these inventions practically useless.

THE PRINTING-PRESS.

The press is the machine whereby impressions are obtained of the type, when set up by the compositor as above described. On the skill and care of the pressman depends the beauty of the work. If the press-work be not good, all the labor of the compositor is thrown away; his work makes no respectable appearance, and the master gets no credit.

It has already been mentioned that very little alteration had been made in the printing-press from the time of the first printers to that of Blaew of Amsterdam, about 1620. Blaew's improvements, although very great, only consisted in alterations in the details, and not in the principle. These

PRINTING.]

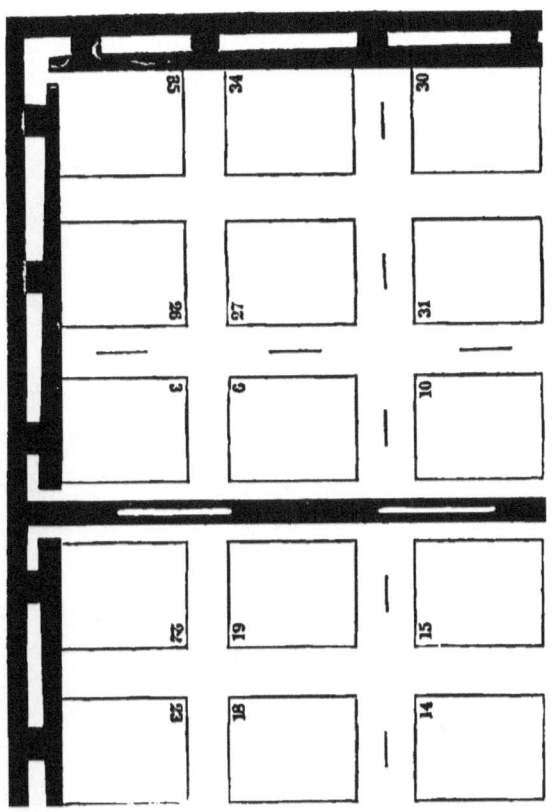

PRINTING.]

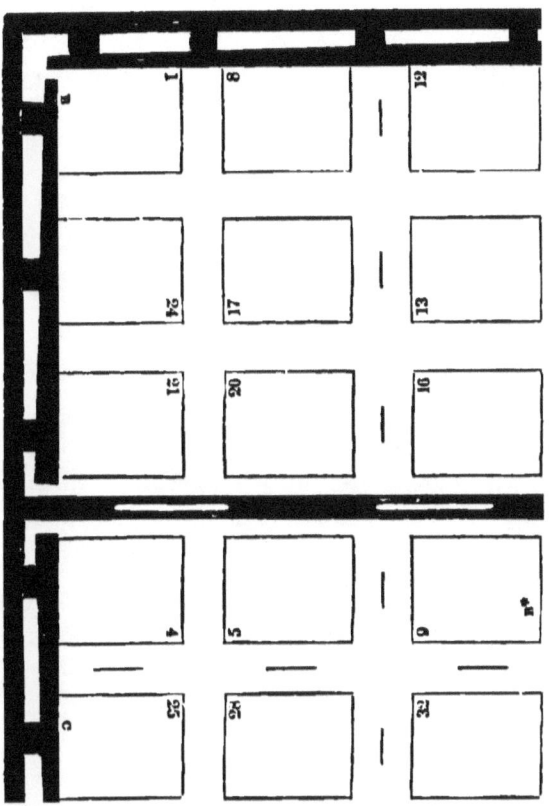

presses have in their turn been superseded by those of Lord Stanhope; and the latter has found successful competitors in the Columbian, Albion, and others of more modern invention. Very few of Blaew's construction are now in existence, save in old offices in England, where they are used as proof-presses, or kept merely as curiosities. As a description of these by-gone pieces of mechanism would be of little utility, the Stanhope press, by which they have been superseded, has been selected for illustration, for which it is best adapted, from the simplicity of its construction and its being easily explained. The novelty of his lordship's invention consists in an improved application of the power to the spindle and screw, whereby it is greatly increased. Upon reference to fig. 6, it will be seen that this press possesses great strength and compactness. The heavy mass of iron AA, somewhat resembling a vase in outline, is called the *staple*. It is united at the top and bottom, but the neck and body are open. The upper part is called the *nut* B, and answers the purpose of the head in the old press; it is in fact a box with a female screw, in which the screw of the spindle C works; the lower portion of the open part, described as the *neck*, is occupied with a piston and cup D, D, in and on which the toe of the spindle works. On the nearer side of the staple is a vertical pillar or *arbor* A (fig. 7), the lower end of which is inserted into the staple at the top of the shoulder; the upper end passes through a top-plate B, which being screwed on to the upper part of the staple, holds it firmly. The extreme upper end of the arbor (which is hexagonal) receives a head C, which is in fact a lever of some inches in length; this head is connected by a *coupling-bar* E to a similar lever or head D, into which the upper end of the spindle is inserted.

The bar or lever F, by which the power is applied by the workman, is inserted into the arbor, and not into the spindle, by which ingenious contrivance—1*st*, the lever is in length the whole width of the press, instead of half, as in Blaew's press, and is, moreover, in a much better situation for the application of the pressman's strength; 2*d*, there is the additional lever of the arbor-head; 3*d*, the additional lever of the spindle-head; and, *lastly*, the screw itself may be so enlarged in diameter as greatly to increase its power. The *platen* L is screwed on to the under surface of the spindle;

the *table* M has slides underneath, which move *in* the *ribs* N, N, instead of *upon* them, as in the old presses, and is run in and out by means of girths affixed to each end, and passing round a drum or wheel O. As the platen is of considerable weight, the workman would have to exert much strength in raising it from the form after the impression has been given, were not a balance-weight P suspended upon a lever and hook at the back of the press, which counterbalances the weight of the platen, raises it from the form, and brings the bar-handle back again, ready for another pull. These are the principal parts of the machinery whereby the impression is given, and are sufficient to give the general reader, with the aid of fig. 7, an idea of the mechanism of the Stanhope press. For the printer there are yet other appliances. At the right-hand end of the table is an iron frame Q, moving freely upon pivots, so as to fall upon the table, or rise until stopped by what is called the *gallows* R; this is covered with parchment very tightly stretched, and is then called the *tympan*; upon the tympan blankets are placed, which are covered by an inner tympan, and fastened by hooks; the whole forming a solid yet elastic and yielding surface, admirably fitted for impressing the paper upon the type (for this is its use), inasmuch as the surface of the parchment is soft and without grain, and readily receives the impression of the type, while the blankets give freely to every projection, without retaining any indentation. To protect those portions of the paper which are not intended to be colored from ink or soil, there is at the upper end of the tympan another iron frame, of much lighter make, and also moving upon pivots, so as to fall upon the face of the tympan. This is covered with a sheet of coarse paper, and after an impression has been taken upon it, the exact size and form of the pages are carefully cut out therefrom, the parts left being an excellent protection of the paper under them. This is called the *frisket*.

Such is the ordinary Stanhope press. A notice of the principle of many other excellent presses which have been since invented, and very extensively introduced, will be found in a subsequent part of this treatise. The manner of working is the same in all.

On the left front of the press stands the inking-table.

[PRINTING.] [PLATE 11.

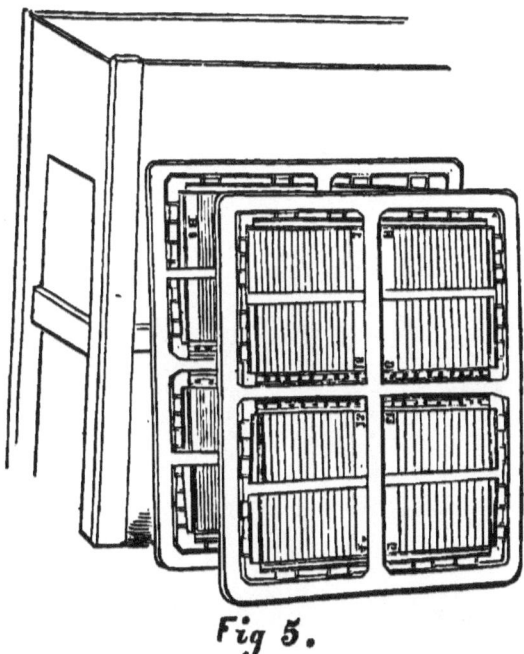

Fig 5.

This is made of iron, about four feet high, and three feet four inches wide; at the back is a solid iron cylinder, turned perfectly true, against which a thin steel straight-edge is made to press by means of levers and weights, thus forming a trough for the ink; of which, when the cylinder is turned round, it becomes covered with a thin film, its thickness being regulated by adjusting the weights on the levers. Against this iron cylinder the inking-roller (which will be hereafter described) is dabbed, and being rolled backward and forward on the table, the ink is evenly distributed over its surface.

It must be fully understood that printers' ink is a very different composition from that used for writing. It is of such consistency that if a small portion be taken up between the finger and thumb, when they are opened it will produce a thread of an inch or an inch and a half in length. Of all the materials used in printing this is the most important, and the most opposite qualities are required in it. It must be of excellent color. Formerly excellence of color was deemed to consist in an exceeding dark hue, not exactly black, but black enriched with a hue of the darkest blue or purple. This gave indescribable effect to the works for which it was used, a richness and intensity which it is impossible to describe, but of which the works of Baskerville and Bulmer, especially the *Milton* of the latter, afford the best specimens. Now we hold perfection to consist in the intensest black, and all the resources of chemistay and the arts have been sought to attain this end. It must stand for ever; but here we have miserably failed. Compare the productions of the old printers with those printed twenty years back. What a difference! The works of the Aldi and Elzevirs, of Plantinus, Caxton, Pynson, and Grafton, preserve their color as intense as on the day they were printed; there is no yellowness or brownness, no foxiness; whilst the books printed from 1810 to 1820 are wretchedly discolored. Where fine printing, however, has been required *and paid for*, the modern ink is no whit inferior to the ancient. Witness the works of Bulmer, Macklin, Ritchie, Bowyer, Baskerville, and others; but certain it is that the ink in general use twenty years ago was of very inferior quality. It must be perfectly mixed, and ground until it is absolutely impalpable, otherwise it will speedily clog the types and inking apparatus; it must adhere to the paper, and not to the type, or it will tear off the face of the former, and clog up the latter; it must be sufficiently thick; it must keep perfectly undried when in large masses, and dry very quickly when it is transferred to the paper. Few printers of the present day make their own ink, although some add ingredients which they believe to improve the color or quality. Ink-making is a distinct business; and by the aid of machinery, capital, and exclusive attention to the manufacture, the ink now supplied is admirable in the qualities of being thoroughly mixed and ground, drying, blackness, etc.; but whether it will stand the test of time, time alone can show. It is an expensive article, the commonest book-ink being one shilling and sixpence per pound,* whilst the usual qualities are two shillings and sixpence, three shillings, and four shillings per pound; those used for superior work are five shillings or six shillings; and those for cuts as high as ten shillings—though it is questionable whether, at the latter price, the consumer is not paying for a mere name.

Every manufacturer has of course his own secrets both of ingredients and process. The universal ingredient is of the finest possible lamp-black; the great secret probably consists in the manner in which, and the material from which, this is made. There are vast buildings appropriated to the sole purpose of burning oil, naphtha, spirits, coal-gas, etc., to produce this black, which is collected from the sides, ceilings, etc., of the buildings; it is brought from Germany and many other countries; and no expense is spared to get the most superior quality. The next most important article is nut or linseed oil boiled and burnt into a varnish; then oil of turpentine, etc. The following receipts have been given. The first is the method used by Baskerville and Bulmer, and nothing can be better than the results:

1. Fine old linseed oil boiled to a thick varnish, and cooled in small quantities, three gallons; a small quantity of black or amber rosin dissolved therein; the mixture then stands for some months, that all impurities may be deposited; after which it is mixed with the finest lamp-black, and carefully ground for use.

* In England. In America, the ordinary price is thirty cents per pound.

Printing—Practical.

2. One hundred pounds of nut or linseed oil are reduced by boiling and burning one-tenth or one-eighth of its bulk, and to the thickness of a syrup, two pounds of coarse bread and several onions being thrown in to purify it from grease. Thirty or thirty-five pounds of turpentine are boiled apart, until, on cooling it on paper, it breaks clean, without pulverizing. The former is poured nearly cold into the latter, and well mixed. The compound is then boiled again. Lampblack is next thoroughly mixed with it, in quantity according to the ink required, and being well ground, the ink is then ready for use. Some add indigo, some Prussian blue, which considerably improves the color; but these inks are so difficult to work, and so clog up the type, that the *improvement* is better let alone. The turpentine is added to give greater varnish, and improve the drying quality; but if the oil be old and fine, the quantity required is proportionally less.

3. Mr. Savage, an admirable artist, denies that any ink can be depended on of the varnish of which oil is the basis; he therefore gives the following receipt: Balsam capivi, 9 oz.; best lamp-black, 3 oz.; Prussian blue, $1\frac{1}{2}$ oz.; Indian red, $\frac{3}{4}$ oz.; turpentine soap dried, 3 oz. This ink is of beautiful color, but appears to work foul.

At the right front of the press stand the *bank* and *horse*. The bank is a deal table of some size; the horse is an inclined plane which stands upon the bank; upon it is laid the white paper properly damped for working; and as each sheet is worked, it is taken off the tympan and laid on the bank. There are two pressmen to each press, one of whom attends to the inking only, to ascertain the excellence of which, whenever he has a moment to spare, he turns to the worked sheets upon the bank, glancing his eye rapidly over each, to see that every part is of its proper color, and that no picks or other imperfections mar the work; the other attends only to the press, and gives the impression. These men are paid by every two hundred and fifty impressions, called a *token*. Thus, if the number be five hundred, and the price $4\frac{1}{2}d.$ per token, each man receives $9d.$ for the five hundred impressions of each form, and the cost therefore is,

Inner form, two men, two token, at $4\frac{1}{2}d$.................1s. 6d.
Outer form,   do   do   do   ....1s. 6d.

                        3s. 0d.

The price varies with the size of the type and the form; with the quality of the paper and the ink; with the number, and the care required. Common work used to be paid for at $4\frac{1}{2}d.$, good at $6d.$, superior at $7d.$, the very best at $8d.$, $9d.$, or even $1s.$ per token.* But now the price is matter of agreement between the master and pressmen.

One of the pressmen, having received the forms after the final correction, lays the inner form, or that one which contains the second signature, upon the table of the press, and secures it in the center by quoins; the other in the meanwhile pastes a stout sheet of paper upon the frisket frame, and then secures it upon the tympan. The form is then inked, and an impression taken upon the frisket, and the printed parts only being cut away, that which is left protects the paper from ink or soil. The puller now carefully folds a sheet of the paper according to the crosses of the chase, and laying it upon the form, opens it carefully, by which the paper is made to lie evenly upon the form, with the same margin with which it is to be afterward worked. Having slightly wetted the tympan, he turns it down upon the form, and takes an impression, when the paper will be found to adhere to the tympan, and thus become a guide whereby to lay all the subsequent sheets, and therefore much care should be taken to lay it properly. They now choose their *points*, which are thin and narrow pieces of iron, having a short point or spur projecting from one end, and a shank at the other made to screw on to the tympan-frame, which must be done in such a manner that the spurs may fall into the grooves in the cross of the chase; because if they did not, they would be battered or broken at the first pull. It is advisable to make the inner form *register*, for it may be very difficult to correct any error in the furniture when the *reiteration*, or outer form, is laid on.

The puller now brings his paper from the wetting-room; for before any good impression can be taken the paper must have been damped, by rapidly passing it, one-fourth or one-fifth of a quire at a time, through water, and then allowing it to soak for two or three days under a heavy weight, until it is evenly and thoroughly damped; and laying a ream upon the horse, he takes a sheet, and placing it carefully over

---

* The price for hand-press work in America is twenty-five cents per token.

the tympan-sheet, closes the frisket over it, shuts both tympan and frisket down upon the form, which in the mean while his companion has inked (a process that will be described below), runs the table in under the platen, pulls the handle of the bar or lever over by his full weight, until brought up by the stop, at which moment the platen descends, and exerts a powerful pressure to the tympan, etc., upon the form, producing upon the paper a perfect fac-simile in reverse of the surface of the pages. The pressman now gradually releases his hold, the balance-weight raises the platen, the bar returns to its first position, the table is run out, the tympan and frisket are raised by the workman, and the frisket thrown up to the catch. The sheet is taken off the spurs of the points, which have been forced through it by the pressure, and the back of the impression is carefully examined, to ascertain that every part of it is just and even, which is the great test of the workman's skill and the excellence of the press. The first impression is, however, invariably defective: the parchment may have been thicker in some parts than in others, the blankets worn, or one of two fonts of type may not have been of equal height, in which respect " the estimation of a hair" would produce a manifest imperfection, but which may be remedied by the thinnest possible tissue paper. The pressman now proceeds to *overlay;* that is, by pasting upon his tympan-sheets portions of paper of the exact size of the defects, thicker or thinner as may be required, to *bring up* the form; he overlays the faint parts of the impression; or if the defect be great, he places a part of a sheet of paper within the tympan, or, which is a much better plan, he raises the form, and pastes the paper under the defective part. If there be any small portion of undue prominence, or that " comes off hard," he rubs down a portion of the tympan-sheet with his wet fingers, or cuts it away altogether. Having, as he supposes, remedied all blemishes, he takes another impression, which he again examines with equal closeness, and carefully removes every remaining defect by the same method; and having at length satisfied himself, and his master or overseer, that the form is well brought up, the work is proceeded with, the inker taking off from the table with the roller or balls even portions of ink, which has been well distributed on its surface, and rolls or beats the form, being very careful that every part is equally inked; the puller taking a sheet and laying it on the tympan as before. They thus proceed until the whole number of the white paper is worked off; when it is a good precaution to count the heap, to ascertain that the number printed is correct. The form is now lifted from the table, and carefully washed with very strong lye. The outer form is then laid on and made ready.

The making ready of this form varies a little from the mode previously described. It has been stated that the spurs of the points penetrate the paper at the first impression. The holes thus made are the guides whereby perfect *register* is obtained; that is, whereby not only the pages, but the lines, are made to fall exactly upon the back of each other, any variation in this respect being a great defect in good book-work. The outer form, therefore, having been placed on the table in precisely the same position which the inner previously occupied, a printed sheet is taken from the heap, and laid upon the tympan with its printed face inward, in such manner that the spurs of the points pass through the holes made by them in the working of the inner form, but of course the opposite way; and an impression is taken. If the pages do not back, the points are shifted until they do; or if the defect arise from the furniture of the form, such alterations are made in it as may be necessary. The impression is then brought up as before, and when all is ready, a thin sheet of white paper, called the *set-off* sheet, is placed over the tympan-sheet and under the points. It must be remembered that one side has been worked, that the ink has not yet dried, that the paper is still damp; therefore at every impression some portion of the ink will be transferred to or impressed upon the set-off sheet. When this has taken place in many impressions, some of the ink of the print will be re-transferred from the set-off sheet to the sheet then working, producing a most unpleasing blurred appearance, very perplexing to the eyes, and utterly destructive of the beauty of the press-work. To obviate this, the puller, after a few impressions, moves the set-off sheet slightly, and when it has become very black, takes it off, and replaces it with another. The pressman should be very attentive to this; and the master should not grudge ample supplies of set-off paper, for it is not destroyed, but, when dried, may be used again for the same purpose, or in other departments as waste paper. The form is now lifted, and carefully washed with lye, and the two are ready for the composing-room, where they are laid up, as previously described. Two good pressmen are supposed to do about one token, or 250 impressions, per hour of fair work. This, however, must depend entirely upon the quality of the work required; with small type, stiff ink, and many rules, the work is more slow, and paid for accordingly. The finest work is seldom paid for by the token, the pressmen being placed upon weekly wages, and allowed as much time as they require, the rapidity being at the discretion of the overseer. Frequently they are limited to a certain number per hour, often as few as fifty, the most careful inspection being given to every sheet by both pressmen, and continual attention by the press-overseer and other chief persons in the establishment. In such work the very best materials are employed. Instead of parchment, the tympans are covered with fine calico, or even silk; instead of blankets the finest broadcloth; picked blotting paper for the thick overlays, the thinnest tissue-paper for the finer. It will readily be understood that in all operations of the press-room, where every thing depends upon the skill of the workmen, there are infinite minutiæ, which it would be tedious, if it were even possible, to enumerate. Seven years' apprenticeship are not more than sufficient to educate a good pressman; it is the accumulated labor of a life to make a first-rate one: and, after all, excellence depends upon the native talent and ingenuity of the man himself.

The ink is distributed over the type either by balls or by rollers. The rollers are of modern use. The balls, which are such prominent objects in the representation of ancient printing-offices, and which form part of the armorial bearings of the printers' guilds on the Continent of Europe, were formerly made of sheep-skins, with the hair taken off by lime, and formed into a ball with wool, gathered at all corners, and nailed upon a wooden handle. One of these was held in each hand; and a small portion of ink being taken, they were well beaten upon the inking-table, and then upon each other, until the ink was so evenly distributed over the whole surface, that if touched gently with the finger, the prominent lines of the skin would be blackened, whilst the channels would be left perfectly clean. The balls were then beaten over every part of the type, so that the whole surface should be evenly covered; an operation requiring much skill and practice. The skins were prepared and softened by the nastiest processes imaginable, which converted a press-room into a stinking cloaca. Thanks, however, to the observation and ingenuity of Mr. Forster, a practical printer, and Mr. Donkin, an engineer, this has been entirely done away, and a press-room now regales the nose with a warm scent of ink and paper, any thing but unpleasant. This invention has been of the greatest consequence to printing. The printing-machine is said to be the handmaid of modern literature; and so it is; but without this, printing machines were mere old iron and brass. Earl Stanhope had attempted to substitute skin rollers for skin balls; but his plan failed owing to the difficulty of preparing the pelts, and the inevitable seam, which left a broad mark upon the type. But the use of rollers, which in the hand-press would have been merely an improvement on a process in use, was a necessity to the printing-machine, and the complete failure of the earliest of these machines was in a great degree owing to the imperfection of their inking appliances. For many years the workmen in the potteries had used a composition of glue and treacle for applying colors to their ware. Mr. Forster observed that this composition possessed every requisite for the use of the printing-office, and he immediately proceeded to form balls of canvas, with a facing of composition. They answered admirably, proved beautifully soft, distributed satisfactorily, kept clean, and were easily washed and purified if soiled. Some opposition was offered by the workmen; but the advantages proved so great that they were readily adopted by the masters, and speedily drove away forever the nasty skins. The next step, however, was more important still. Mr. Donkin observing the adaptability of the composition to casting rollers for printing-machines, devised moulds, by which he was able to cast cylinders without seam, and of somewhat greater tenacity than the original compound. The rollers answered perfectly for printing-machines; and there was little difficulty in perceiving that at the hand-press the roller might be advantageously substituted for beating by balls. They were

[PRINTING.] [PLATE 12.

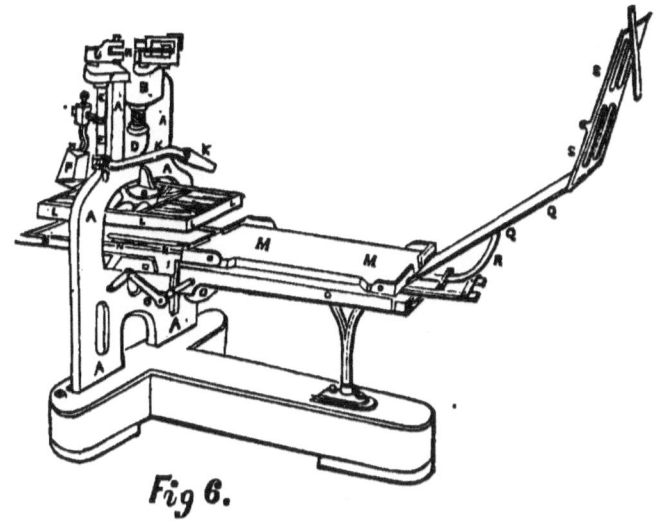

Fig 6.

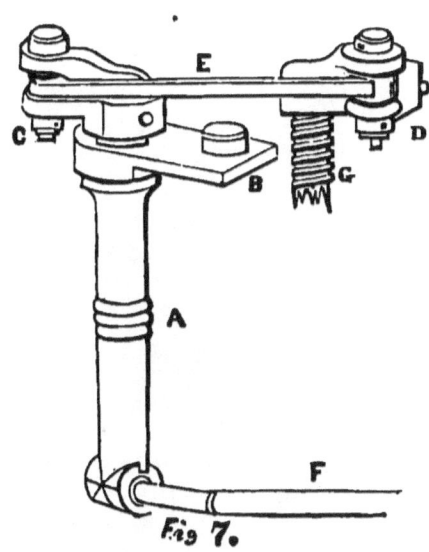

Fig 7.

accordingly introduced, and after meeting with some opposition, are now in universal use. They consist of a solid wooden cylinder, with a thick coating of composition cast in a metal mould perfectly true; through the middle of the cylinder passes an iron rod attached to a curved bar, upon which are fixed two handles; the roller revolving freely upon the rod. The pressman regulates the quantity of ink to be taken by adjusting the pressure of the straight edge against the cylinder at the back of the table, as above described; and according as that pressure is greater or less, the cuticle of ink on its surface is proportionately diminished or increased in thickness. Having taken off upon the inking-roller a line of ink, he distributes it carefully upon the table until the entire face is evenly covered, and then rolls the form, taking care that the whole surface receives its due proportion. If he does this lightly and steadily, there is no fear of the result; he cannot in rolling leave any part without ink; but it nevertheless requires some judgment. If there be any heavy titles or large type, he must roll that portion several times; if there be blank pages, he must take care that the roller does not sink, and so leave the pages in line with them slightly touched. The greatest judgment, however, is displayed in choosing the exact quantity of ink required for the form. If the type be small, the quantity taken must also be small; it must be very carefully distributed, and the form rolled many times; for if the quantity be too great the type will become clogged, and if too little, the color will become faint. The pressman must from time to time examine the sheets as they are printed, and in working the reiteration, turn up the corners of the sheets to see that the color corresponds with that of the inner form, detecting with quick eye every defect; and he must be particularly careful that for every sheet of the same work he takes the same quantity of ink, so that the book when bound may present an even and beautiful color, every bold line being perfectly covered, and yet every fine stroke clear and distinct. This can only be effected by careful distribution and repeated rolling, with nice judgment as to the quantity of ink to be taken.

The sheet having thus been worked off, the printed paper is taken away by the warehouseman, and hung by the boys upon poles stretched under the ceiling, by means of a *peel*, which is a handle with a broad end, upon which a quire or two is hung at a time, thence transferred to the poles, and distributed in portions of four or five sheets. Here they hang a day or two, until the ink and paper are perfectly dry. This should be a gradual process, for if by artificial heat the drying is hurried, a skin will be formed upon the surface of the ink, which will prevent that underneath from drying; the work will look very well until it is pressed or bound, when the skin breaks, the ink spreads, and the sharpness of the impression is entirely destroyed. When perfectly dry the sheets are taken down and laid in heaps upon the gathering-board, each signature separately; thus, first, a heap, say 1000, of B, then C, D, E, F, and, lastly, the title-sheet A. The boys then take one sheet from each heap; consequently, when they have got to the last signature, each boy has gathered one complete copy of the work. These are laid upon one another at the end of the gathering-board in such a manner that each book is perfectly distinct. The warehouseman then takes away this heap, and with a collator (a needle inserted in a handle) goes over the whole with great rapidity, ascertaining that no sheet has been carelessly omitted, and that more than one of each signature has not been taken. The books are then folded down the middle, counted out in tens, thirteens, or twenty-fives, and tied up in bundles of convenient size. The process of printing is thus complete, and the work is ready for the binder.

Works of finer description, indeed most works of the present day, are submitted to another process after they have been taken down from the poles, viz., hot or cold pressing, which very much improves their appearance. In cold pressing the sheets are placed one by one between glazed boards, which are sheets of coarse material pressed and glazed on both surfaces by burnishing on a steel plate with a steel ball. The heaps are then placed in a hydraulic press, with cold iron plates at small intervals, and the whole is subjected to considerable pressure for some hours; they are then taken out, and the sheets extracted from the boards, when the indentations consequent upon the working will have been all pressed out, the roughnesses of the paper smoothed out, a slight gloss given to the ink, and the whole will present a very agreeable smoothness to the eye and the touch. Hot-pressing is used when the paper is very stout and the ink strong. The sole difference is, that the iron plates are heated until they can hardly be touched. The effect produced is much greater than that by cold pressing; the whole surface of the paper is perfectly glazed, and the ink absolutely shines; but the effect is not so agreeable to the eye; it is too glossy. A machine of great power has been invented for superseding the use of glazed boards and the hydraulic: in this machine the sheets are placed between two plates of copper or zinc, and passed in rapid succession between two hard steel rollers, and come out more perfectly smoothed than by the ordinary hot or cold pressing. As these processes set the ink and also make the books lie perfectly flat, they render much beating by the binder unnecessary, which is a great advantage, as the beating causes the ink to set-off upon the opposite pages when the work is recently printed. The glazed boards must be often cleaned by rubbing with waste paper, or they will soil the sheets placed between them. Every printing-office of credit should have an hydraulic press and glazed boards; for it is incredible how much smartness pressing gives to the work, and how greatly the warehouse work is facilitated by the readiness with which the hydraulic is pumped up, and by its great power. A press of eight-inch ram will be found sufficient for most purposes; but where much hot and cold pressing are required, one of ten-inch ram will prove cheapest, because, from its immense power, a few hours are sufficient to give the requisite surface, and the press may therefore be filled twice or thrice a day.

Wood-blocks are very often worked along with the common type. The block, having been carefully reduced by the engraver to the exact height of the type, is placed in the composing-stick, and justified to the width of the page; it is then made up along with the other matter in its proper place. When laid upon the press for working, and an impression of the form has been taken, the pressman examines with great minuteness whether it stands well with the type; if not, the form is unlocked, and paper placed under it if it be too low, or under any corner that may be lower than the rest; if the block be too high, it must be scraped or filed at the bottom. The artist in wood contents himself with producing his lights and shades by cutting his lines in greater or less degrees of fineness upon a plane, leaving to the printer the task of producing the required effects by a tedious process of overlaying; so that the pressman becomes to a certain extent an artist, and must have a good eye for perspective and for the proper adjustment of tints. These effects he produces by careful and skillful overlaying. But Bewick and some other eminent engravers, instead of imposing this tedious process upon the pressman, used to cut away the parts of the block intended to appear light before engraving them; and thus, by repeated lowering and rounding, they so regulated the lights and shades that the cut left their hands in a fit state to be worked. This process was, however, very costly, and has been discontinued by modern artists. In machine-printing, to prevent the loss that would be incurred if the machine were to stand still during the operation of bringing-up, the machiner, some time before the sheet is laid on, takes an impression of the cuts, and by overlaying and other processes, so prepares them that they require very little additional work when the forms are laid on. Where it can be managed, the cuts should be worked in the outer form, to prevent setting-off and the impression of reiteration upon them. The cuts may then be worked with the type without any other care than that of keeping them clear from clogging or picks. When done with, they must be very carefully cleaned with spirits of turpentine and a brush.

The working of wood-cuts by themselves, as illustrations of works, differs from type-printing in no other respect than in the superior materials and skill required. The wood-cut must be imposed in a chase, and locked upon the table of the press, which is generally a smaller one than that used for ordinary printing, of most excellent construction, and in good order. The tympans are, as before stated, often of silk or cambric. For the inking, balls are preferred to rollers. The greater opportunity for manual skill offered by the former enables the pressman to exercise an artistic judgment which is not possible when rollers are used. The ink is generally brayed out by a muller on a slab.

There are in London, and probably in the larger provincial cities, parties who make an especial business of the manufacture of composition balls and rollers, which they supply to printers upon payment of a rent. The skill and experience of these persons enable them, as must be the case in every instance where a manufacture engages exclusive attention, to supply a much better and cheaper article than could be manufactured by any individual whose engagements are varied; consequently there are not many printers, either in town or country, who do not avail themselves of these opportunities. The rent is paid for each roller required, and by the quarter; that is to say, if a printer employs six presses, and consequently six rollers, he pays for six rollers, the manufacturer engaging to supply him with as many changes as he may require from their getting out of order or being injured; in fact, to keep him supplied with six rollers in good condition. The rent for a common press-roller is the moderate sum of six shillings per quarter; they are sent into the country in boxes fitted for the purpose. There are, of course, situations in which it is not easy to obtain a regular supply of the necessary article, and in this case the printer may very easily make them for himself; but the expense of the utensils is so great as to exceed the usual rent for years. They consist of the following: For rollers, a hollow cylinder of iron, the bore of which must be most accurately turned and well polished; this mould consists of two semi-cylinders closely fitted, and brought into contact by screws along the sides and collars at the end, and a head is made to fit into the lower end. The core, a wooden or iron cylinder, upon which the composition is cast, is held in the center of the bore by means of a star, through the radii of which the composition flows. For balls are required a concave mirror of about half an inch cavity, and a board of the same size and of a quarter of an inch convexity. A kettle for melting and mixing the composition is also required. This is made double like a glue-pot, fitting exceedingly close, and with a small orifice for the escape of the steam from the hot water between the two; and the inner vessel should have a large lip. The recipes for making the composition vary, and this appears to arise from the different circumstances under which it is made. The ingredients are but three, and these easily purchasable, viz., fine glue, treacle (not that procured from the sugar-bakers, which is adulterated, but the best from the sugar-refiners), and a small quantity of carbonate of barytes, called in commerce Paris white, or of carbonate of soda. The first

two ingredients are quite sufficient with a little skill. The following are good recipes: *
1. Two pounds of glue to one pound of treacle.
2. Two pounds of glue to three pounds of treacle.

* An approved method of composing and casting rollers, in America, is described by A. E. Senter, pressman in Follett, Foster & Co.'s Printing and Publishing House, Columbus, Ohio, as follows:
Take seven pounds of Upton's frozen glue, put it into hard water, and let it soak until the water has struck half way through it. In good frozen glue, this will be in ten minutes—in ordinary glue, considerably longer. Then take it out of the water, and let it lay long enough so that it will bend easily; it is then ready for the kettle.
The kettle for melting and mixing, should be so set as to heat and boil the composition by steam or a hot water bath, in the manner in use by cabinet-makers. Let the glue heat in the kettle until it is all dissolved, or if there should be any pieces that do not readily melt, take them out, or they will make the roller lumpy. When the glue is all melted evenly, take four quarts of good sugar-house molasses or sorghum syrup, stir it in, and continue to stir occasionally for three or four hours, during which time the heat under the kettle should be kept up so as to give the composition a gentle boil. To try the composition, take a little out on a piece of paper, and when cool, if it is tough so as to resist the action of the finger without feeling tacky, it is ready to cast. A person can generally tell when it is done, by taking out the stirring stick and holding it up, when, if the composition will hang in strings, it is done.
A very important feature of roller-making, is in preparing the core. Strip off the old composition with a knife, and scrape the core. Keep water away from it, and also sweaty hands. If water is used at all, let it be hard water, and let the core dry thoroughly before casting. If the core is likely to give the composition the slip, brush it over with lime water newly made with quick lime, and let it dry well, and the composition will stick fast.
Have the mould carefully cleaned and oiled on the inside, set it upright, with the core in its place in the center, then pour in the composition hot from the kettle, carefully, upon the end of the core, so as to run down the core, and not down the inner surface of the mould, as that would be likely to take off the oil from the mould, and by flowing it against the core, would make it peel off when cast.
When the composition is cold in the mould, and ready to be drawn out, draw it steadily; trim the ends with a sharp knife, beveled toward the core, so the ends will not be so likely to get started loose; take a hot iron and run it around the ends of the composition, soldering it to the core, which operation will prevent water, lye or oil from getting in between the composition and the wood, and making it peel at the ends. Do not wash a roller when it is taken from the mould; it will be all the better for two or three days' seasoning, with the oil on the surface. It is always good economy to have enough rollers cast ahead, so as not to be obliged to use new ones until they are seasoned. In washing, use lye just strong enough to start the ink, and rinse off with water immediately, and carefully wipe dry with a sponge. Rollers should always be kept in an air-tight box, without water, and in the room where they are worked. Sudden changes of temperature, as from a cold cellar to a warm press-room, will soon use them up.

3. One pound of glue to three pounds of treacle and a quarter of a pound of Paris white.

(Sugar is sometimes used in lieu of treacle, and is said to make the composition firmer.)

Soak the glue in water until it is soft; then place it in the inner vessel, and boil quickly, until the glue is thoroughly dissolved; add the treacle, mixing it well, and let it boil for an hour or more; then sift in the Paris white, but do not stir it violently, or the mixture will be full of air-bubbles, which are destructive to the roller or ball. Rub the mould slightly with a rag dipped in thin oil, taking care that no globules and streaks remain upon the surface. When the mixture is ready, pour it gently between the radii of the star, so that no air be detained within the cylinder, until the mould be filled; allow it to set, and then take it from the mould, cutting off the superfluous portion with a string. When the roller has been hung up twenty-four hours it will be fit for use. Owing to the rapidity of the printing-machines recently introduced, the ordinary rollers have proved inadequate to the work; but improvements have been introduced into the manufacture which remedy the defect. The excellence of the new rollers is said to depend entirely on skillful manipulation. The ingredients are the same, but great experience is required in the choice of the glue, the proportion of the ingredients, the mixture, and the heat applied. In making balls, having oiled the mirror, pour the composition upon the center, and having allowed it to spread itself, lay over it a piece of coarse canvas, place the board upon it, and lay weights upon it to press it down; it will consequently be found that the composition face of the ball will be slightly thicker in the center than at the edges, which, besides being a convenience in the working, will allow it to be *knocked up* with much facility, which is done in the ordinary manner. These balls and rollers are very easily kept in order: if they are too soft, cold water will harden them; if too hard, warm water will soften them. When not in use they should be covered with refuse ink, and hung up in a room of even temperature, and carefully scraped with a pallet-knife before use. They should not be cleaned with spirits of turpentine, as that will give them a hard surface. These rollers will be fit for use for a long while if attention be paid to them; and when spoiled, the composition may be repeatedly melted down, and, with an addition of new materials, will make as good rollers as before. When the proper apparatus is wanting, small balls for wood-cuts or single pages may be made upon an earthern pallet, or even upon a smooth dinner-plate.

A new process has recently been patented by Messrs. Harrild for the manufacture of composition rollers, which enables them to resist the friction of the fastest machines even in the warmest weather, and to continue in working order for a much longer period than those at present in use. They are also but slightly affected by atmospheric changes. These are great advantages for the fast newspaper machines, and for country printers who have not the same facilities as the printers in the metropolis for changing their rollers when out of order. The principal difference in the new process is, that the glue is liquified without any admixture of moisture, the condensed steam which floats on the surface of the glue being entirely drawn off by a syringe.

STEREOTYPING.

Stereotyping is a mode of making perfect fac-similes in type-metal of the face of pages composed of movable types. Letter-press printing being a very expensive process, the price of books consequently high, and the heaviest expense consisting in the composition, the printers of the Continent very soon set up the entire of such small works as were in constant demand, and thus were enabled to sell them at little more than the cost of paper and press-work. Some works of very great extent, especially Bibles and prayer-books, were kept standing by the privileged printers. This, however, was exceedingly expensive, as the cost of the type would be very great; the forms would occupy much space in storing, and be liable to continual damage from the dropping out of letters, from batters, and other accidents to which they would be unavoidably exposed. Some method, therefore, by which all or some of these disadvantages might be remedied, became desirable. About the beginning of the eighteenth century, Van der Mey, in Holland, sought to avoid this liability to accidents, by immersing the bottom of his pages in melted lead or solder, and thus rendering them solid masses: "c'est une réunion des caractères ordinaires par le pied, avec de la matière fondue, de l'épaisseur d'environ trois mains de papier à écrire;" therefore the mass together would be somewhat less than the height of our type. It is not very easy to imagine how they contrived to make the backs of these blocks of such evenness as to produce any thing like a good impression; but Dibdin says that *the book* is very handsome. The same process was followed by a Jew of Amsterdam, in printing an English Bible; but he was utterly ruined by his speculation.

Some time before the year 1735 there is sufficient evidence that the French used casts of the calendars placed at the commencement of church books. These plates are thus described by Camus: "It (one of the plates) is formed of copper, and is three inches and a half long by two inches broad and one-seventh of an inch thick. From the roughness of the casting, it has evidently been made in a mould formed of sand or clay." After the plate had been cast, the back of it had been dressed with a file, in order that it might bear equally upon a block of wood to which it had been attached.

Who really invented the art of stereotyping as at present practiced (and after all, he who finds out the efficient *modus operandi* is the inventor of the art, though he may not be of the principle) is, like the inventor of the parent art, a matter of some controversy, which has been carried on with more vigor than the subject merited. It seems, however, most probable, when all assertions are weighed, that William Ged, a goldsmith of Edinburgh, deserves the credit. According to his statement, being in 1725 in company with a printer, they lamented the want of a good letter-founder in Scotland; and the printer asked him whether he could do any thing to remedy the inconvenience. He immediately answered that it would be more easy to cast plates from pages when composed in movable type; and he undertook to produce, and very shortly did so, a specimen cast on his new plan, and not long afterward made arrangements with a capitalist for the advance of the requisite funds. The latter failing to perform his part of the engagement, Ged made a similar contract with a London stationer, in conjunction with whom he made many attempts; but being repeatedly thwarted in perfecting his plans, he separated from his partner, and made proposals to the universities and the king's printers for the stereotyping of Bibles and prayer-books. These all entered into the scheme with eagerness, and some works were produced from plates quite equal to the ordinary printing of the day. Nevertheless, so much ignorance and prejudice prevailed amongst the workmen and other interested persons that Ged was obliged to abandon the undertaking. He entered into several subsequent arrangements, in which he was equally unsuccessful; a type-founder, in particular, causing so much opposition that the invention made no progress. Ged died before he had met with much encouragement; and his son was equally unsuccessful, although, as the practicability was made more manifest, the very parties who had rejected his plans, subsequently made extensive use of his plates. What was Ged's method of stereotyping is unknown, as he kept it private; nor did he fully communicate the secret to his partners.

Fifty years afterward Mr. Tilloch made a similar invention; but from private circumstances the design was laid aside, not, however, before several volumes had been printed from his stereotype plates at the press of Mr. Foulis. Some years after this, Lord Stanhope engaged an ingenious London printer, Mr. Wilson, to prosecute the invention; and after many trials, the noble lord's ingenuity succeeded in bringing the invention to practical use.

When a work is expressly intended to be stereotyped, the spaces, quadrats, and leads generally used are somewhat different from those commonly employed, being cast of the same height as the stem of the letter, in order that the base of the plate may be more solid and of uniform thickness. When low spaces, etc., are used, plaster is poured upon the face of the type to fill up the interstices, and just before it sets the superfluous plaster above the stem of the letter is removed by a brush, which damages the face of the type not a little. The page is composed in the ordinary manner, and very carefully corrected; it is then imposed in a small chase with metal furniture, and the whole is placed within a moulding-frame, somewhat less than half an inch higher than the type. The surface of the type is then rubbed with a soft brush holding a small quantity of very thin oil.

The plaster of Paris (gypsum) of which the mould is formed is of the finest quality, and may be purchased ready prepared. Having been carefully mixed with water to the thickness of cream, a small portion is gently poured upon the surface of the page, and softly worked in with a brush, care being taken that every part is fully covered, and that no air-bubbles remain. Then a larger quantity is poured on, and spread over the previous layer without disturbing it; a straight-edge is then passed over the moulding-frame, clearing away the superfluous plaster, and leaving that within the frame of uniform thickness. It is then left to set. When sufficiently dry, the moulding-frame is raised, and the mould with it, from off the face of the page; the mould is then dressed, and placed in a heated oven until it be perfectly dry, and raised to an adequate temperature for the casting. The oil with which the page is rubbed prevents the plaster from adhering to the type.

The melting-pot is a square vessel of iron about two inches and a half deep, having a separate lid, of which the four corners are cut off, the inner face being turned true, but the outer face hollow toward the center. A floating plate, of which the upper surface is turned, is placed at the bottom of the pot. Over the melting-pit is a crane with a rack, upon which a pair of nippers are made to run. These lay hold of ears upon the melting-pot, closing with its weight, and opening when relieved. The metal does not differ from type-metal, and must be sufficiently fluxed to flow easily, but not made too hot, or it will prove brittle. The melting-pot having been heated in the same oven with the mould, and consequently to the same temperature, the latter is placed within it, the face being turned down upon the floating-plate. A bar or other piece of iron is screwed down upon that part of the lid which is turned hollow; and the whole being suspended by the rack and crane, is swung over the melting-pit, and gradually let down into the metal, which flows gently into the pot through the openings left at the corners. The metal flowing slowly in gradually dispels all the air; the mould immediately rises to the inner surface of the lid; the floating-plate being specifically lighter than the metal, rises also to the edge of the mould; consequently the metal which has run in between is of the exact thickness of the depth of the mould, the upper surface being the field upon which are the casts of the type, the under surface the smooth surface of the floating-plate, and the rest of the melting-pot being filled with metal. The pot is allowed to remain immersed ten minutes or a quarter of an hour, that is, until the air is supposed to be perfectly expelled. It is then drawn up, and swung to a board resting upon a trough of water, and there allowed to cool. The cooling is a process requiring much care and attention. It is obvious that unless the whole mass cool equally, the plate will be warped, and consequently spoiled; it is equally clear that the heat will more readily radiate at the corners, and consequently that the center will remain fluid after the other parts are set, and that the contraction must be unequal. This is provided against by the lid having been turned hollow in the center, and it will therefore allow the metal under it to cool more rapidly. The mass having been turned out from the pot, the metal under the plate is separated by a smart blow or two of the mallet; the floating-plate will be readily disengaged, and the mould be removed from the cast. Some defects will invariably be found in a new plate; but these are removed by the picker, who goes carefully over it, clearing away the picks from the face of the letter, and deepening the larger white lines with a graver, that they may not blacken in working at press; for it must be remembered that the quadrats and spaces used in stereotyping are higher than those in movable-type printing. If the face of the plate has cooled evenly, and it is in other respects a successful cast, it is placed, the face inward, in a turning-lathe or planing-machine, and the back rendered a plane parallel to the face; the margins are then squared, and the edges flanched. The plate is now ready for use. If any errors or batters occur in the plates, they are cut out, and the corrections made with movable type let in and soldered at the back.

 A great improvement in the stereotype art was a number of years ago introduced by Mr. Thomas Allen, printer in Edinburgh, into his establishment, by which a number of plates are cast at once, whilst the risk of broken casts is considerably lessened. This is effected by means of a pot sufficiently deep to contain moulds placed in a perpendicular position. The pot is an oblong square cast-iron box, widening toward the mouth, and having placed inside, at each end, a wedge-like block, of which one face is parallel to the side, while the other is perfectly vertical. On the vertical side are perpendicular grooves, at distances rather greater than the thickness of a stereotype mould. Into these grooves are inserted plates of malleable iron, by which the interior of the box or pot is partitioned into spaces sufficiently wide to admit with ease the plaster moulds. The moulds, when baked, being inserted into these spaces, a cross bar of metal is placed over the top, instead of a cover, which serves to prevent the moulds from being raised by the liquid metal flowing beneath them; and it is then suspended upon the crane, and dipped into the metal-pit in the usual way. By this method not only the moulds are saved from all risk of breaking by being placed horizontally and pressed between the two broad surfaces of a float-block and cover, as in the method of single-page casting, but a number of plates are produced at one cast, and thus additional celerity is combined with greater certainty of sound plates. The plates of the *Encyclopædia Britannica*, which is the most extensive work ever stereotyped, were for the most part produced by this process, in pots containing each five moulds; and it is especially advantageous for large plates, the risk of breakage by the old method increasing in a greater ratio than the increase in the size of the page.

The plates are sometimes screwed down at the corners upon blocks of wood, the height of which is the difference between the thickness of the plate and the height of the type. This answers very well for jobs and standing advertisements; but for ordinary book-work it is usual to have the blocks formed of several separate pieces of mahogany furnished on one side and at one end with brass or iron catches (let in and screwed to the blocks), the upper part of which is turned over so as to take hold of the flange of the plate. But as wood is liable to warp and to other accidents, a plan has recently been devised of making hollow blocks of type-metal of the requisite height and of different sizes, by means of which pages may be easily composed to any required size, the plates being fastened on by brass holders. At a small expense, once incurred, the stereotype printer may furnish himself with blocks capable of being made up to suit works of any measure.

There are many smaller instruments requisite, which it is unnecessary to mention. The founder requires some practical skill, which, however, it is not difficult to acquire; and the excellence of the casts will depend upon his personal knack and observation. The best metal for stereotyping is composed of new metal and old type in moieties. The price of prepared metal is about 28s. per cwt.* The following, however, are proportions which may be used when the prepared metal cannot be procured:

1. From five to eight parts lead, one of regulus, one fiftieth of block-tin.

2. One seventh of pure regulus, six sevenths of lead. The best lead is that which comes from China, in the lining of tea-chests.

The mixing of the metals is exceedingly injurious to the workman, and should be avoided whenever it is possible. The foundery should be thoroughly ventilated, as the fumes from the melting-pit, and the moisture and smell of the drying oven, are very noxious.

In some cases stereotyping is of great advantage; but chiefly in books of numbers, in which it is of the utmost importance that every figure should be correct. In this case the proofs must be read again and again, until the correctness is unquestionable; when once stereotyped, there is no fear of alteration from the error of compositors or carelessness of readers, but the book remains the same for ever. Such works also are most expensive in getting up, and the cost of composition very much exceeds that of stereotyping. Books of logarithms may be especially mentioned, tables of longitude, indexes to maps, and other works, which being once written, remain unchangeably the same, such as ready reckoners, interest tables, etc.; or when it is found expedient to have duplicates of the work where large numbers are required, and it is necessary for speed to work on double-sized paper, the cast and the movable types are imposed together, and are worked side by side at the same moment, producing two copies instead of one. There is also another advantage, for

---

* In the United States, about $10 per cwt.

the stereotype remains without further expense for another edition; again, where it is expedient to send duplicate plates to other countries to be worked.

Wood-cuts may be stereotyped with great advantage; for a small cut which has cost several guineas to engrave may be multiplied indefinitely, and at a cost of only a few shillings.

No printer should stereotype by the common process who wishes his type to be a credit to his house. The wear of the type in casting is very great, especially when low spaces, etc., are used; the gypsum is at best a fine powder, and grinds away the edge and face of the letter when rubbed in with the brush, in a frightful manner. The letter can never be entirely freed from the plaster, and will present a very dirty appearance ever after. The wear of a font of 1000 lbs. weight, returned six times from the foundery, is greater than would occur in six years' constant fair usage; besides which, the high spaces, quadrats and leads, are all extra expenses, for which the economical bookseller makes no remuneration whatever.

The plan of stereotyping Bibles and prayer-books has been nearly abandoned, and the entire sheets are kept standing, in movable types, at a great expense, by the Queen's printer, and the universities of Oxford and Cambridge. Before every edition, however, is worked, each sheet must undergo a careful reading, in order to guard against accidents which may have occurred since the last edition.

Such is the process of stereotyping at this time in common use, and which will probably continue in practice in provincial and colonial printing-offices, by reason of the readiness of the materials and the knowledge now acquired by the workmen.

A greatly improved method has, however, been recently introduced by Messrs. Dellagana, by which all the inconveniences incident to the existing system are obviated. The page is composed with the ordinary spaces, leads, etc., and there is therefore no additional charge for composition; the destructive tampering with the face of the type is avoided; the plaster-mould is not required; and there is no necessity for reimposition, as the new moulds can be taken from the pages as they are imposed in the chases; and the forms can be returned to the printer within an hour from the time of their being sent to the foundery. So great are the resources of this invention that the largest or the smallest pages can be cast with equal facility, and either plane or curved to suit the periphery of cylinder machines. The pages, for instance, of *The London Times* newspaper are each cast in a single plate, in a curved form to fit the cylinders of the great machines used in that establishment. The following is a brief account of the process:

A page of a newspaper or a sheet of book-work (as imposed), carefully cleaned and perfectly dry, is laid on an iron chest previously filled with hot water. A fine brush, having the whole of its surface slightly anointed with olive oil, is rubbed over the face of the type to remove any picks or other impurities from the pages, which are then ready for moulding. A substance, in appearance resembling two or three sheets of wrapper-paper pasted together, of a soft and pulpy nature (the matrix), understood to be composed of an earthy material very finely ground, and afterward felted together, and which is not affected by heat, in a damp state, is laid smoothly on the face of the type, and carefully beaten in with a brush until every letter is indented into this substance, and the matrix is thus formed. The type, with the matrix unremoved, is taken to a press and subjected to a steady pressure, continued for two or three minutes. The matrix is then removed from the type, which may now be returned to the printer. Not more than ten minutes is required for these operations. The matrix is next laid upon a plate heated to $200°$ or $300°$, and covered with a piece of flannel (as a non-conductor of heat and an absorbent of the moisture generated in drying) upon which is placed a thin metal plate of the dimensions of the page or form, to keep the matrix flat. It remains on this hot plate about two minutes, and is then ready for casting. The matrix, with its face upward, is now placed in a "register" flat or curved, as the plates are required to be plane or convex. The register is formed of two iron plates, the inner surfaces of which are accurately planed; these plates are joined together by hinges at the further end. The matrix is placed, face uppermost, on the lower of these plates, and is secured on three sides by an iron gauge, which varies in height according to the intended thickness of the plate about to be cast. The upper plate is closed over, and the two, inclosing the matrix, are firmly clamped together by an iron bar which passes over, with a screw in the center, which presses the two plates upon the gauge. The register swings upon trunnions; and thus prepared, is turned into a vertical position, and the metal, at a temperature of 500°, is poured in through a mouth. In one minute the metal is set sufficiently hard to bear removal, the register is brought back to a horizontal position, the upper plate is thrown back, and the cast and matrix are taken out and placed (the matrix uppermost) on an iron table, which is flat or curved like the register, otherwise the cast in cooling would contract or spring, and its flatness or curvature would not be preserved. The matrix may now be carefully lifted off, and, if required, again placed in the register for another cast.

The curved casts for newspapers are fixed on the cooling table by four screws, and the dressing is performed by a tool on the lever principle, which cuts off the flange or waste piece of metal at the top of the page, and bevels it at the same time. For book-work, the under surface of the cast is planed, as in the ordinary mode. A little chiseling is required to lower the white and break lines, to prevent their blacking the paper when worked. The casts obtained by this process are remarkably true, and require little "bringing up." The matrix is uninjured by the casting, and may be used again for any number of casts, or preserved for future use. The power of multiplying casts from the same matrix is of immense advantage where large numbers are required to be printed in a short space of time. As before stated, a matrix and the first cast may be obtained in less than a quarter of an hour, and several subsequent casts will not require more than five or six minutes each. In half an hour, therefore, several machines may be at work simultaneously.

It is of course not necessary that any cast should be taken from the matrix; and therefore when a second edition of a book is doubtful, the matrix only need be made, and may be kept until required, at a cost of not more than one-third of a casting; and when used, may be put by without inconvenience, and another cast taken when the first is worked out or injured.

In book-work also this process will be found of great advantage, as compared with the charge of recomposition. The matrices of a work of 500 pages would occupy no more space than a ream of demy, and not weigh more than 10 lbs. They will remain unchanged for years if preserved free from damp or water.

The cost of casts by this process is about 10 per cent. less than by the ordinary mode; and the proportions of lead and regulus used in the composition of the metal are those given above in recipe No. 2.

The great excellence of the imperial Austrian printing establishment in the art of stereotyping should not escape mention. In the Exhibition of 1851 were some magnificent moulds taken from type by the electrotyping or galvanoplastic process. From these moulds other copies in relief were obtained by doubling the process, which are stated to produce beautiful work; or casts in type-metal could be taken of great perfection. A curious specimen was also exhibited, the work of the Rubeland ducal foundery, of a stereotype-plate of cast-iron.

OF POLYTYPAGE, AND OTHER METHODS OF PRODUCING PRINTING SURFACES ON METAL PLATES.

Many considerable improvements in stereotyping are to be ascribed to French artists; but stereotyping has never been a favorite with them, and they have rather exerted their inventive talents in a series of experiments which may be classed under the general name of *polytypage*.

In 1780 Hoffman, a German residing in France, not satisfied with his success in stereotyping, made many ingenious experiments in polytypage. Whilst he was thus engaged, a practical printer named Carez discovered a method which Hoffman afterward pursued. The page, after being composed in the ordinary manner, was attached, with the face downward, to the under side of a heavy block of wood, suspended from a long beam. Immediately under the page was an anvil, whereon was a tray of oiled paper into which the workmen poured a portion of type-metal, attentively watching the cooling. When the metal was on the point of setting, the page, block, and beam, were brought down with a very smart blow, forcing the face of the type into the setting metal, and producing a very sharp matrix; which again was made to take the place of the type upon the block, was struck in a similar manner upon the fused metal, and thus produced a perfect and excellent polytype plate. This having been properly dressed at the edges and back, was affixed to the usual wooden raiser and made type height, and might be printed separately or in conjunction with movable type. Several casts might be made from the same mould. This process was designated *cliche*.

Ign, a native of Alsace, who settled in Paris as a printer in 1784, availing himself of the discoveries made in the art of stereotyping, endeavored to extend them by inventing logotypy, or the art of uniting several characters into a single type. He printed on solid plates several sheets of his *Journal Polytype*, and advertised Father Chenier's *Recherches sur les Maures*, 3 vols. 8vo, as a polytyped book; but being deprived of his printing-office in 1787 by a decree of the council, he was prevented from executing his design.

In 1791 M. Gegembre made considerable improvements in the art of polytyping in printing the fifty-sous notes of the Caisse Patriotique. He caused the whole print of the notes to be engraved in relief upon a plate of steel, and this engraving he pressed into a plate of copper, from which polytype casts were taken. Any number of these casts could be taken from the copper mould, and if by chance the copper mould became injured, a new one could be readily made from the steel engraving.

When the revolutionary government commenced issuing assignats, it became necessary to have an immense number of plates to work the enormous quantity required of these documents. A design having been approved of, artists were employed to engrave *three hundred* fac-similes. Of course, if three hundred so-called fac-similes could be engraved, other artists would find no difficulty in engraving another hundred, nor could even the bank-officers tell which document was printed from a forged fac-simile and which from the plates engraved by their authority. The consequence was an utter want of confidence in the government paper. To remedy this, the committee of assignats caused many experiments to be instituted for the production of plates which should be not only imitative and similar, but *pro re* identical. The plan adopted was the engraving a plate *in intaglio* on steel, from which copper matrices were obtained *in relief*. From these perfect fac-similes of the original engraving were struck and were worked by the roller-press in the manner of copper-plates. But it was a great defect in this process, that the air compressed within the hollows of the letters frequently destroyed the form in the reproduction. Upon the suppression of assignats this establishment was broken up; but some of the plates and matrices are preserved in the public repositories of France.

Polytyping, as now practiced in England, is confined to the production of casts from metal plates in intaglio and from wood-cuts. Instead of the cumbrous machinery employed by Carez, a fly-press is used, the wood-cut is fixed upon what may be called the platen, and a tray containing semi-fluid metal is placed upon the table of the press immediately under the cut to be matriced. By a slow motion the cut is impressed into the metal, and an intaglio matrix is produced. The matrix is then attached to a drop stamp to perform the *cliche* process, and by the rapid descent of the stamp with the matrix attached into a tray of molten metal, a polytype in relief is obtained. The type-founders have adopted this process for the production of casts for their ornamental designs; and Mr. Bramston has practiced this mode so successfully that he is able to take fac-simile polytype casts of the most elaborately engraved wood-cuts, without in the slightest degree injuring the original.

A method of producing raised surfaces for the purposes of printing has of late years been extensively used in Paris and London, chiefly for forming maps and rough designs for the cheap illustrated press. The art is of French origin, but has been patented in England. In a patent granted in 1853 to Mr. Vizetelly it is described for "improvements for producing plates for printing surfaces, by which the manipulatory process of engraving is superseded."

A plate of highly-polished zinc, copper, or steel, is thoroughly rubbed over with very fine pounce powder moistened with water, and then with a soft dry piece of linen it is again rubbed until no greasy appearance remains on the surface, which is now in a fit state to receive the transfer.

Where the engraving has been recently printed, say within a month, the transfer is thus effected: The print is soaked for five minutes in a flat dish containing a liquid composed of seven parts of water, one of azotic acid, and six drops of phosphoric acid. It is then taken out and placed between two sheets of blotting-paper, to absorb the superfluous moisture, after which it is laid on the prepared plate and covered with a sheet of soft paper, and subjected to the strong pressure of the lithographic press. When the transfer is thus effected, the plate is washed with a sponge moistened in a solution of gum-arabic, slightly acidulated with nitric acid; this preparation having remained on the plate for five minutes, is sponged off with clean water. While the plate is still wet, a lithographic roller charged with ink composed of bitumen of Judæa, powdered very fine with a muller and mixed with linseed oil, is passed over it. The linseed oil must be of the purest quality, and be boiled for at least an hour, and afterward filtered through a felt bag containing some animal black. For zinc plates, lithographic transfer-ink and melted virgin wax, well mixed and ground together, must be substituted. When the plate is well rolled over with this ink, it will be observed that the transfer only has taken up the ink, the parts of the plate where the lines of the print do not occur having no power to take it up. While the ink is still wet, some resin, ground to an impalpable powder, is distributed over the plate with a piece of cotton wool or a camel's hair brush, care being taken that it adheres to the inked transfer only, and not to the other parts of the plate. The plate is now placed over a spirit-lamp, and gradually heated until it becomes luke-warm, in which state it is allowed to remain undisturbed for at least two hours; if expedition is not required, it will be better not to disturb the plate for twelve hours, as the resin and ink will then have thoroughly combined, and more completely protect the portions of the plate covered by the transfer from the corroding action of the acid, by which the surface in relief is produced. Before the plate is subjected to this "biting" process, it is necessary to cover its back with a varnish or other substance, to protect it from the action of the acid. When this is done, it is placed in a slanting position, and a liquid composed of nitric acid, diluted to about $4°$ Reaumur, for zinc and steel plates, and to about $12°$ for coppers, to which is added a table-spoonful of spirits of wine to every half-pint of acidulated water, is applied with a clean sponge to the surface of the plate. This bathing is continued for a quarter of an hour, and pure water is then poured over the plate until the acid is entirely washed off. The plate is then again sponged over with the slightly-acidulated gum-water, reinked, submitted to the action of the acidulated water, and washed with pure water as above described; and these operations are repeated four or five times, until the exposed portions of the plate are so much bitten away by the acid as to leave the transfer sufficiently in relief to be printed from.

The "whites," *i. e.* the blank spaces in the engraving, must be lowered or removed to prevent their receiving the ink in process of printing, and blacking the paper. This is effected by covering the surface of the raised lines of the transfer, and the sides also where practicable, with engraver's varnish, which is composed of bitumen of Judæa dissolved in essence of turpentine, with the addition of lamp-black to make it of a proper consistency, and allowed to stand two hours before it is used. The plate is then bathed with the solution of acidulated water and spirits of wine, and washed as before described; but in this operation a stronger solution is used, being 8° instead of 4°. Where the whites are very large, essence of spikenard (aspic) is substituted for essence of turpentine, or they may be lowered by scrapers or gouges, or cut out with a fine saw. Great care must be taken that the bitumen is entirely dissolved, and that the varnish is made of the proper consistency.

A raised printing surface being now produced, the plate is cleaned with turpentine and well rubbed over with charcoal, after which it may be mounted on raisers to type height, and used as a stereotype cast.

When an *old* print is to be transferred, it is treated in the manner commonly employed by lithographic printers prior to making a transfer.

Anastatic printing is a process by which a print, whether from type or a copperplate, may be reproduced without drawing or engraving. The print is saturated with a strong solution of nitric acid; it is then placed between sheets of blotting-paper, and the superfluous fluid absorbed; after which it is laid, face downward, upon a polished plate of zinc, and another placed over it. The plates are then passed between iron rollers, and subjected to great pressure. The nitric acid is thus squeezed out upon the zinc, except in those parts which are protected by the ink of the old print. The acid bites away the zinc, and a rough surface is produced, the protected parts continuing bright and unaffected. The plate is then wetted with a solution of gum in water. The corroded surfaces retain the fluid, while the unaffected portions remain dry. A roller charged with the ink used by copperplate printers is then rolled over the plate: the ink covering the dry and being repelled by the wet surfaces, This is repeated until the lines of the print are well covered with the ink— a process which is rapidly effected if the ink of the original print is fresh, and has parted with a portion of its oil under the pressure of the rollers. Impressions may now be readily taken in the same manner as lithographic prints.

PRINTING FOR THE BLIND.

The invention of printing for the blind forms a new era in the history of literature. In European countries, one individual in every 1200 or 1400 of the entire population is blind, and in America one in every 2000. To open up to this large and unfortunate class such a source of profit and pleasure as reading could afford was long considered very desirable, and also very doubtful; but while, of late years, embossed books have very rapidly increased, it is exceedingly gratifying to find that blind readers have far more rapidly multiplied. The credit of this invention belongs to France. In 1784 Valentine Haüy printed the first book at Paris with raised letters, and proved to the world that those for whom such books were intended could easily be taught to read with their fingers. He seems to have caught the hint from a blind pianist of Vienna, who distinguished the keys of her instrument by the sense of touch. After many experiments as to the form of his raised letters, he at last chose a character a little approaching the Italic. A new institution was at once established—*Institution Royale des Jeunes Aveugles*—and Haüy was placed at the head of it. Twenty-four of his pupils exhibited their attainments in reading, writing, arithmetic, music, and geography, before the king and the royal family at Versailles, on the 26th December, 1786, to the very great delight of those high personages. In 1814, when Haüy was pensioned off, Dr. Guillié was chosen in his stead. This enterprising *directeur-general* modified Haüy's letters, and prosecuted the publication of embossed books with renewed vigor. Still, however, very little progress was made toward the extension of Haüy's system; and their books could only be read by those possessing a very delicate touch. In 1806 M. Haüy established schools for the blind in Germany and St. Petersburg, but they have made very slow progress. It was in Scotland and the United States that improvements were first made in embossed typography. To Mr. James Gall of Edinburgh belongs the merit of reviving and improving this very useful art. After canvassing every form of letter, he at last adopted his angular alphabet. Before 1826, when Mr. Gall began his experiments, not a single blind person using the English language could read by embossed printing. On the 28th September, 1827, he published *A First Book for Teaching the Art of Reading to the Blind*, the first book printed for the blind in the English language. In October, 1834, this zealous individual published in a perfected alphabet *The Gospel by St. John, for the Blind*. The text, which was embossed, and, unlike his former effort, printed not with wooden but with metallic types, consisted of 141 pages, with 27 lines on a page of 70 square inches. This book was counted a great improvement, but it was objected that the types were too angular. He afterward printed a number of books with serrated edges. It is unquestionably to Mr. Gall, more than to any other man, that the interest in the education of the blind was awakened throughout Great Britain and America. While Mr. Gall was engaged in perfecting his plan in this country, Dr. S. G. Howe, of the Perkins Institution, Boston, United States, was busily engaged in developing his system. In 1833 Dr. Howe began, like Gall, by taking Haüy's invention as the basis of his system, and soon effected those improvements upon it which have given so wide a fame to the Boston press. He chose the common Roman letter of the lower case, reducing it by cutting off the flourishes, etc., until it occupied but a space and a half instead of three. This alphabet remains unchanged. So rapid was his progress, that in 1836 he printed in relief the whole of the New Testament for the first time in any language, in 4 small quarto volumes, comprising 624 pages, for four dollars. More than twelve times this amount has now been printed, and seventeen of the American States have adopted Dr. Howe's method.

The Society of Arts in Edinburgh awarded a medal, on the 31st of May, 1837, to Dr. Fry of London for the *invention* of an alphabet, which seems, however, to have been in use in Philadelphia since 1833. Mr. Alston of Glasgow improved upon Fry's alphabet, by reducing the size of the letters, and sharpening the embossing. In 1840 Mr. Alston published the entire Old Testament in 15 quarto volumes, of 2535 pages, and 37 lines to a page, in double pica type. Alston, in his just pride, designated this "the first Bible ever printed for the blind;" in which he was wrong, however, for Boston had claimed the honor years before. Some 70 distinct volumes have been printed by the Glasgow press; but since the death of Alston, on the 20th of August, 1846, it has almost ceased to work. Since 1837 it has supplied England, Ireland, and Scotland with embossed books in Roman type. The best of all the arbitrary systems is that of T. M. Lucas of Bristol, who set it on foot about 1835, and which "The London Society for Teaching the Blind to Read" has been gradually improving since its establishment in 1839. In May, 1838, "The London and Blackheath Association for Embossing the Scriptures" adopted the phonetic method of James Hartley Frere. A cheap plan of embossing or stereotyping was devised by Mr. Frere in 1839. His books read from left to right, and back, after the ancient Greek βου-στροΦηδόν writing. Mr. Moon, of the Brighton Blind Asylum, has slightly improved on Mr. Frere's method. Dr. Howe's typography is judged, however, to be superior to the British both in cheapness and in size. There are at present no less than five different systems of typography in use in Great Britain.

The following table shows the results of the six systems of printing for the blind used in the English language, taking the New Testament as a standard of comparison:

Systems.	No. of Vols.	Size.	No. of Pages.	No. of Lines in a Page.	No. of Square Inch. in a Page.	Price.
The New Testament—						*L. s. d.*
Howe's	2	4to.	430	......	117	0 16 0
Alston's	4	"	623	42	90	2 0 0
Gall's	8	"	......	28	70	2 0 0
Lucas's	9	"	841	27	70	2 0 0
Frere's	8	Ob. 4to.	723	......	110	2 10 0
Moon's	9	"	......	25	110	4 10 0

OTHER PROCESSES.

To the magnificent establishment of the imperial printing-office at Vienna we owe the introduction of several processes, which, though not founded on the use of type, belong to the art of printing. The description of these new arts is derived from the Reports of Jurors of the Exhibition of 1851.*

Galvano-plastic Process.—The Austrian department contained some extraordinary prints of fossil fishes, which were produced by the following process: By means of successive layers of gutta percha applied to the stone inclosing the petrified fish a mould is obtained, which being afterward submitted to the action of a galvanic battery, is quickly covered with coatings of copper, forming a plate upon which all the marks of the fish are reproduced in relief, and which, when printed at the common press, gives a result upon the paper identical with the object itself.

Galvanography.—The artist covers a plate of silvered copper with several coats of a paint composed of any oxide—such as that of iron, burnt terra sienna, or black-lead—ground with linseed oil. The substance of these coats is thick or thin according to the intensity to be given to the lights or shades. The plate is then submitted to the action of the galvanic battery, from which another plate is obtained reproducing an intaglio copy, with all the unevenness of the original painting. This is an actual copperplate resembling an aquatint engraving. It may be touched up by the engraving-tool. This process has been improved upon by outlines etched in the

* At London.

usual manner, and the tones laid on with a roulette. A galvano-plastic copy of this sunk plate is obtained. On this second raised plate the artist completes his picture by means of chalks and Indian ink, and puts in the lights and shades; from this a second galvano-plastic copy is produced. This second copy or sunk plate, the third in the order of procedure, serves, after being touched up, for printing from in the copperplate-press.

Galvanoglyphy.—Upon a plate of zinc coated with varnish a drawing is etched; then ink or varnish is rolled over. The ink adheres only to the parts it touches, every application when dry raising the coating and consequently deepening the etched lines—a galvanic battery produces a plate in relief, which is printed at the common press.

Chemitypy.—A polished zinc plate is covered with an etching-ground. The etching is bitten in with diluted aquafortis. Remove the etching-ground, and carefully wash out the aquafortis. Heat the plate thus cleansed over a spirit-lamp, after covering with filings of a fusible metal, until fusible metal has filled all the lines of the engraving. When cold, scrape down to level of zinc plate until none of the metal remains but what has entered into the engraving. Place compound plate in solution of muriatic acid; and as of the two metals one is positive the other negative, the zinc alone is eaten away by the acid, and the fusible metal which had filled the lines of the engraving is left in relief, and may be printed by the common press.

Paneiconography.—On a polished plate of zinc draw with lithographic crayon or ink, or transfer impressions from lithography, wood engraving, or copperplates. The thickness of the drawn lines is increased by repeated rollings or powdered resin. For relief-block, place plate in trough of very dilute sulphuric or hydrochloric acid. The acid eats away the unprotected parts of the plate, and leaves raised lines of the protected parts.

NATURE-PRINTING.

Mr. Henry Bradbury, who has had a principal share in introducing this beautiful process into England, describes it as a method of producing impressions of plants and other natural objects, in a manner so truthful that only a close inspection reveals the fact of their being copies. So deeply sensible to the touch are the impressions, that it is difficult to persuade those who are unacquainted with the manipulation that they are the production of the printing-press. The process, in its application to the reproduction of botanical subjects, represents the size, form, and color of the plant, and all its most minute details, even to the smallest fibers of the roots. The distinguishing feature of the process, compared with other modes of producing engraved surfaces for printing purposes, consists, firstly, in imprinting natural objects—such as plants, mosses, sea-weeds, feathers, and embroideries—into plates of metal, causing, as it were, the objects to engrave themselves by pressure; and, secondly, in being able to take such casts or copies of the impressed plates as can be printed from at the ordinary copperplate-press.

The art is by no means new in idea, many persons having attempted something analogous to the present process, and produced results which were imperfect, merely because science had not yet discovered an art necessary to its practical development. It is to the discovery of electrotyping that the existing art of nature-printing is due.

The progress of the art, and the persons to whose ingenuity the steps were severally due, are stated by Mr. Bradbury thus:

Professor Kniphof of Erfurt took impressions from leaves, etc., which had been colored with lamp-black, printers' ink, etc., 1728–57.

Kyhl, a goldsmith of Copenhagen, took copies of natural objects in plates of metal between two steel rollers. These were not for the purposes of printing, but for reproduction of embossing and ornamentation in metal. 1833.

In 1851 Dr. Ferguson Branson of Sheffield read a paper before the Society of Arts, in which he detailed some experiments in nature-printing. He had taken impressions from plants, etc., in gutta percha, for the purpose of having them printed. The experiment failed through the softness of the material. Dr. Branson then bethought himself of the electrotype process; but appears to have found it too tedious and costly, and he abandoned the idea.

In 1849 Professor Leydolt of Vienna availed himself of

the facilities afforded by the imperial printing-office to carry out experiments in the representation of flat objects of mineralogy,—such as agates, fossils, and petrifactions,—and obtained great results. Soon after, Haidinger and Abbate suggested, the former the reproduction of plants, etc., and the latter the representation by this means of different sorts of ornamental woods on woven fabrics, paper, and plain wood; and lastly, Andrew Worring, of the imperial printing-office, Vienna, perfected the application of these processes to printing, 1853.

These circumstances are dwelt upon at some length, because nature-printing is yet in its infancy, and appears capable of development to a degree at which it will be an impressorial art of greater importance than any which has been invented since the art of printing itself. Worring's services were so highly estimated that the emperor rewarded him with a munificent gift, and with the Order of Merit.

The plant, perfectly dry, or any other suitable subject, is placed on a plate of fine rolled lead, the surface of which has been polished by planing. The plate and subject are then passed between rollers, by the pressure of which the subject is forced into the surface of the lead. The leaden plate is then subjected to a moderate heat, by the action of which the subject is loosened from its bed and easily removed. This mould is then subjected to the galvano-plastic process, the second cast being a perfect fac-simile of the leaden mould. When the subject to be printed is of one color only, that pigment is rubbed in, and any superfluity removed; but when it is of two or more colors, the process is simple, but, it is believed, perfectly novel in any process of printing heretofore practiced. In the case, for instance, of flowering plants, having stems, roots, leaves, and flowers, the plan adopted in the inking of the plate is to apply the darkest color, which generally happens to be that of the roots, first; the superfluous color is cleaned off; the next darkest color, such, perhaps, as that of the stems, is then applied, the superfluous color of which is also cleaned off; this mode is continued until every part of the plant in the copperplate has received the right tint. In this state, before the plate is printed, the color in the different parts of the copper looks as if the plant were imbedded in the metal. The plate thus charged, with the paper laid over it, is placed upon a copper-plate-press, the upper roller of which is covered with five or six layers of blanket of compact fine texture. The effect of the pressure is, that all the colors are printed by one impression; for when the paper is removed the plant is seen quite perfect, highly embossed, with the roots, stems, and other parts, each of its proper tint.

The great national work which the Austrian establishment has produced as the exemplar of the new art is truly imperial. The *Physiotypia Plantarum Austriacarum* consists of five volumes large folio, containing 500 plates (about 600 plants), with a quarto volume of plates and text. The first production of the English press, though it will bear no comparison in extent with the imperial magnificence of the Austrian work, fully equals it in beauty of execution. It is *The Ferns of Great Britain and Ireland*, by Thomas Moore, edited by Dr. Lindley, imperial folio, with 51 plates. It is printed by Mr. Bradbury.

PRINTING IN COLORS.

One of the most beautiful aids to typography, the art of printing in colors, has been unduly neglected in this country; at least so far as relates to the embellishing works of ordinary excellence with vignettes, capitals, tail-pieces, and other devices of fancy, in beautiful tints, in the manner of the early typographers. It is true that some very beautiful works, illustrated with remarkable richness of design and color, have been produced; but these have been executed rather as examples of the beautiful in art than as books,—the work of the artist has been the principal object, and the work of the author the occasion and vehicle. In other works, chiefly ecclesiastical, the object has been to reproduce in *fac-simile* the rich illuminations of the monkish scribes. But as regards the average printing—the literature of the day—the art of printing in colors has been very much neglected. This may very easily be accounted for. To print in two colors occupies more than twice the time necessary to print in one; and it also requires more skill and ingenuity. These unfortunately must be paid for; and this pecuniary consideration is sufficient to banish from our pages this lovely art. So did not our forefathers; *they* took pride in choosing the most tasteful designs, the most harmonious colors, to illuminate their productions, and beguile the reader into study by the illusive charms of gold, and blue, and crimson. Fortunately, either time was of little value, or the exclusive possession of the market enabled them to demand remunerating prices for the time thus well bestowed; but in the bustle and competition of our more mercantile days, time is money, and blue and gold, scarlet and green, give way to the equally useful but infinitely less beautiful uniformity of unredeemed black. To a country printer, however, some knowledge of color-printing would be of advantage, because, as his fonts of type are more limited, he can create unlimited variety by a judicious use of colors in job-work: moreover, as he has usually much more time upon his hands, his ingenuity would have ample scope for the production of small works of *vertu*, in a taste which cannot be indulged by the denizens of a busy metropolis.

Except in the execution of works of a very high order, and the imitation of intricate and delicate patterns, printing in colors requires no addition to the ordinary accomplishment of printing, other than considerable ingenuity and a little practice in preparing the colors. The latter may, it is true, be purchased of the ink-maker, prepared for use; but the charge for them is enormous, and they require constant replacement, whilst it is not possible to have on hand every variety of tint. By the purchase of the most simple materials from the oil-shop, the ingenious printer has at his hand every color that fancy can require, at the most moderate cost, without waste or delay. The appliances are few and cheap: a muller, a marble slab, and the pallet-knife; the materials, a can of printers' varnish, to be purchased of the ink-maker, which will keep any length of time, and the raw colors hereafter given, which may be purchased from time to time; care, however, being taken that they are of the best quality, or they will fade and turn rusty in a short time, and be a deformity instead of an ornament to the work.

Useful tints of *red* may be prepared of orange lead, vermilion, burnt sienna, Venetian red, Indian red, and lake. Vermilion is the most brilliant of these reds; but its beauty depends very much upon the particular parcel used. The pale vermilion is best for a bright tint, as the dark, when mixed with the varnish, produces a dull red. Orange lead and vermilion ground together produces a very bright tint, which is more permanent than vermilion alone.

Yellows are prepared with yellow ocher, gamboge, and chromate of lead. Of these, the brightest is the chrome; yellow ocher, when mixed with the varnish, produces a very dull tint.

Blues are made from indigo, Prussian blue, and Antwerp blue. Of these, indigo is exceedingly dark, and not very easily lightened. Prussian blue is a very useful color; Antwerp blue is very light.

Greens may be produced from a mixture of any of the blues and yellows, as gamboge and Prussian blue, chromate of lead and Prussian blue. These may be mixed in any proportions until the required tint is produced; but it must be remembered that the varnish has a considerable yellow tinge, and will produce a decided effect upon the mixture. With a slight portion of Antwerp blue it will, without the mixture of any of the yellows, produce a decidedly greenish tinge.

Purples of any degree of richness are made by judiciously mixing reds and blues.

Sæpia produces a nice brown tint, burnt umber a very hot brown, raw umber a much lighter brown, bister a brighter still. Neutral tints may be obtained by mixing Prussian blue, lake, and gamboge. In fact, every pigment that painters use can also be used in printing, avoiding, as much as possible, all heavy colors. In truth, if the printer is desirous of imitating any particular color, or of producing any particular tint, he cannot do better than consult the nearest artist in oil or water colors (oil in preference), or in default of that, the neighboring house-painter.

The necessary colors having been procured, the method of preparing them is very simple. Each must first be well ground by the muller upon the slab, even although they may have been purchased well powdered. The color should then be well mixed with the pallet-knife with the varnish, until the pigment has attained the required consistency, which will vary with the quality of the work to be executed; for if it be a posting-bill or coarse job, the ink should be very thin, and consequently a much larger proportion of varnish should be used. If, however, the work be a wood-cut, or in small type, the pigment should be made as thick as possible. If the color required be a compound, the predominant tint should be first mixed with the varnish, and the lighter tint added in small quantities, until the exact shade required be produced. Thus, if the color be a dark green, the blue should be mixed up first, and the yellow added; but if it be a very light green, then the yellow should be first applied, and the blue added. If the tint desired be exceedingly light, it will be found that the quantity of raw material to be employed will not make the mixture sufficiently thick to be applied to the type or wood-block: in this case whitening is added to thin colors, and dry white-lead to the heavier, in considerable quantities, which must be adjusted in the course of mixing. To insure thorough combination, the mixture should be scraped into a corner of the slab, and a very small portion of it spread with the pallet-knife, and well ground with the muller until no specks or lumps appear, then scraped up and placed in another corner. This should especially be done when white-lead is used, as it will be found that every little lump when crushed will produce a white streak upon the slab. If this be not carefully done, independently of its tendency to clog the type, it will very materially alter the tint. When the pigment seems sufficiently mixed, it is better to bray it out with the muller instead of the usual brayer, and grind again each particular portion immediately before it is used. Colors may be worked either with a ball or a roller. If the job be large and coarse, and the ink consequently thin, the roller will answer every purpose; but if it be small, and requiring much nicety in the manipulation, decidedly with a ball; but in either case the ink should be well distributed, and the form well beaten or rolled. When two or more colors are employed, they must be worked at as many different times. In this case extreme nicety in the register and justification is required, in order that every color may fall in its just place, without overlaying any other tint employed in the print. This would be a great dis-sight in any case, but most especially where the combination of colors would produce a third; as, for instance, if any part of a blue line should unfortunately fall upon a yellow, a green outline would be the result. The simplest way to guard against this is to have the wood-blocks all cut to precisely the same size, with the print in the proper place upon each; when, therefore, the first color has been worked, the form is unlocked, the block taken out, and the second block inserted; it then falls at once into its proper position. If the form consist of type, each line should be carefully composed in its proper body; that is, if three colors be employed for as many different lines in *pica*, *small pica*, and *long primer*, the one to be first worked should be composed in pica *letters*, the other lines in small pica and long primer *quadrats*. When the second line is to be worked, its quadrats should be taken out and letters inserted, while the type of the first line should be removed and quadrats substituted; and so of the third line. The points on the tympan must never be moved. It is clear, therefore, that if the paper be placed upon the same pointholes as before, and if the form has never been moved, the new line cannot fail to fall in its proper place. In these cases the paper must never be suffered to dry; indeed the sooner each color succeeds the other the better. If it be covered with a wet blanket, and the edges well sprinkled, the danger will be little; but if it should dry and shrink in the slightest degree, it will be impossible to obtain register. For printing *red-letter days* in almanacs and the rubrics in prayer-books (an almost extinct practice), an especial type is used called rubrical; it is cast about an m higher than ordinary type. The black is first worked, quadrats having been inserted in the places of the red-letter, which are subsequently withdrawn and the rubrical type inserted. But as, in so small an insertion in so large a body this process does not attain any very good register, and is expensive withal, the red-letter days have been abandoned, and some other distinguishing type (generally old English or black) has been substituted, which sufficiently indicates the day. It would not be possible here to give sufficient instructions to enable the printer to execute landscapes, portraits, and other delicate subjects, in various colors and shades. The difference between this and other color-printing consists mainly in the superior individual skill and ingenuity of the artist, the excellence and truth of his engravings, and the superiority of his appliances. In truth, before the printer can produce any great effect, he must be excellently qualified as a painter, which it is not the province of an article on printing to teach. It will be sufficient to state that the lighter and more extensive tints, and especially those in which transparent colors are used, are worked first; that the color is gradually deepened by successive blocks until the required effects are produced; and that the outline is printed last, which has the effect of giving sharpness and finish to the design.

The curious reader is referred to Mr. Savage's beautiful book on Decorative Printing, and to the many admirable productions of Mr. Baxter and Mr. Vizetelly. Nor should the accurate work and beautiful colors of Mr. Delarue's playing-cards be passed over without notice. To Mr. Delarue, indeed, the revival of color-printing in England as a practical art is greatly due.

The lottery system and the stamp duties gave extensive employment to the color-printer, and also gave occasion to a process which is denominated "compound plate-printing." The effects are produced by an ingenious system of mechanism, by which several plates are made to separate for the purpose of receiving the colors, and to combine with perfect accuracy, for the purpose of transferring these colors to the paper by a single impression. This process is in daily use at the stamp and excise offices, and the most familiar examples are to be seen in the intricate patterns printed on the labels of reams of paper, or those of patent medicines. The printing is effected by the cylinder printing-machine with the greatest rapidity.

There is no difficulty in printing in gold; it is within the power of any typographer. The type is composed and made ready at press in the usual manner. Take the best printer's varnish, grind it to a thick consistency with burnt sienna or brown umber; reduce this with gold-size, the same as that used by gilders and japanners. The first admixture is necessary because it has been found that the umber will not combine with the size. The type is then rolled with this compound in the same manner that ordinary ink is applied, and the impression is taken upon the paper. Leaf-gold is then laid over it with a piece of cotton-wool, and pressed lightly upon it. When the varnish has had time to set, a piece of cotton-wool is rubbed steadily over the part printed, and the superfluous leaf is thereby removed, leaving the gold adhering to the varnish. The print should then be passed between steel rollers, or hot-pressed—care being taken in the latter process that the plates be not too hot, or a dull drossy surface will be produced. The sharpness of the print will vary with the judgment of the printer in the quantity of sizing applied to the type; for if the press-work be bad, the print will be bad also. For inferior gold-printing bronze-powder is extensively used. For this the varnish is made very much thicker than for gold; the method of printing is the same. After the impression has been given, the powder is brushed over the print, and adheres thereto, whilst the superfluity is easily removed. In printing the golden "Coronation Sun" with this powder, a very distressing disease arose,—the hair became perfectly green, and the men were very seriously affected; great care should therefore be taken that particles of the powder be not allowed to fly about the room. Dutch gold cannot be used as a substitute for gold-leaf. When all these appliances cannot readily be obtained, very fair gold-printing may be produced by the following process: Let the surface of the type be heated by any convenient means—as by laying upon it for a space a heated metal plate—and then cover it carefully with leaf-gold by a ball of cotton-wool. Having carefully sifted dry white-of-egg or resin, finely pulverized, over the surface of the paper, place it on the tympan, and bring it gently down upon the type. Dwell upon the pull. The leaf-gold will be found perfectly adherent to the impression on the paper, and the superfluous part may be brushed off. The sheet, after drying, should then be hot-pressed. Some observation is required to ascertain the proper heat to be given to the type: if it be insufficient, the gold transfer will be imperfect and the tint light; if too great (of which there should be no danger) the color will be dull.

BANK-NOTE PRINTING.

The Bank of England notes were formerly printed from steel-plates; but in 1853 the Bank adopted the surface or letter-press mode of printing. The plates are produced by the electrotype process. An original is first engraved in metal *in relief*. This original is subjected to the galvano-plastic process, by which a matrix is obtained, and from this matrix a second cast is obtained in relief, a perfect *fac-simile* of the original engraved plate. From this plate the bank-notes are printed. The metal of which these plates are formed is exceedingly hard, frequently yielding nearly one million impressions without being worn out. The original engraving is never used for printing, but only for the production of matrices; consequently it always remains unimpaired, and thus perfect identity is maintained in the appearance of the notes.

The notes are printed at platen-machines possessing great advantages over the ordinary printing-machines, more particularly in the distribution of the ink. Three machines are employed, two of which were manufactured by Messrs. Napier & Sons, and the other by Messrs. Hopkinson & Cope. A *tell-tale*, or register, is attached to each machine, which marks the number of impressions. These registers are set by a clerk before the printing commences, and are checked by him at the close of the day, when the printer must account for (either in bank-notes or "spoils") the number of impressions registered by the dial. The notes are printed upon dry paper, a process which has been very greatly accelerated by the recent improvements introduced into the ink by Mr. Winstone, who manufactures for the bank.

The number and dates of the bank-notes are added in an after-printing. This is effected at Messrs. Napier & Sons' cylinder machines: a very ingenious mechanism being attached to these machines which makes it impossible to commit any fraud by printing two notes of the same number. The apparatus consists of a series of brass discs, of which the rim is divided by channels into projecting compartments, each containing a figure. The numbers 1 to 9 having been printed in the course of the revolution of the first disc, the second disc then presents the figure 1, which, by combining with the 0 of the first disc, the number 10 is formed. The second disc now remains stationary until, in the course of the revolution of the first disc, the numbers 1 to 19 have been printed, when it presents the figure 2, and does not again move until another revolution of the first disc completes the numbers 20 to 29. Thus the two discs proceed until 99 notes have been numbered, when the third disc comes into operation, and with the first two, produces 100, consequently the first disc performs one hundred revolutions to ten of the second and one of the third. The notes may be numbered indefinitely by this process, without the possibility of error, the machine, meanwhile, being its own check.

PRINTING-MACHINES.

As long as the thirst of literature was confined to books and a few periodicals of limited sale and size, the ordinary printing-presses sufficed to supply the demand : nor was it discovered that any further speed was requisite, until the increased facility of conveyance, and the important events at the close of the last century, created a demand for news which the utmost exertions of the printers were unable to supply ; for the attempt to increase the speed by the composition of two distinct forms of type would avail little, so long as the presses could turn out only 250 or 300 impressions each per hour. Accordingly for this branch of the art were the first machines projected. Many schemes were proposed for accelerating the movements of the press; but the first attempts at any thing like the machine afterward introduced were made by William Nicholson, a gentleman connected with periodical literature, who took out a patent about 1790 for a printing-machine, of which the chief points were the following : The type being rubbed or scraped narrower toward the bottom, was to be fixed upon a cylinder, in order, as it were, to radiate from the center of it. This cylinder, with its type, was to revolve in gear with another cylinder covered with soft leather (the impression-cylinder); and the type received its ink from another cylinder, to which inking apparatus was applied. The paper was impressed by passing between the type and impression-cylinders. Most of these plans were, when modified, adopted by after-constructors. This machine was never brought into use.

König, an ingenious German, was the next who undertook to construct a machine; and having made considerable advance in his plans, obtained a contract with Mr. Walters, the proprietor of *The London Times* newspaper, for manufacturing two for that journal. His machine was successful, and the number for the 28th November, 1814, was worked by it at the rate of 1100 impressions per hour. In this Nicholson's plan was so far altered, that the ordinary type was used and laid upon a flat surface, and the impression was given by the form passing under a cylinder of great size. König afterward invented a machine in which the sheet was printed on both sides before it left the machine; but his arrangements for the equal distribution of the ink were so complicated and clumsy (consisting of not less than forty wheels) and the works of every part of the machine so intricate, that it never came into practical use.

The first really useful machine was constructed by Messrs. Applegath and Cowper, being an extensive modification of that of König; its principal improvement consisting in the application of two drums between the impression-cylinders, one of which reverses the sheet, and the other secures the register, by retaining it, after the impression of the first form, just so long that it may pass on to the second cylinder in exact time to be impressed thereby upon the second form; and of the distribution of the ink upon a plane surface, instead of by a number of rollers, by which König's complicated machinery was got rid of. These machines, with numerous modifications, according to the plans of different makers, are now in general use.

For newspapers, machines are generally made to work but one side at a time. It is manifest that a machine will work a much greater number (more than double) of one form than of two, and that the machinery will be lighter and less expensive, and of course require less motive power. One form, therefore, of a newspaper, containing advertisements and the less important matter, is worked at leisure; and the second form, containing the leading article, important news, and other matter of consequence, is reserved until the last moment, and is then thrown off with immense rapidity. For the usual description of book-work, machines (perfecting-machines) are constructed to work both forms at a time. In these, perfect register, and the exact and even distribution of the ink, are of the greatest consequence, and such immense rapidity is not necessary. These machines, therefore, differ very much in construction, though not in principle, from those used for newspapers.

The machine constructed by Messrs. Applegath and Cowper in 1827 for *The Times*, two of which are still used for printing the supplements and advertising pages, has four impression-cylinders, which are so arranged that two are in contact with the type as the table passes to the right, and two as it passes to the left. It will print from 4000 to 5000 impressions per hour.

One of the principal impediments to great speed in this form of printing-machines is the necessity for a reciprocating motion in the type, table, and inking-table,—a great weight, the *vis motus* of which has to be neutralized, and then the *vis inertiæ* overcome, at each end of the traverse. This not only occasions a great waste of motive power, but also causes breakages and serious accidents. Mr. Applegath, finding these and other difficulties insuperable, abandoned the principle of placing the type on a plane table and the reciprocating motion, and constructed a machine in which the type is placed on the surface of a cylinder of large dimensions, which revolves on a vertical axis, with a continuous rotatory motion. *The Times* has the credit of being first in adopting this great improvement in newspaper printing.

The following is a careful description of this vast and complicated piece of machinery:

In the center of the machine is a vertical cylinder or drum, 5 feet 4 inches in diameter. In contact with it, and revolving each on its own vertical axis, are eight impression-cylinders, 13 inches in diameter, each of which has a set of inking-rollers working in advance of it. The cylinders move with the same velocity as the surface of the drum. The columns of type are placed in a kind of iron galley, or *turtle*, curved to fit the surface of the drum. The outer surface of these galleys is not formed into a segment of a circle, but into facets, each the width of a column; the wedge-shaped interval, which is left between the top and bottom of the types of every two adjoining columns, is compensated by column-rules, made thicker at the top than at the bottom in the same proportion. The middle column-rule is fixed. The columns are locked up in the galleys by means of screws, and the column-rules press the types together like key-stones in an arch. The fixed rule in the center prevents the types from rising. The galleys are then screwed on to the drum, the columns vertical. The outer face of the forms is now, it must

PRINTING.] [PLATE 13.

THE ADAMS PRINTING PRESS.

be remembered, a series of facets, sides as it were of a polygon; the surfaces of the impression-cylinders are made to conform to these facets, with sufficient accuracy, by paper overlays. When stereotype plates are used, they are cast by Dellagana's process, in accurate segments of a circle, and the overlaying is unnecessary. The forms of types do not, of course, occupy the whole circumference of the central drum: a large part of the remainder is made the inking-table. The ink-box, which is also vertical, supplies ink to a ductor-roller, which works between two straight edges. As the drum revolves, a portion of ink is taken from the ductor by two vibrating rollers, and distributed on to the inking-table. The inking-table precedes the type-forms, and as it passes the inking rollers attached to each impression-cylinder come into contact with it, and receive ink from its surface. The type-forms, following next, come into contact with these inking-rollers, and take from them the ink they have just received. The inking-table passes under the impression-cylinders without touching them; but the type is brought into contact with the paper upon them, and the impression is given. Therefore, at every revolution of the drum, the type is inked eight times, comes into contact with eight impression-cylinders, and prints eight sheets of paper.

It is most difficult to convey, by any verbal description, the singularly ingenious mechanism by which the sheets of paper are conveyed to and around the impression-cylinders. It must be remembered that the sheets are necessarily laid on the feeding-table *horizontally*, and that they pass around the cylinder *vertically*. The task will be rendered somewhat simpler by reminding the reader that each impression-cylinder is a complete machine within itself, acting with the drum, but independent of the other cylinders; and that, as each has its own system of inking-rollers, so each has its own system of feeding-drums and tapes. The white paper is laid on the feeding-table at the top; each sheet is placed by the layer-on to the center of a feeding-drum. At the right moment, the sheet is advanced by finger-rollers until its forward edge is brought between two small rollers, each connected with a series of endless tapes, between which it is passed vertically downward. At the right moment its further progress is arrested by two vertical slips of wood called "stoppers," which start forward and press the sheet against two fixed stoppers; and, at the same moment, the two rollers and their tapes separate, and leave the sheet extended vertically between the two pairs of stoppers. Observe that, up to this moment, the travel of the sheet has been vertically downward, and that its plane surface is part of a radius from the axis of the central drum. The problem now to be solved is, to give it a horizontal movement toward the center, preserving its vertical position. The instant the sheet is arrested vertically between the stoppers, its top edge is caught by two pairs of small finger or suspending rollers; at the same instant the stoppers separate, and the sheet is suspended for a moment between these rollers; a slight inward motion is then given to the suspenders, sufficient to bring the inner edge of the sheet into the mouth of two sets of horizontal tapes, by which it is carried around the impression-cylinder and printed. As the sheet, after being printed, issues from the horizontal tapes, it is delivered to other sets, by which it is conveyed outward, under the laying-on board; arrived at the proper point, it is again caught at the top edge between suspending rollers, the tapes separate, and it hangs for a moment; when the taker-off, who sits below the layer-on, releases it by a slight jerk, and lays it on his board.

No description can give any adequate idea of the scene presented by one of these machines in full work,—the maze of wheels and rollers, the intricate lines of swift-moving tapes, the flight of sheets, and the din of machinery. The central drum moves at the rate of six feet per second, or one revolution in three seconds; the impression-cylinders make five revolutions in the same time. The layer-on delivers two sheets every five seconds, consequently, sixteen sheets are printed in that brief space. The diameter of an eight-feeder, including the galleries for the layers-on, is twenty-five feet. *The Times* employs two of these eight-cylinder machines, each of which averages 12,000 impressions per hour; and one nine-cylinder, which prints 16,000.

These vast machines, however, are only useful when the necessity of working a very large number with the utmost rapidity overrides all considerations of cost and space. An excellent machine, in which considerable speed is obtained with comparative economy of expense and room, the invention of Messrs. Hoe of New York, has been lately used for newspapers and periodicals of long numbers. In principle, it does not differ from Applegath's vertical, inasmuch as the type is fixed upon a central cylinder or drum, which has a continuous rotatory motion, in contact with impression-cylinders set around it. The chief difference is, that the drum and impression-cylinders are not vertical, but horizontal. The machines are manufactured of different sizes, according to the number of impression-cylinders required. Those more generally made have six cylinders, some have eight, and *The Times* has recently constructed one with ten. This last machine is calculated to produce 20,000 impressions per hour. The following is a description of a six-cylinder machine:

A horizontal central cylinder is mounted on a shaft with appropriate bearings, and around it, arrayed at proper distances, are six horizontal impression-cylinders. The movable types or stereo-casts are secured on a portion of the central cylinder, about a quarter of its circumference, and compensated by a balance-weight on the opposite side; the remainder of the cylinder is used as a distributing-table for the ink. This portion of the cylinder is lower than the face of the type, in order that it may pass under the impression-cylinders without being touched by them. The ink is contained in an ink-box placed beneath the central cylinder, and supplies the ink to the ductor-roller, from which it is transferred by a vibrating distributing-roller to the distributing-table. The ductor-roller receives a slow and continuous rotary motion, so that it always presents a uniform line of ink to the vibrating roller. The machine being put in motion, the form of type on the central cylinder is brought into contact with each of the six impression-cylinders in succession; and six sheets of paper, which have been introduced, one to each impression-cylinder, are printed in one revolution of the central cylinder. For each impression-cylinder there are two inking-rollers, which roll over the distributing surface and take a supply of ink; at the proper time they rise, pass over the type, and then fall on to the distributing surface.

Each page is locked up upon a detached segment of the large cylinder called a "turtle," which constitutes the bed and chase. The column-rules, like those for the vertical machine, are wedge-shaped, and are held down to the turtle by tongues projecting at intervals along their length, and sliding in rebated grooves cut crosswise in the face of the turtle, the space in the grooves between the column-rules being filled with sliding blocks of metal, accurately fitted, the outer surface level with the surface of the turtle, the ends next the column-rules being cut away underneath to receive a projection in the sides of the tongues. The head and cross rules are segments of a circle of the same curvature as the turtle. The types are secured by screws and wedges.

Six persons, one to each impression-cylinder, are required to supply the paper,—three on each side of the machine. The paper is conveyed from the laying-on board to the impression-cylinders by gripers. The sheets when printed are carried by tapes to six self-acting fly-frames, which lay them regularly in piles.

Another American, M. S. Beach, has improved upon Hoe's machine, by converting it into a perfecting-machine. His improvement consists in placing the second form upon the central type-drum, superseding the necessity for the balance-weight: the sheet, after being printed on one side, is immediately drawn back and printed on the other side from the second form, without checking or changing the uniform revolution of the cylinder; and thus the work done by it is doubled. The diameter of the type-drum in this machine, which is calculated for eight impression-cylinders, is only four feet; the type has therefore to travel a less distance in one revolution of the drum; and the consequence is, that in traveling the same distance in this machine, and at the same speed, 22,000 *double* impressions would be produced in an hour. This account is taken from the *New York Sun*.

A horizontal cylinder-machine, on the same system as Hoe's, made by Middleton, capable of printing 20,000 impressions per hour, is now used for printing *The London Morning Herald*. The type is secured on the central cylinder, $2\frac{1}{2}$ feet in diameter, in the same way as in *The Times* vertical machine; the ink is supplied from a ductor below the type-cylinder, and distributed upon an inking-table attached to the type-cylinder, to which a slight lateral motion is communicated by two straps, one on each side of the machine. There are five impression-cylinders at equal distances around the central cylinder, to which the paper is supplied from ten feeders, on the same principle as in the other horizontal machines, four on one end of the machine and six on the other; the printed sheets are delivered on to five taking-off boards, one to each two feeders, and received by five lads. The machine is 26¼ feet long, 5 feet wide, and 17½ high.

The machines of Mr. Napier, intended for book-work, are in good repute. They have the advantage of being easily worked by two men, thus rendering steam-power unnecessary. They stand in a very small compass, and do beautiful work. As far as regards motion and impression, they do not greatly vary from the cylinder machines already described; but in the method of conveying the paper, obtaining register, and inking, they are altogether different. The paper is laid to a certain gauge, when, in the revolution of the cylinder, gripers are made to compress the edge of the paper upon it, very much in the manner in which the fore-finger closes on the thumb. It is by these means conveyed with it during one revolution, in the course of which it is printed on one side. At the commencement of the second revolution these gripers open at the precise moment, when the gripers attached to the second cylinder close, and thus convey the sheet over the second form. Tapes pass under the second cylinder, *between* the blanket and the paper, and over a pulley upon a bar, by the mere friction of which the sheet is thrown out upon a board. These gripers are made to act with such perfect certainty that the best possible register is obtained. The inking apparatus consists of a trough with a ductor and vibrating roller, which communicates the ink to composition-rollers, by the revolution of which in contact with each other the ink is perfectly distributed, and from these to the type. A cross motion is communicated to the distributing-roller by means of a worm in the elongated spindle. As but one impression is given during the traverse of the table in each direction, the cylinder which does not at the moment hold the paper would be in contact with the type, had not Mr. Napier added a beautiful adjustment, whereby the cylinders rise and fall alternately, so that the one not in use passes over the form intact. This machine will work from 1000 to 1200 perfect sheets per hour, and requires but two boys. Mr. Napier has constructed several other machines of great merit, one of which, for newspapers, will perfect 2000 sheets per hour by the labor of two men.

Messrs. Hopkinson and Cope have also produced a double-cylinder perfecting griper machine adapted for book-work or newspapers. The peculiarity of this machine is, that it is supplied with a set-off sheet apparatus, by which a "set-off sheet" is fed in with each sheet to be printed, which it meets as the latter enters on the second cylinder, and, passing round with it, prevents the ink on the printed side of the paper setting off on the blanket of the cylinder, and being thence transferred to the following sheet. This apparatus can be easily dispensed with when ordinary work is being printed. They have also made a single cylinder griper machine called a "Desideratum." It is supplied with a pointing apparatus, which renders it available for book-work.

Before the invention of cylinder machines, the desire to obtain increased speed led to many ingenious contrivances for accelerating the action and economising the expense of the ordinary printing-press; all of which, however, either failed, or were superseded by the steam machine. There are now in general use, for book-work of a quality superior to that produced by the cylinder, several machine-presses which are in every respect satisfactory. They generally consist of two tables, on each of which a form is laid; these pass alternately under a self-acting platen: while one form is receiving the impression, the other is delivering its printed sheet to the taker-off, and receiving its white sheet from the layer-on. This double operation is effected at the same time, by the frisket being attached to the tympan at the bottom (not at the top as in the common press). When the tympan opens, it falls back inward; the white paper is laid on the frisket, the tympan closes upon it, it is printed; but when the tympan opens, the printed sheet is made to rise with it, and is taken off while the layer-on is placing another sheet on the frisket. The ink is conveyed to the type by a similar apparatus to that used in cylinder machines. These machine-presses do excellent work at the rate of 600 or 700 impressions per hour, and are made by the same firms as supply cylinder-machines.

The "Scandinavian" machine-press differs from all others in respect that the form of type is stationary, and that the tympan and inking-roller are passed between the form and the platen. As the power required to set this press in motion is much less than that required where the form and table travel, manual labor is sufficient; but only one form can be worked at a time.

These are by no means all the machines that have been devised or brought into use. They are, however, all that it is necessary to mention, as the same principle is common to all. Every maker is at liberty to manufacture almost all of them, with such modifications as his own talents may suggest, the patents, where any were taken out, having, with few exceptions, expired.

POTTERY AND PORCELAIN.

BY CHARLES TOMLINSON.

POTTERY AND PORCELAIN.

THE word *pottery* is said to be derived from the low Latin term *potus*, a pot, which is from the classical Latin *potus*, drink;* but the etymology of porcelain is more uncertain. Some writers derive it from *porcellana*, the Portuguese for a drinking-cup; others from a similar word in Italian, which is applied to a univalve shell of the genus *Cypræidæ*, or cowries, having a high arched back resembling that of a hog (*porco*, Ital.), and a white, smooth, vitreous glossiness of surface similar to that of fine porcelain. The essential ingredients of every article in pottery and porcelain are silica and alumina. The pure chemical compound, silicate of alumina, must, however, be regarded as an ideal type, unattainable even in the finest porcelain; while in the coarser varieties, and in pottery, impurities, such as iron, lime, potash, etc., give character to the resulting wares. Even if it were possible to obtain pure silica and alumina in sufficient quantities for manufacturing purposes, it would still be necessary to add certain substances to increase somewhat the fusibility of these refractory materials. Pottery is also distinguished by being opaque, while porcelain is translucent. Wares of either kind are further distinguished by the terms *soft* and *hard*, or, as the French term them, *tendre* and *dur*—distinctions which relate as well to the composition of the ware as to the temperature at which it is made solid. Common bricks and earthenware vessels, pipkins, pans, etc., are soft; while fire-brick and crockery, such as queen's-ware, stone-ware, etc., are hard. Soft pottery, consisting of silica, alumina, and lime, admits of being scratched with a knife or file, and is usually fusible at the heat required merely for baking por-

---

**Pot is said by Tooke to be the past tense of the verb to *pit*—*i. e.*, to excavate or sink into a hollow.

celain. Stone-ware is composed of silica, alumina, and baryta, and may be regarded as a coarse kind of porcelain. Hard porcelain contains more of alumina and less of silica than the soft; it is baked at a stronger heat, and is more dense. Soft porcelain contains more silica than the hard, and is also combined with alkaline fluxes, so that its softness is manifested in being easily scratched and less able to resist a strong heat.

HISTORICAL SKETCH.

Articles of fictile ware are at once the most fragile and the most enduring of human monuments. A piece of common pottery, liable to be shivered to pieces by a slight blow, is more enduring than epitaphs in brass and effigies in bronze. These yield to the varying action of the weather; stone crumbles away, ink fades, and paper decays; but the earthen vase, deposited in some quiet but forgotten receptacle, survives the changes of time, and even when broken at the moment of its discovery by the pick of the laborer, affords instruction in its fragments. In their power of traversing accumulated ages, and affording glimpses of ancient times and people, fictile articles have been compared to the fossils of animals and plants, which reveal to the educated eye the former conditions of our globe.

Clay is so generally diffused, and its plastic nature is so obvious, that the art of working it cannot be considered as above the intelligence of a savage; hence the production of articles in clay may be said to belong to every people and to all time. The first drinking-vessels would be sun-baked, and consequently very destructible; so that few articles would survive a single winter. A considerable period must have elapsed before the method of giving permanence to these articles by the action of fire was discovered; but it is chiefly to this discovery that we owe the preservation of so many ancient relics of the fictile art. The sun-dried bricks of Egypt, Assyria, and Babylonia, have, however, been preserved to this day, and "not only afford testimony to the truth of Scripture by their composition of straw and clay, but also by the hieroglyphs impressed upon them, transmit the names of a series of kings, and testify the existence of edifices, all knowledge of which, except for these relics, would have utterly perished. Those of Assyria and Babylon, in addition to the same information, have, by their cuneiform inscriptions, which mention the locality of the edifices for which they were made, afforded the means of tracing the sites of ancient Mesopotamia and Assyria with an accuracy unattainable by any other means. When the brick was ornamented, as in Assyria, with glazed representations, this apparently insignificant but imperishable object has confirmed the descriptions of the walls of Babylon, which critical skepticism had denounced as fabulous. The Roman bricks have also borne their testimony to history. A large number of them present a series of the names of consuls of imperial Rome; while others show that the proud nobility of the Eternal City partly derived their revenues from the kilns of their Campanian and Sabine farms." *

The excellent authority just quoted refers to the next step in the progress of manufacture, namely, that of modeling in clay the forms of the physical world, the origin of the plastic art, " to which the symbolical pantheism of the old world gave an extension almost universal." When stone and metal came to be used as materials for sculpture, clay was still employed for the elaboration of the model, and also for the multiplication of copies for popular use of celebrated pieces of sculpture. The invention of the mould caused the terra cottas of antiquity to be as widely diffused as the plaster casts of modern times. Among the Assyrians and Babylonians clay was used as a material for writing on. The traveler Layard discovered in the palace of Sennacherib a whole library of clay books, consisting of histories, deeds, almanacs, spelling-books, vocabularies, inventories, horoscopes, receipts, letters, etc. About 2000 of these clay books of the Assyrians have been discovered : they are in the form of tablets, cylinders, and hexagonal prisms of terra cotta.

Before the invention of the potter's wheel, clay vessels could have had but little symmetry of shape. The necessity for some such contrivance must have been early felt, and it was probably invented by several nations. It is represented on the Egyptian sculptures; it is mentioned in Holy Scripture; and was in use at an early period in Assyria. Mr.

*History of Ancient Pottery, by Samuel Birch, F.S.A., London, 1858.

Birch states, that "the very oldest vases of Greece, some of which are supposed to have been made in the heroic ages, bear marks of having been turned upon the wheel." The art of firing the ware is also of the highest antiquity. Remains of baked earthenware are common in Egypt in the tombs of the first dynasties, and the oldest bricks and tablets of Assyria and Babylon bear evidence of having passed through the fire. The oldest remains of Hellenic pottery owe their preservation to their having been fired. As the clay by this process is rendered porous and incapable of holding liquids, the necessity for some kind of glaze must have been early felt. Opaque glasses or enamels have been found in Egypt as old as the fourth dynasty, and both the Egyptians and the Assyrians seem to have preferred an opaque enamel to a transparent glaze, somewhat after the fashion of the modern *faience*. Numerous fragments testify to the use of glazing amongst the ancient Greeks and Romans. With respect to form, the Greek vases, by their beauty and simplicity, have become models for various kinds of earthenware; while the application of painting to vases has transmitted to us much information respecting the mythology, manners, customs and literature of ancient Greece. Even the Roman lamps and red ware illustrate in their ornaments many customs, manners, and historical events. As the pottery of different modern nations has its characteristic features, so the ancient pottery has its distinctions of time and place. It is impossible not to distinguish between the rude and simple urns fashioned by the early inhabitants of Great Britain and the more carefully finished specimens of the Roman conquerors of these islands. Then, again, the simple unglazed earthenware of Greece contrasts with the more elaborate Etruscan forms, the finest of which, however, are probably by Greek artists. Then, again, the red and black potteries of India contrast with the black and white potteries of North America, the latter being interspersed with fragments of bivalve shells. On the discovery of the extraordinary ruins in Central America, specimens of pottery were found which showed considerable advance in the art compared with the date assigned to these ruins, namely, 1000 B.C. The specimens had been formed without the assistance of the potter's wheel; but they are well baked, the ornaments are in different colors, and they are coated with a fine vitreous glaze, such as was unknown in Europe until within about ten centuries. The religious employment of earthen vessels in early times, and the custom of placing them in tombs as receptacles for medals, trophies, insignia, money, rings, and votive offerings, has greatly assisted the studies of archæologists in modern times, and we can do no more in this brief sketch than refer to their useful labors.

Porcelain is of modern introduction into Europe, but it was known in China more than a century before the Christian era. The Chinese appear to have improved their art during four or five centuries, and then, supposing themselves to have attained perfection, they allowed the art to remain stationary. So completely was the manufacture identified with that nation, that on the introduction of porcelain into Europe by the Portuguese in 1518, it received the name of *china*, which it still partially retains. The Chinese continued to supply us with porcelain during many years. It was supposed that the fine clay or *kaolin* used in its production was peculiar to China, and that it was consequently hopeless to attempt to manufacture porcelain in Europe. The porcelain of Japan is only a variety of the Chinese.

While the Chinese were improving their manufacture, the art of making decorative pottery became lost in Europe amid the darkness which followed the overthrow of the Western Empire. The first symptoms of revival were due to the Mohammedan invaders of Spain, whose tiles of enameled earthenware are to be seen in the Moorish buildings of Seville, Toledo, Granada, and the Alhambra. They are of a pale clay, "the surface of which is coated over with a white opaque enamel, upon which the elaborate designs are executed in colors."* The Spaniards acquired from the Moors the art of manufacturing enameled tiles, or *azulejos* as they are called, and they still continue to be made in Valencia. The Moors also adorned their pottery with Arabic inscriptions, and with arabesque patterns resembling a lace vail in richness. The vase known as that of the Alhambra is of earthenware; the ground is white, the ornaments are either blue of two shades, or of gold or copper luster.† The Moors continued to manu-

---

* *A History of Pottery and Porcelain, mediæval and modern*, by Joseph Marryat, 2d edition, London, 1857.
† This vase is figured in Owen Jones's work on the Alhambra.

facture ornamental pottery until the time of their final expulsion from Spain at the beginning of the seventeenth century. This Hispano-Arabic pottery, as it is called, is the prototype of the Italian majolica, and was long confounded with it. Specimens of it are to be seen in several celebrated collections. The majolica, or enameled ware of Italy, probably dates from the twelfth century. It is related that a pirate king of Majorca, about 1115, was besieged in his stronghold by an armament from Pisa, and being vanquished, the expedition returned to Italy laden with spoil, among which, it is supposed, were a number of plates of painted Moorish pottery, such specimens being found incrusted in the walls of the most ancient churches of Pisa. They appear to have been regarded as religious trophies. No attempt, however, was made to imitate them until the fourteenth century, when specimens of majolica, so called from the island of Majorca, were produced; they resemble the Moorish examples in having arabesque patterns in yellow and green, upon a blue ground. About the year 1451 the manufacture had become celebrated at Pesaro, the birthplace of Luca della Robbia, who is regarded by persons who set aside the foregoing origin of majolica as the inventor of this ware. He appears to have earned distinction as a sculptor when he took to working in terra cotta, and gave permanence to his productions by the invention of a white enamel. His Madonnas, Scripture subjects, figures and architectural pieces are still prized by collectors. Mr. Marryat refers to them as "by far the finest works of art ever executed in pottery." He is also "the founder of a school which produced works not much inferior to his own." Existing specimens are of a dazzling whiteness, and the glaze, after so great a lapse of time, continues to be quite perfect. The manufacture of majolica flourished during two centuries under the patronage of the House of Urbino. The first duke, Frederick of Montefeltro (1444), took a lively interest in the manufacture; his son established a manufacture at Pesaro, and the most eminent artists were employed in furnishing designs, a system of patronage which was maintained by succeeding dukes. There is a tradition that Raffaelle was employed in furnishing designs; whence majolica sometimes passes by the name of *Raffaelle ware*. But as the finest specimens do not date earlier than 1540, or twenty years after the death of that great artist, he was probably not directly concerned in the manufacture. But it is admitted that his scholars used his drawings in composing designs for the finest specimens. In the middle of the fifteenth and during part of the sixteenth century, many towns of Italy had become renowned for their majolica ware, of which the coarser specimens were named *mezza-majolica*, and the finer, however inappropriately, *porcelana*. The manufacture had attained its greatest celebrity between 1540 and 1560. After the last-named date the art began to decline, and the introduction of porcelain, properly so called, helped to complete its downfall. The caprices of fashion cannot be alone charged with the destruction of this beautiful art, since, so far as utility is concerned, a hard paste covered with a vitreous glaze, as in porcelain, must be very superior to a soft paste coated with a metallic glaze, as in the case of majolica. The best examples of mezza-majolica are distinguished by the beauty of their color, and the perfection of their enamel glaze; the latter imparting to the yellow and white tints the metallic luster of gold and silver. There is also a remarkable mother-of-pearl luster, together with an iridescent ruby, peculiar to Pesaro and Gubbio. The most general colors used in the painting were blue and yellow, with their mixtures. The drawing is not so good as the coloring, until the so-called porcelana raised the art to its zenith. After the year 1560 the designs became more fanciful and grotesque, and the colors inferior. It must not, however, be supposed that the articles manufactured were ornamental only. During the whole reign of majolica ware, all kinds of common articles were produced, such as pilgrim's bottles, with holes in the bottom rim for the strap or cord by which the vessel was carried; various forms of vases, adorned with paintings, with handles in the form of serpents, and rims surmounted by grotesque figures of animals and fishes; fruit dishes, with embossed patterns in high relief; small plates for ices and sweetmeats; vases, for holding different kinds of wine, which could be poured out from one spout; small flasks, in the shape of lemons and apples; cups covered with tendrils or quaint devices; small figures of saints; jocose figures; birds colored after nature; painted tiles for walls and floors, etc. Some of the most interesting specimens of majolica are known as *amato-*

rii, and consist of vessels, plates, or deep saucers, containing the portrait and name of a lady; these were filled with fruits or sweetmeats, and presented as pledges of affection. The portraits not only perpetuate the female beauty of a former age, but also the costume by which it was sought to make that beauty more attractive. Some of the amatorii represent hands united, hearts a-flame, or pierced with darts, after the fashion of the modern valentine. The painters who executed the designs were usually copyists, the design itself being furnished by an eminent artist. In some cases, however, the painters themselves were the artists, and are known by certain monograms and marks. Occasionally the painters bought the pieces ready prepared for painting, executed them at home, and took them to the kiln to be fired. In such cases, the piece is often marked with the name of the potter, as well as that of the artist. The custom of attaching signatures to the pieces is peculiar to some manufactories: those with names and monograms for the most part belong to Gubbio and Albino. Different towns had their distinguishing marks, and it was common to mark in blue characters on the back of the dish the subject of the design; but when a complete service was painted, only the principal piece was marked: it was also customary to introduce the arms of the family for whom the service was prepared.

Majolica was introduced into Germany in 1507 by Hirschvögel of Nuremberg, but the manufacture does not appear to have survived him. It prospered better in France, where, under the name of *faience*,* it flourished under the patronage of Catherine de Medici and her kinsman Louis Gonzaga. The latter established Italian artists in his dukedom of Nevers, and they were successful in producing enameled pottery from native materials. Gradually as native artists succeeded the Italian ones, the classical designs of the latter were degraded, and the enameled ware of Italy was represented only by the common faience of France. In the eighteenth century Nevers recovered her reputation, and became celebrated for the brilliancy of a dark blue enamel with white patterns upon common ware. A variety of enameled pottery was also

---

* This term is supposed to be derived from the small town, now a village, of Faience, in the department of Var, which, as early as the sixth century, appears to have been celebrated for glazed pottery.

produced at Rouen: this attracted some notice; but the kind of ware which may be said to be peculiar to France is that known as Palissy ware. There is a good deal of romance mixed up with the life of the inventor of this ware. Bernard Palissy and his adventures, real or imaginary, have assisted in multiplying the number of those dangerous books which ascribe imaginary events to real characters. Palissy was born at the commencement of the sixteenth century, of poor parents; but nature had implanted within him a love of the beautiful, which became his teacher. He managed to acquire a knowledge of reading, writing, and land-surveying, by which last-named art he earned his livelihood. In the intervals of employment he was much given to the study of the Italian masters, and he was delighted to obtain work in painting images and designs on glass. This enabled him to gratify his taste for travel, and for studying natural objects. He became master of the chemistry and mineralogy of his day, such as it was. In 1539 he settled at Saintes as an artist, where he married. His attention was directed to pottery by being shown a beautiful enameled cup, and on proceeding to inquire into its mode of manufacture, he found that there were secrets connected with it, and especially with the composition of the enamel. He at once undertook a course of experiments on the subject, but without success. The desire to master the subject had, however, taken such possession of him, that during several years he devoted nearly all his time and means to this pursuit, in spite of the claims of his wife and family and the remonstrances of his friends. He borrowed money to enable him to construct a new furnace; and when too poor to buy fuel, he used his furniture instead. When unable to pay his assistant's wages, he gave him the coat from off his back. Thus becoming every year more wretched than the preceding, the folly of sixteen years (as it would have been called had he failed) ended in a triumph. His *figulines* or rustic pottery became the fashion of the day, and his beautiful patterns were every where admired. The general style of his ware is marked by quaintness and singularity; his figures are usually chaste in form: the ornaments and subjects of an historical, mythological, and allegorical character are in relief, and colored. His natural objects, with the exception of certain leaves, were all moulded from nature. His shells are those of the tertiary formation of the Paris basin; his fish are those of the Seine; the reptiles and plants are from the neighborhood of Paris; and he made use of no foreign natural production. The colors are usually bright, and mostly confined to yellows, blues, and grays; sometimes extending to green, violet, and brown. Mr. Marryat says that Palissy never succeeded in attaining the purity of the white enamel of Luca della Robbia, or even that of the faience of Nevers. The *pieces rustiques* of this artist, intended to adorn the large sideboards of the dining-rooms of the period, are loaded with objects in relief. A favorite subject with him was a flat kind of basin or dish, representing the bottom of the sea, covered with fishes, shells, sea-weeds, pebbles, snakes, etc. We have also from the hand of this artist, ewers and vases with grotesque ornaments, boars' heads, curiously-formed salt-cellars, figures of saints, wall and floor tiles, etc. Mr. Baring Wall speaks of Palissy as " a great master of the power and effect of neutral tints." *

France is also celebrated for a fine ware known as *faience fine* and *gres cerame*. Some of the earliest specimens are known under the name of *renaissance*, or fine faience of Henri II. There are only thirty-seven pieces of this manufacture extant; and as twenty-seven of them have been traced to Touraine and La Vendée, it has been conjectured that the manufactory was at Thouars in Touraine. The material is a fine white pipe-clay, the texture of which is seen through the thin transparent yellow varnish. The patterns are engraved on the paste, and the hollows filled up with colored pastes, so as to resemble fine inlaying, or chiseled silver works in *niello;* whence this ware has also been termed *faience a niello*. There are also beautifully-modeled raised ornaments: the articles are for the most part small and light, consisting of cups, ewers, and a vase with a spout for pouring, called a *biberon*. A single candlestick of this ware was sold a few years ago for 220*l*.

Germany had its enameled wares as early as the thirteenth century, the secret of success being of course the discovery of a fine glaze. Ratisbon, Landschut, and Nuremberg thus became formidable rivals of the Arabs and the Italians. The

---

* Lecture delivered before the Literary and Scientific Society of Salisbury, January, 1853. Printed for private circulation.

distinctive characters of this ware are the fine green glaze, the complex form, the number and variety of ornaments, lightness, and good workmanship. Nuremberg also became famous for its large enameled tiles used for covering stoves.

Holland, from its exclusive trade with Japan, was induced to imitate the Japanese porcelain. The chief seat of the manufacture was Delft; and the ware was known and esteemed in the sixteenth century by its fantastic design, good color, and beautiful enamel—the latter being smooth and even, and slightly tinged with blue. The Japanese origin was seen in the monstrous animals of the *chimera* class, the three-ringed bottle, the tall shapeless beaker, and the large circular dish, which were long regarded in Europe as favorite ornaments; while the common articles were so generally distributed as to obtain for Delft, the title of the "parent of pottery." The fine English wares introduced by Wedgwood and others were the means of injurng the trade of Delft.

In England, the first manufactory of fine earthenware is said to have been erected in the reign of Elizabeth at Stratford-le-Bow. The well-known Shakspeare jug is cited as a good specimen of Elizabethan pottery. It is of cream-colored earthenware, about 9 inches in height and 16 in circumference in the largest part. Its shape resembles that of a modern coffee-pot. It is divided lengthwise into eight compartments, each containing a mythological subject in high relief and of considerable merit. The silver top is a modern addition. The Elizabethan pottery nearly approaches in hardness that of fine stone-ware; it is of a dingy white, and its ornaments in relief consist mostly of quaint figures and foliage. In the reign of Elizabeth the Staffordshire potteries came into notice, of which some of the earliest specimens consist of butter-pots of native brick earth, glazed with powdered lead-ore, which was dusted on while the ware was in a green state; the *tig*, or drinking-cup, with three handles; and the parting-cup, with two handles. In 1684 a manufactory of earthenware was established at Fulham, some of the products of which, under the name of *Fulham-ware*, are still valued by collectors. They consist of white *gorges* or pitchers, marbled porcelain vessels, statues, and figures. The proprietor, Mr. John Dwight, attempted to produce the transparent porcelain of China, but his success was not such as to turn him from the more profitable manufacture of earthenware. About the time of the Revolution, ale-jugs of native marl, ornamented with figures in white pipe-clay, were introduced. During the reigns of Anne and George I. an improved ware was made of sand and pipe-clay colored with oxide of copper and manganese, forming the well-known *agate-ware* and *tortoiseshell-ware*, conferring on the pottery the character of a hard paste, which was subsequently so much improved by Wedgwood, and introduced under the name of *Queen's-ware*.

The proceedings of Wedgwood form an epoch in the history of the art. Josiah Wedgwood was the son of a potter at Burslem in Staffordshire. He was born about the year 1730, and can scarcely be said to have received any formal education. At the age of eleven he entered his brother's pottery as a thrower; but he had not been long so engaged before he was attacked by small-pox, which left him with a lame leg, and rendered amputation necessary. His first attempts to settle in life were not fortunate; he became partner for a short time in 1752 with a man named Harrison, at Stoke, where he is said to have first felt a strong desire to manufacture ornamental pottery. His next partner was named Wheildon, and his employment consisted in manufacturing knife-handles in imitation of agate and tortoise-shell, melon table-plates, green pickle-leaves, etc.; but he could not induce his partner to embark largely in the production of ornamental wares, nor was there much encouragement to do so. The upper classes of Great Britain obtained their porcelain from China; the great bulk of the earthenware in domestic use was supplied by France, Germany, and Holland; and even the trade in tobacco-pipes, in which England had attained some success, was becoming monopolized by the Dutch. To compete with these formidable rivals required the courage and persistence of genius; and Wedgwood was not slow in bringing them to bear upon the native materials which surrounded him. Accordingly, in 1759 he established a small factory on his own account at Burslem. Here he must have been successful, for he soon undertook a second manufactory, where he produced a white stone-ware, and afterward a third, where he manufactured his celebrated *cream-colored ware*. Some specimens of the latter having been shown to Queen Charlotte, her Majesty was so pleased with them that she appointed Wedgwood the royal potter, and gave permission for calling the ware Queen's-ware. Wedgwood had now no longer reason to complain of want of taste or of prtronage on the part of the public, and nobly did he use his best exertions to encourage the one and respond worthily to the other. He studied the chemistry of his day, and courted the society of scientific men, with a view to improve the composition, glaze, and color of his wares. He invited good artists to furnish him with designs, among whom was the celebrated Flaxman. Among Wedgwood's inventions may be mentioned a *terra cotta*, resembling porphyry; *basalts*, or black ware, which would strike sparks like a flint; *white porcelain biscuit*, with properties similar to basalt; *bamboo*, or *cane-colored biscuit; jasper*, a white biscuit, of exquisite delicacy and beauty, well adapted for cameos, portraits, etc.; also *blue jasper* and *green jasper*, and a *porcelain biscuit* little inferior to agate in hardness, and used for pestles and mortars in the laboratories of chemists. He also succeeded in imparting to hard pottery the vivid colors and brilliant glaze of porcelain. About the year 1762 Wedgwood opened a warehouse in London, and intrusted it to the care of Mr. Bentley, a gentleman of recognized taste, who succeeded in attracting attention to the rising Staffordshire works, and also in obtaining the loan of vases, cameos, oriental porcelain, etc., which at that time were difficult to procure, especially for the purposes of the manufacturer; but such was the sympathy of persons of taste with Wedgwood's pursuits, that they freely lent their fictile treasures either to be copied or to suggest new designs. Even the Barbarini vase, which was purchased by the Duchess of Portland for 1800 guineas, was lent to Wedgwood, who, after executing fifty copies, destroyed the mould. Wedgwood's wares now became so deservedly popular that the extension of his works in Staffordshire led to the formation of a new village near Newcastle-under-Lyne, which was named "Etruria," from the resemblance which the clay dug there had to the ancient Etrurian earth, and also probably to mark the success with which Wedgwood had imitated the ancient Etruscan ware. This village long continued to be a center of attraction for travelers from all parts of Europe, and we may still trace that celebrity in many noted collections of the ceramic art, Wedgwood's finest productions taking rank with the choicest specimens of Dresden and Sèvres. Wedgwood died at his mansion in Etruria in 1795.

The stone-ware which Wedgwood so greatly improved had long existed under various forms in different potteries of the world. In some cases it was common, and in others fine—the difference consisting in the composition of the paste. The Chinese were acquainted with this ware, and were accustomed to use it as the basis for a surface of porcelain paste. The stone pottery of the Rhine of the sixteenth century is esteemed by collectors for its quaintness of form, richness of ornament, and the color of its enamel. *Gres Flamand*, or Flemish stone-ware, of the period between 1540 and 1620 is remarkable for its beautiful blue color, quaint forms, and rich ornaments. France also appears to have manufactured stone-ware before the sixteenth century. In England, Dutch and German workmen were engaged in the manufacture at an early period. In 1690 the mode of glazing by means of common salt enabled the stone-ware manufacturers to compete successfully with delft and soft paste fabrics. Toward the end of the seventeenth century a very fine unglazed stone-ware, with raised ornaments, known as red Japan ware, was made in England, after the failure of many previous attempts. It appears that two brothers named Elers, from Nuremberg, discovered at Bradwell, about two miles from Burslem, a bed of fine red clay, which they worked at a small factory erected on the bed itself. They endeavored to conceal their discovery, as well as their mode of working, for which purpose they employed the most ignorant assistants that they could meet with; but no sooner did their ware attract attention than a potter named Astbury, feigning to be an idiot, entered the service of the two brothers, and having learnt all their secrets, established a factory for himself; the processes soon became known, and others followed the example. In 1720 the two brothers closed their establishment, and entered the porcelain manufactory at Chelsea. Mr. Marryat characterizes their ware as being fine in material and sharp in execution, the ornaments being formed in copper moulds.

Regarding stone-ware as a connecting-link between earthenware and porcelain, we come now to the history of the latter article. China, Japan, and Persia are the earliest nations which produced this beautiful material. Bottles of Chinese manufacture have been found in the tombs of Thebes; and from an inscription on one of them, the date of the manufacture would appear to be between 1575 B.C. and 1289 B.C. The workmanship, however, is inferior. Porcelain seems to have been common in the Chinese empire in the year 163 B.C., and to have attained its greatest perfection in the year 1000 A.D. The porcelain tower near Nankin was erected in 1277. Marco Polo describes the manufacture in China during the thirteenth century. Specimens of the ware had gradually found their way to Europe, but were not generally known until the Cape of Good Hope had been doubled by the Portuguese. The latter were so struck with the resemblance between the texture of this fine ware and that of cowrie-shells or "porcellana," as they were called, that they imagined that the ware might be made of such shells, or of a composition resembling them, and named it accordingly. They imported numerous and splendid collections of the ware into Europe, where it was also named from the country which produced it; and, from its ringing sound, "China metal." It was also called "China earth." On the expulsion of the Portuguese, the Dutch succeeded in establishing a traffic with India and Japan; and Europe was for a long time supplied with porcelain through Holland. The English shared in the trade somewhat later, through the medium of the East India Company; but the taste for collecting china had become very general, and about the middle of the seventeenth century had amounted to a passion. The writers of the day frequently refer to it, especially in Queen Anne's reign. The French, who had established missions in China, succeeded in obtaining, from time to time, information respecting the manufacture. Fokien was represented as the seat of manufacture of the pure white porcelain of China, some of which consists of small cups and similar articles, with inscriptions, devices, etc., under the glaze, so that they can only be seen by holding the article up to the light. Nankin produced the blue and white porcelain, as also the pale buff on the necks of bottles and backs of plates. King-te-tching was named as the origin of the old sea-green and crackle porcelain. To the former the term *celadon* has been applied; but the French extend the term to porcelain of any tint in which the colors are mixed with the glaze, and burnt in at the first firing. In some cases two or more colors are blended so as to give the appearance of shot-silk; a variety, known as *marbled*, belongs to this class, and resembles marble in its coloring and veining. Crackle china, in which an immense number of cracks occur on the surface in small regular figures, is due to the unequal expansion of the glaze on the paste. The crackled "tsouï-khi" are produced by combining steatite with the glaze; and this when fired, splits into a net-work over the surface. A similar effect can be produced by plunging the heated porcelain into cold water; the cracks are then filled in with a thick ink or red-ocher. The ancient crackle is so much esteemed in Japan that as much as 300*l*. has been paid for a single specimen. The Chinese call this ware *snake-porcelain;* and the French apply to it the term *porcelaine truitee*. But the perfection of the ceramic art among the Chinese is exhibited in their egg-shell porcelain, which is thin and transparent, and resembles an egg-shell in appearance. This ware is colored citron-yellow for the exclusive use of the emperor, and ruby for the use of the imperial family. The porcelain in common use in China is brown, the inside being white, and white medallions outside. There is also an inferior and modern porcelain, manufactured at Canton, and known as Indian china. But all the specimens of Chinese porcelain, however beautiful may be the material and delicate the texture, however brilliant the color and pure the glaze, the form and the design are hideous. It has been remarked that the vase of the humblest Greek potter of the best period has an æsthetic value far surpassing the most costly productions of the Celestial Empire. The porcelain of Japan is in better taste than that of China, the dragons being less monstrous and the flowers more natural.

After the introduction of Chinese porcelain into Europe, many attempts were made during two centuries to imitate it. The first successful experiment was the result of one of those accidents which are doubtless of frequent occurrence, although the quality of mind required to take advantage of them is rare. John Frederick Bottcher was an apothecary's assistant at Berlin: he was fond of chemistry, and conducted his experiments with so much ardor that the authorities could not resist the conclusion that he was practicing the black art. He found it convenient to make his escape from Berlin and to visit Dresden, where the Elector of Saxony, Augustus II., patronized chemistry, not from the love of science, but from that of gold. Bottcher claimed the protection of the elector, who eagerly inquired of him respecting the transmutation of the baser metals. With the natural frankness of his character, Bottcher confessed his ignorance, but was disbelieved. Why should a man study chemistry except to enrich himself? it was argued; and as the elector was already patronizing the alchemist Tschirnhaus in his endeavors to discover the art of transmuting old age into youth, by means of the *elixir vitæ*, he associated Bottcher with him, with strict orders not to let him out of his sight. Bottcher was employed to seek after the philosopher's stone; and in the course of his experiments he made some crucibles, which, on being fired, possessed many of the characters of oriental porcelain. The vessels were made from a brown clay found near Meissen, and they were of a reddish tint. When the result was brought before the elector he appreciated its importance; and in order that Bottcher might pursue the inquiry in secret, he sent him to the castle of Albrechtsburg, near Meissen, where he was magnificently entertained, but restrained in his personal liberty. So much importance was attached to the secret, that during the troubles consequent on the invasion of Saxony by Charles XII. of Sweden, Bottcher, Tschirnhaus, and three workmen, were sent to the fortress of Königstein on the Elbe, where a laboratory was prepared for them. Bottcher's fellow-prisoners formed a plan of escape, which he communicated to the commandant, whereby he gained favor and a little more personal liberty. In 1707 he returned to Meissen, where he continued to prosecute his experiments, delighting every one around him with his active cheerfulness, and keeping up the spirits of the workmen during the furnace operations, which sometimes lasted sixty hours consecutively. Tschirnhaus died in the following year, and Bottcher enlarged the scale of his operations; he caused a new furnace to be erected, and extended the time of firing to five days and five nights. The elector was present at the opening of the furnace, and expressed his satisfaction at the progress which was being made. Up to this time, however, the only result was a kind of red and white stone-ware; and when, in 1709, Bottcher succeeded in producing a white porcelain, it became bent, and cracked in the fire. The progress, however, was deemed to be sufficient to determine Augustine to establish a manufactory at Meissen, and to appoint Bottcher the director. In 1715 the new factory produced a beautiful description of porcelain by means of the kaolin of Aue in the Erzgebirge, the discovery of which was made by an ironmaster of the district named Schnorr. This man had observed, while riding near the place, that his horse's feet stuck in a soft white tenacious earth, and it occurred to him that if this earth were dried and reduced to powder, it would make a good substitute for hair-powder, which the fashion of the day required, all except the poor, to use. Accordingly he manufactured the powder in large quantities, and found a ready sale for it in Dresden and elsewhere. Bottcher's valet used it, and so increased the weight of his master's wig as to lead to inquiry; and finding that the new hair-powder was of mineral origin, the idea flashed across his mind that this white powder might be useful in his experiments. He made the attempt, and was delighted to find that he had at length discovered the long wished-for material for making white porcelain. The secret so curiously obtained was for a long time as carefully guarded. The powder was made to retain its commercial name of "Schnorr's white earth" (*Snorrische weisse Erde*), its export was forbidden, and it was introduced into the factory in sealed barrels by persons sworn to secrecy. All persons connected with the factory were obliged to take a similar oath; no visitor was admitted; and the factory was regulated after the manner of a fortress. The motto in large letters, "Be secret unto death" (*Geheim bis ins Grab*), was set up in each room; the oath to the workmen was renewed every month; and when the king or any distinguished visitor was allowed to enter the factory, a similar obligation was imposed on him.

But all this parade of secrecy would make it clear to the most ill-informed workmen that the secret had a high marketable value, and we cannot wonder that it should have been sold to one or other of the monarchs of Europe, most of whom were ambitious to manufacture oriental porcelain. Bottcher died in 1719, at the age of thirty-seven, but before his premature death, a foreman had escaped from the factory, and proceeding to Vienna, submitted to be bribed, and it was not long before rival factories sprang up in different parts of Germany. A few years ago the writer visited the Meissen factory, which is pleasantly situated on the banks of the Elbe ; it still retains something of its fortress character, although the workshops are light and cheerful. The principal room is adorned with the bust of Bottcher. The factory, however, has lost its former vigor : an air of lassitude seems to pervade the place, and neither there nor at Sèvres are we impressed with the idea that the work is being done in earnest, as it is at such an establishment as Minton's at Stoke upon-Trent. There can be no doubt that private enterprise, unshackled by state restrictions, is the only healthy condition of the useful arts. A royal factory, which can neither become bankrupt nor meet with the wholesome stimulus of competition, is not likely to be worked at a profit, nor to inspire activity in its attendants.

The temporary success of the Meissen factory depended on the singularity of its position. There was a great demand in Europe for fine porcelain, and Meissen was in a condition to supply it. The first productions of the factory were mostly imitations of oriental patterns, but they were deficient in grace and lightness. There was a marked improvement when Kändler, a professional sculptor, was appointed in 1731 to superintend the modeling. He introduced wreaths and bouquets, animals and groups of figures, with the feeling of an artist. The works were arrested by the Seven Years' War ; but after this calamity Meissen became celebrated for its exquisite miniature copies of the best works of the Flemish school, together with birds and insects, painted by Lindenir, and flowers and animals by the best artists. In 1745, when Frederick of Prussia took possession of Dresden, he obtained among the spoils of war enormous quantities of porcelain. He also removed to Berlin some of the workmen, together with the models and moulds of the finest pieces. Again, in 1759, the factory was plundered and its archives destroyed : it revived somewhat under Dietrich the painter, Lüch the modeler, Breicheisen, and the sculptor François Acier. Gradually, however, the factory ceased to be profitable, and was for many years maintained at a loss ; when some years ago the king gave it up to the finance department of the state. The finest works of art are no longer produced; and it is also stated that the beds of fine clay in the neighborhood are nearly exhausted, and that an inferior material from Zittau is used instead.* Various marks were placed on different periods; the first mark consisted of the letters A. R. (Augustus Rex), and was placed on all pieces not intended for sale. The well-known mark of the electoral swords, crossed, also distinguishes Dresden china. Fac-similes of these marks, and of the marks and monograms of other celebrated European potteries, are given in Mr. Marryat's work.

Among the best of the Dresden works are groups from antique models; lace figures, so called from the fineness of the lace-work in the dress; flowers, evidently studied from nature; and vases richly adorned and incrusted, forming what is called *honey-comb* china. But even during the palmy time of this manufacture, namely, from 1731 to 1756, the productions were sometimes disfigured by the highly artificial taste of the age.

The first rival of Meissen was the porcelain factory of Vienna, which originated in 1720, in consequence of the perjury of a Meissen workman, as already noticed. The factory does not, however, appear to have flourished until warmed into life by the patronizing smiles of Maria Theresa in 1744, and of the Emperor Joseph. The porcelain of Vienna holds a lower rank than that of Dresden or of Berlin. It is not so light as that of Dresden, and the glazing has a grayish tint. Its chief feature is its raised and gilded work, which are in good taste, and of late years the application in relief of solid platinum and gold. The works are now in private hands, and the chief markets for the sale of the ware are in Turkey, Russia, and Italy.

As the Vienna works were based on treachery, so was the next important establishment based on the defection of a Vi-

---

* This statement is made on the authority of Mr. Marryat; but at the time we are writing an account is given in the German papers of an order from Paris having been executed at Meissen, consisting of portraits of the Emperor and Empress of the French, of a medallion shape, and inclosed in a rich porcelain frame. According to the German critics, " these are the finest works of art which porcelain painting has yet produced." If this criticism be true, or even partially true, the Meissen works must have experienced an extraordinary revival.

ennese workman. A celebrated pottery was already in existence at the village of Höchst on the Nidda, when in 1740 a man named Ringler undertook to superintend the manufacture of porcelain if the proprietors would introduce it. This man appears to have been simply a knave without skill or invention; he had committed to writing the various processes of the Vienna establishment, and concealing his manuscript about his person, consulted it every time he had to give out materials to the workmen. As knavery propagates itself, the workmen, taking advantage of Ringler's fondness for wine, invited him to a feast, where they made him helplessly drunk —when they robbed him of his papers, carefully copied his recipes, and then decamped to other parts of Germany, where they sold the secrets to those who were anxious for their possession. Hence originated from one source the porcelain factories of Switzerland, of the Lower Rhine, of Cassel, and even of Berlin. The Fürstenburg works, in the duchy of Brunswick, originated in a bribe offered by one of the dukes to a Höchst workman. The works at Frankenthal in Bavaria originated in a pottery which was visited by Ringler after he had been plundered of his papers. The factory of Nymphenburg in Bavaria had a similar origin. The porcelain of this factory is much esteemed, many of the designs having been furnished by the celebrated picture-gallery of Munich. A factory at Baden was conducted by some of the Höchst workmen until 1778. The factory of Ludwigsburg, begun in 1758 under the patronage of the Duke of Wirtemberg, has executed some beautiful works, which are known as *Cronenburg* porcelain, from the town of that name, and the mark CC on its wares. The distance from which the clay and the fuel had to be procured prevented the success of this establishment. The porcelain factory of Berlin was first undertaken in consequence of the information supplied by the men who robbed Ringler; but it was not very successful until a more magnificent fraud had been perpetrated, namely, the transference of the best of the workpeople, and the material of the Meissen factory, as already referred to. The Berlin porcelain was, of course, only an imitation of the Dresden, but the factory was carried on with such vigor as to yield to the king an annual revenue of 200,000 crowns. In 1790 a second royal porcelain factory was established about two miles from Berlin. To one of Ringler's fraudulent comrades is also due the factory established at Fulda, about 1763. The prince-bishop of Fulda established another factory in a house adjoining the episcopal palace; but it is said to have failed in consequence of the taste for porcelain extending to the dignitaries of the church, who claimed the privilege of carrying off specimens without paying for them. The porcelain factories of Thuringia originated about 1758, when an old woman having sold some sand at the house of the chemist Macheleid, his son, struck by its appearance, experimented on it, and obtained by its means a porcelain-looking substance, whereupon the Prince of Schwartzburg sanctioned the erection of a factory at Sitzerode, which was afterward removed to Volkstadt. The abundance of fuel supplied by the Thuringian forest led to the erection of other factories, such as that of Wallendorf in Saxe-Coburg, Limbach in Saxe-Meinengen, the director of which succeeded so well as to be able to purchase the factory of Grosbreitenbach in Rudelstadt, and also that of Kloster Veilsdorf. Factories were also founded at Gotha in 1780, at Hildburghaus, at Anspach, at Ilmenau, at Britenbach, and at Gera. All these factories had their periods of prosperity, and produced porcelain which is still esteemed by collectors. Some of them have degenerated into potteries, and some produce pipe-bowls as their only article in porcelain. Nor will our list approach completeness without mentioning a factory established by the Empress Elizabeth in 1756, near St. Petersburg, which still continues to produce good porcelain from native materials. Denmark has a factory at Copenhagen; it is supported by the government, but is said to be, commercially, a failure. The factory at Zurich in Switzerland was established on the information supplied by one of Ringler's workmen. A factory at Nyons, in the Canton de Vaud, has also produced some good porcelain.

During all this active rivalry on the Continent it will not be supposed that England had escaped the porcelain-making mania. Bow and Chelsea produced the first porcelain works. They made a soft ware from a mixture of white clay, white sand from Alum Bay, and pounded glass. The Chelsea works do not appear to have been in a very flourishing condition until George II. imported workmen, models, and materials from Brunswick and Saxony. Chelsea porcelain then became the rage, and such was the eagerness to obtain it, that it was sold by auction to the highest bidders, the dealers rushing in crowds to compete for it. Some of the best works were produced between 1750 and 1755: they are in the style of the best German; the colors are fine and vivid, and the claret color is peculiar. Bow china, made at Stratford-le-Bow, has some resemblance to that of Chelsea, but the material is not so good. Its principal productions were tea-services and dessert-sets. In 1750 was established the factory at Derby, which became important in consequence of the introduction of the Chelsea artists, workmen, and models, the junction of the two factories being notified by the anchor and the letter D, the monograms of each manufacture. Flaxman furnished designs for the establishment; but the union did not continue long; the partners quarreled, and one of them destroyed the models. Mr. Marryat describes the Derby porcelain as being very transparent, of fine quality, and distinguished by a beautiful bright blue, often introduced on the border or edge of the tea-services, the ground being generally plain; the white-biscuit figures are said to equal those of Sèvres. The Worcester works were established in 1751 by Dr. Wall and some others, under the name of the Worcester Porcelain Company. The company first imitated the blue and white Nankin china; they afterward adopted the Sèvres style, with the Dresden method of painting. These works are remarkable as being the first to make use of the Cornish stone or kaolin, discovered by Cookworthy in 1768. They are still carried on with distinguished success by Messrs. Kerr and Binns. In 1772 a factory was established at Caughley, near Broseley, Colebrook Dale, the productions of which are known as *Salopian ware*. Early in the present century some good porcelain was made at Nantgarrow and Swansea; it is also stated that the Bristol china, a white ware formerly common in the west of England, was made in Wales, and sold in Bristol.

France regarded with impatience during sixty years the progress of porcelain in Europe, and although eminently qualified in point of taste, skill, and science to contribute to the ceramic treasures of the world, she was unable to compete with other nations for want of a suitable raw material. It is true that as early as 1695 a soft porcelain had been manufactured at St. Cloud, and that some of the scientific men of France had endeavored, under royal patronage, to discover the secrets of the art, but no great success was attained. The company had been established at Vincennes, but in 1756 they removed to a large building which they had erected at Sèvres. In 1760 Louis XV. bought up the establishment, probably at the instigation of Madame de Pompadour, who seems to have shared with her sex the passion for china. The factory became celebrated for its porcelain, or *pate tendre*, but the great point aimed at was to produce the hard porcelain which had rendered Saxony the envy of Europe. But kaolin was not known in France, nor was its presence even suspected, until about 1768, when the wife of a surgeon named Darnet of St. Yrieix, near Limoges, having noticed in a ravine near the town a white unctuous earth, thought that she might relieve her husband's poverty somewhat by using it in her house instead of soap. The surgeon showed a portion of the substance to an apothecary of Bordeaux, who being aware of the search that was being made for porcelain earth, forwarded a specimen to the chemist Macquer, who recognized it as the much-desired kaolin. Assuring himself that an abundant supply could be had, he established the manufacture of hard porcelain at Sèvres in 1769. At first some difficulty was experienced in managing the colors upon the more compact and less absorbent material, so that the soft porcelain continued to be made until the year 1804.

Such, in few words, is the origin of the hard porcelain of Sèvres. The *pate tendre* was not considered as real porcelain, but the taste and skill of the French are remarkable in carrying it to the highest pitch of perfection under many difficulties, arising from its complicated and expensive composition, and from its liability to collapse during the firing. Mr. Marryat speaks of it as being "remarkable for its creamy and pearly softness of color, the beauty of its painting, and its depth of glaze." The ware for common or domestic use had generally a plain ground, painted with flowers in patterns or medallions; articles *de luxe*, and pieces intended for royal use, had commonly grounds of various colors, such as *bleu de roi*, *bleu turquoise*, *jonquille*, or yellow, *vert pres*, or green, and a lively pink or rose color, named after Madame Dubarry.

Skillful artists were employed upon the finest porcelain, which is adorned with landscapes, flowers, birds, boys, and cupids gracefully arranged in medallions. Some of the specimens are painted with subjects after Watteau, and other known masters. The jeweled cups, with the *blue de roi* ground are celebrated. The best period of the soft porcelain was from 1740 to 1769, and the tests which Mr. Marryat gives to distinguish it form its highest praise, namely, " the beauty of the painting, the richness of the gilding, and the depth of color." In point of form the Sèvres china is not equal to that of Dresden. A law was passed in 1766, and renewed in 1784, limiting the use of gold in the decoration of porcelain to the royal manufactory of Sèvres, which accounts for the rarity of old French gilded porcelain.

At the time of the Revolution many fine specimens of Sèvres porcelain in the royal palaces and mansions of the nobility were destroyed. The establishment of Sèvres, however, was supported by the revolutionary government, who appointed three commissioners to manage it. In the year 1800 the first consul appointed M. Brongniart as director. He held the appointment during forty-seven years, and originated the celebrated *Musee Ceramique*, consisting of a historical series of specimens illustrative of the ceramic art in all times and among all people, together with a collection of raw materials, tools implements, trial-pieces, models of furnaces, etc. On our visit to this museum, we were particularly struck with a collection of *failures*, or specimens showing what had been done to overcome faulty results, and what it was hopeless to attempt. M. Brongniart is also the author of a classical work on the art to which he devoted his life with such distinguished success.* M. Ebelman succeeded Brongniart as director, and held the appointment for a year or two. The present director, M. Regnault, was appointed by the Emperor Napoleon III.

The following is a list of the more celebrated porcelain manufactures of France : Chantilly, which owed its origin in 1735 to a workman from St. Cloud; Menecy, founded in 1735 under the patronage of the Duc de Villeroi; Sceaux-

---

* Traité des Arts Céramiques ou des Poteries considerées dans leur histoire, leur pratique, et leur theorie, par Alexandre Brongniart, etc,. etc. 2 vols. 8vo, with an Atlas of plates. Paris, 1844.

penthièvre, established in 1751; Clignancourt, 1750, under the patronage of the Duke of Orleans; Etiolles, near Corbeil, 1766; Bourg la Reine, Paris, 1733. Lille, established, it is supposed, in 1708, when the Dutch were masters of the town; Arras, 1782; Tournay, 1750. At St. Amand les Eaux, near Valenciennes, and at Tournay in Belgium, are two factories, the only two in Europe where the old *pate tendre* of Sèvres is still produced.

As respects Italy, a factory was established at Doccia, near Florence, at the beginning of the eighteenth century. Venice also manufactured porcelain until 1812. There was also a factory at Vineuf, near Turin; but the most famous factory in Italy is the Capo di Monti at Naples, founded by Charles III. in 1736. This sovereign appears to have excelled the other royal amateurs of Europe in the ardor with which he cultivated the ceramic art, and he even surpassed Augustus III., who was nicknamed by Frederick of Prussia "the Porcelain King," and who exchanged a whole regiment of dragoons for some huge useless china vases. Charles III. even worked in the factory with his own hands, and held an annual fair in front of the royal palace at Naples, where there there was a shop for the sale of the royal productions; and there was no more certain road to the king's favor than to become a purchaser. When Charles became king of Spain he founded a factory at Madrid, and that at Naples declined. His successor Ferdinand sanctioned the erection of other porcelain works, and allowed the royal workmen to assist in their formation; and they appear not only to have assisted but to have robbed the parent factory of its gold and silver models and other valuables. The royal factory was closed in 1821. The porcelain of Capo di Monti is not, as is commonly the case, an imitation of that of some rival factory. Its beauty and excellence are due to the design from shells, corals, embossed figures, etc , artistically moulded in high relief. Mr. Marryat regards the tea and coffee services of this ware as perhaps the most beautiful porcelain articles ever produced in Europe, for transparency, thinness of the paste, elegance of form, and gracefully-twisted serpent handles, as also for the delicate modeling of the ornamental groups in high relief, painted and gilt, contrasting well with the plain ground. The factory at Madrid was conducted with the utmost secrecy during several reigns, but was destroyed by the French in 1812. Portugal has a factory of hard porcelain near Oporto.

The prices paid for porcelain are high. As much as 150*l.* has been paid for a single specimen of majolica; while a service of Chelsea ware has cost 1200*l.* One of Sèvres, of a good period, 30,000 livres; while the Dresden ware was equally costly. Although our modern manufacturers have produced porcelain rivaling that of the best periods of celebrated works, the price still continues to be necessarily high, where the materials require to be treated with the precision of a chemical process, and the design and ornamentation require high artistic skill. Mr. Minton received 1000*l.* for his service of turquoise and Parian; Lord Hertford gave 1000*l.* for two vases; Mr. Mills the same; one of the Queen's vases has been valued at 1000*l.*, and Lord Ward gave 1500*l.* for a dessert service of Sèvres. Such works as these, however, belong rather to the fine arts than the useful arts, to be preserved in cabinets and museums. Formerly it was customary on great occasions to serve the guests on porcelain, which gave to wealth a real distinction. In those days the transition from porcelain to earthenware was abrupt; but through the exertions of Wedgwood and others, porcelain now descends through numerous varieties of material, style, taste, and decoration; so that every class of consumer may suit his own taste and means. Our trade in earthenware has of late years gone on increasing. In the year 1835 the declared value of earthenware exported from the United Kingdom was 540,421*l.*; in the year 1857 it amounted to 1,488,668*l.* Our exports extend to most parts of the world, including Russia, Austria, Turkey, and even France. The United States of America take nearly the half of our exports in earthenware, so little has the potter's art been encouraged in the New World.* Our exports to foreign countries would doubtless be larger if the restrictions were fewer and less clumsy. In Germany and Italy the duty is levied on the weight; so that Wedgwood, on account of the lightness of his ware, was long able to command the market in those states. In France the duty on common English china of one color, without gilding or ornament, is 104 francs per 1000 kilo-

---

* Stone-ware is extensively manufactured in Northern Ohio.

grammes (200 lbs.); for fine china, 327 francs for the same quantity. The most whimsical of all tariffs is that of Portugal, where the charge is according to the number of colors; so that, as Mr. Wall remarks, "no man's pocket could stand the choice of a rainbow pattern."

THE MATERIALS.

Clay, which forms the basis of pottery and earthenware, is not only abundant and widely diffused, but presents so many varieties that much experience and judgment are required in adapting the kind of clay to the article to be manufactured. Brongniart enumerates 167 varieties of clay, and states their physical and chemical characters, composition, locality, and application. Some of the commonest varieties of clay consist of—1. *Pipe clay.* It has a grayish-white color, a smooth greasy feel, an earthy fracture; it adheres to the tongue, and is plastic, tenacious, and infusible. It becomes of a cream color when fired; and is used for tobacco-pipes and white pottery. It is found near Poole in Dorsetshire.—2. *Potter's clay.* This is of various colors; those used in the Staffordshire potteries are the *brown* and *blue* clays from Dorsetshire,* and black and cracking clays from Devonshire. The color of the black clay is due to bitumen or coaly matter, which disappears in passing through the kiln; so that the wares formed of it are almost white. Cracking clay is esteemed on account of its whiteness, but as it is liable to crack during the firing, it must be mixed with other clays which are free from this defect. Brown clay when passed through the gloss oven sometimes causes the glaze to crack, or *craze,* as it is called. For ordinary purposes blue clay is preferred; it can be mixed with a larger proportion of flint than the other varieties, and thus produces a white ware. Potter's clay, mixed with sand, is formed into bricks and tiles.—3. *Stourbridge clay.* This is of a dark color, from the presence of carbonaceous matter, and from its being more refractory than potter's clay, it is largely employed for glass pots, crucibles, etc. 4. *Brick clay* or *loam* is abundantly met with on the London clay, and is often found on an interposed bed of sand. Its

---

* In the year 1855 there were exported from Poole in Dorsetshire 53,702 tons of Poole clay, and 582 tons were sent to London by railway.

appearance, texture, and composition vary greatly; and the color depends on the proportion of oxide of iron contained in it.—5. *London clay.* This is an extensive deposit of bluish clay: although near the surface, it frequently has the usual clay color. It extends over the greater part of Middlesex, a portion of Norfolk, and the whole of Essex and Suffolk. It is often found near the surface; but the lower beds are sometimes yellowish, white, or variegated. Organic remains are found in it.—6. *Plastic clay.* This skirts the London clay within the London chalk basin, and is also found in the Isle of Wight. This formation comprises a number of sand, clay, and pebble beds, alternating irregularly, and lying immediately on the chalk.

The above varieties of clay are mixed with such substances as carbonate of lime, magnesia, protoxide of iron, manganese, finely-divided quartz, felspar, mica, organic matter, etc., which greatly modify its properties and applications. Pure clay is soft, more or less unctuous to the touch, white and opaque, and has a characteristic odor when breathed upon. It is a compound, or perhaps only a mixture, of the two earths, alumina and silica, with water. Silicate of alumina enters largely into the composition of many crystalized minerals, among which is felspar, so abundant an ingredient in granite, porphyry, and other ancient unstratified rocks. Under certain circumstances the felspar undergoes decomposition, and is converted into a soft friable mass. In certain districts of Devonshire and Cornwall the felspar of the white granite is often disintegrated to a great depth, and the rock becomes converted into a substance resembling soft mortar. This being collected, is thrown into a stream of running water, which washes off the argillaceous portions, and holds them suspended while the heavier quartz and mica subside. At the extremity of these streams the water is dammed up, forming *catchpools*, where the pure clay sinks and forms a solid mass, which, when the water has been drawn off, is dug out in blocks, and placed on shelves called *linnees* to dry. It is next stove-dried, crushed, packed in casks, and sent to the potteries, under the name of *china clay*, or *kaolin*. It consists of 80 parts alumina and 20 silica; a proportion of undecomposed felspar, under the name of *china-stone*, is sometimes added to the ingredients for porcelain. In the year 1855 as much as 60,188 tons of china clay was shipped from Cornwall, and 19,961 of china-stone; while Devonshire shipped 20,000 tons of pipe clay, and 1100 of china clay. Of late years improved methods have been adopted for getting out the china clay in Cornwall. At the Lee Moor clay-works, for example, Mr. Phillips, the managing director, has introduced the following arrangements: The decomposed felspar is transferred directly from the quarry to the works, where it is thrown into hoppers, and passes into a trough under the action of a full stream of water, encountering on its way a series of knives and iron arms furnished with teeth, which thoroughly beat up the clay in its passage along the trough. Pure spring-water is used in the operation, and great care is taken to exclude the surface drainage from the peat soil of Dartmoor. As the water leaves the trough it flows through sieves which separate the coarser fragments of quartz, and the fluid, charged with clay and mica, passes on; the mica breaking up into thin scales, has a tendency to float, but being heavier than the suspended alumina, it gradually subsides under the regulation of the current, which is now not sufficiently rapid to carry on the mica, nor sufficiently sluggish to allow of the deposition of the clay. When at length the stream holds nothing but pure clay, it is allowed to flow into deep V-shaped channels, which terminate in large covered reservoirs, in which the clay is deposited. Warm-air pipes circulate beneath the reservoirs, so as produce a temperature of 90°. The fine clay soon subsides, so as to allow of the clear water above it being drawn off. The mineral *pegmatite* is also valuable, as containing all the ingredients for hard porcelain. It consists of felspar, kaolin, and a small proportion of prismatic quartz. The mineral must, however, be in the state of decomposition already referred to. The quartz gives whiteness and transparency to hard ware; but for soft porcelain bones are substituted. These melt into a kind of semi-transparent enamel, which imparts transparency to the ware. *Steatite*, or *soap-stone*, is also an ingredient in porcelain. The *statuary* porcelain known as *Parian* or *Carrara*, from its similarity to those beautiful marbles, owes its effect chiefly to the use of a soft felspar instead of Cornish stone; while its agreeable yellowish-white tint is due to the presence of a small portion of oxide of iron contained in the clays and the felspar.

The property possessed by clay of forming a perfectly plastic mass with water, and of being permanently fixed by heat, has led to its employment in the manufacture of bricks and vessels of various kinds, but it undergoes a large amount of contraction in drying and burning, to diminish which the clay is usually mixed with a considerable proportion of quartz-sand, or with the powder of previously-burnt clay. The quartz in pottery ware is in the form of flints; these are obtained from the chalk districts of Gravesend and Newhaven; they are white outside, but dark and clear within. Such flints should be selected the fracture of which is free from yellow or iron stains.

The preparation of the clay for such coarse articles as tiles consists first in *weathering*, or spreading it out to the action of the air, so that by absorbing water the articles may separate, and the clay work freely. It should be exposed to at least one night's frost, or to one day's sun, before a second layer is added to the first. The weather-clay is cast into pits, and left for some time covered with water to *mellow* or *ripen*. Before being used it is tempered by grinding in a pug-mill. If the clay be *foul*, or contain many stones, it is *slung*, or cut into lengths of about two feet with a *sling* or wire-knife, and then further divided into slices of three-quarters of an inch in thickness, during which operation the stones fall out, or are picked out. The clay goes once more through the pug-mill, and is then ready for the moulder. For chimney-pots and such articles the clay is slung once or twice, and pugged or ground two or three times.

The clay for fine pottery undergoes a number of preparatory processes. Two or more kinds of clay being put together in proportions according to the judgment of the manufacturer, they are thrown into a trough with water and left for some hours. They are then well worked with a long blade of ash furnished with a cross-handle, named a *blunger*, until a smooth pulp is formed, a pint of which weighs 24 ounces, or, in the case of china clay, 26 ounces. The operation of *blunging*, as it is called, may be assisted by pugging the clay in an iron cylinder furnished with knives on the inside, and a moving vertical axis also containing knives, which by their joint action divide the clay, and by their position force it downward, and out through an opening at the bottom. It is then removed to a vat, mixed with water, and blunged by means of cross-arms attached to a perpendicular shaft. In this operation stony particles sink to the bottom.

The flints having been heated in a kiln, and plunged in cold water to increase their brittleness, are crushed into fragments by means of stampers, and are next reduced to powder in a *flint-pan*. This is a circular vat ten or twelve feet in diameter, the bottom of which consists of masonry of quartz or felspar. In the center is a vertical axis, from which radiate four arms for moving the *runners:* these are masses of *chert*, a hard siliceous stone found near Bakewell in Derbyshire. The broken flints are thus ground with water, and in the course of some hours are reduced to powder, which forms with the water a creamy mixture. Felspar, broken porcelain, etc., is sometimes ground up in the same manner in smaller vats. The creamy mixture is transferred to another vat furnished with a vertical shaft and arms, and being diluted with water, the arms are set rotating, the effect of which is to keep the finer siliceous particles suspended, while the coarser ones sink to the bottom. The former are drawn off with the water, and the latter are sent back to the flint-pan. The water thus drawn off is received into a reservoir, in which the finer particles subside. The creamy mixture of flint and water is fit to mix with the clay when a wine pint of it weighs 32 ounces. The proportions, however, in which the clay and the flint are mingled vary greatly with the kind of ware intended to be made, and the experience of the manufacturer.

These proportions being determined, the ingredients are first mingled by being agitated together, after which the mixture is passed through sieves of fine hard-spun silk, arranged on different levels, so as to run through comparatively coarse into finer sieves, and thus produce a smooth, uniform mixture of *slip*, as it is called. To assist the easy passage of the mixture a jigging motion is given to the sieves. The water which has thus far served as a vehicle for the ingredients, is next got rid of by evaporation in the *slip-kiln*. This is a long brick trough, heated by flues underneath, and capable of raising the water to the boiling point. During the heating the slip is diligently stirred to prevent the heavier flint from subsiding, and also to prevent the flint and clay from forming a kind of mortar with the water. When bubbles of steam cease to form, the operation is at an end. In countries where fuel is not so abundant as in England, the water is got rid of by filtration, assisted by mechanical pressure, or by rarefying the air beneath the filter by atmospheric pressure.

When the stuff is of uniform texture and sufficiently hard, it is cut up into wedges which are dashed down upon each other, in order to get rid of vesicles and air-bubbles, which might afterward form blisters in the ware. To obtain a fine grain the clay should be *wedged* at intervals during several months. It is stated that in China the stuff is prepared many years in advance. The French missionaries were informed that it was customary to prepare the stuff for a hundred years (*pour cent annees*), whence arose a fanciful derivation of the word porcelain. However this may be, there is no doubt that newly-made stuff produces bad ware, and that *ageing* greatly improves it. During the last-named process a kind of fermentation sets in, carbonic acid and sulphide of hydrogen are liberated, and the mass improves in texture and color. These gases are doubtless formed at the expense of the carbonaceous and organic impurities of the clay or of the water, whence the improvement in color; while the disengagement of the gas accounts for the improvement in texture. The next process is *slapping*, in which the workman takes up a mass of the paste and dashes it down with violence, then dividing the mass with a wire, he dashes the top on the lower; this is done many times, care being taken to preserve the *grain*—that is, to slap the layers parallel to each other, and not obliquely, otherwise the paste would be liable to fall apart during the firing.

THE MANUFACTURE.

There are three processes by which fictile articles are shaped—namely, *throwing, pressing*, and *casting*. Of these throwing is the most common, and by far the most ancient. It is performed by means of the potter's wheel or lathe, which is a disc of wood fastened to the top of a vertical spindle, and made to rotate by being connected by means of a strap with a multiplying wheel driven by an attendant. The paste, as it is received from the slapper, is of the consistence of dough. The thrower's attendant cuts it up into portions, weighs each, according to the quantity required for the intended article, and rolls each portion up into a ball. The thrower, seated before his lathe, dashes one of the balls down upon the rotating board, and with the fingers, which are frequently dipped in water, raises the lump into a conical form, presses down the mass to get rid of air-bubbles, and with one hand, or finger and thumb, in the mass, gives shape to the intended article; he is also furnished with a piece of horn or porcelain called a *rib*, the edge of which accurately represents the curve of the vessel. With this he smooths the inner surface, and gives it shape. During this operation the assistant turns the wheel with varying rates of speed, so that the centrifugal force may act differently in different conditions of the growing vessel. The thrower is furnished with a rude kind of fixed gauge, consisting of an upright stick, from which projects a horizontal rod at such a height above the whirling table as to enable the thrower to make all the articles of one kind very nearly of the same size. When one article is finished, it is removed by passing a wire beneath it, and is set aside in an airy or a warm room until sufficiently consolidated for the next operation, which is *turning*. As it would not be possible for the thrower to produce articles sufficiently thin, they are reduced in size by being put on the chuck of a lathe, and turned to shape by means of cutting tools, the material flying off in long, broad shavings just as if it were wood. When it has thus been properly thinned and brought to shape, the vessel is smoothed and solidified by the pressure of a broad tool upon its surface. Handles, spouts, etc., are formed separately, and are attached to the articles by means of slips. Flowers, leaves, etc., are formed partly in moulds and partly by hand, and are stuck on separately. The article is lastly trimmed with a knife, and cleaned with a damp sponge, and is ready for the kiln.

By the process of pressing, such articles as plates, dishes, saucers, etc., are formed. The exact pattern, say of a plate, having been determined by means of a model, a number of plaster casts are taken, one of which the plate-maker places on a whirling table, bats out a sufficient quantity of paste by means of a plaster mallet, and when sufficiently extended, places it on the mould, much in the same way as a housewife would cover a pie with paste. The table is then set whirling, and a profile or shape in earthenware being brought down upon the paste, gives the required form to the bottom of the intended plate. When the plate-maker is satisfied with his work, the mould, with the plate in its *green* state, as it is called, upon it, is conveyed by a boy to a warm room, and he brings back an empty mould, which has been drying, for another plate. In about two hours the plate is sufficiently dry to be removed from the mould, but the mould itself is left to dry before it is used again. One man and two boys can produce from sixty to seventy dozen of common plates in a day of ten hours, the same mould being used some five or six times during the day.

The above operation is called *flat-ware pressing*. Deep vessels are formed by what is called *hollow-ware pressing* or *squeezing*, for which purpose the mould consists of several parts, which fit accurately together by means of projecting pins and cavities. The clay having been batted out, the several parts of the mould carefully lined with it, and the points of junction well worked and wetted with slip, are brought together and secured by a cord, when the joints are further well worked and pressed, thin rolls of clay being sometimes inserted, and the whole worked and smoothed with moist leather and a cow's lip. The interior is then washed with a sponge, set aside for a time, and when somewhat solidified, is worked or polished with a flexible plate of horn; it is next put into a warm room, and when the plaster has absorbed sufficient moisture, the article is removed from the mould and *fettled* or trimmed with proper tools to get rid of seam marks. The outside is also cleaned with a moist sponge, and the handles, etc., having been added, and the horn again used, it is set aside for baking. For elaborate works, models are formed by experienced artists in clay, and the moulds for the separate parts may be numerous. Works of a comparatively simple character are formed by the united agency of throwing and moulding.

By the third process, called *casting*, such delicate articles as egg-shell china are formed. The paste having been reduced to a creamy state, is poured into a plaster-mould, which, absorbing water from that portion of the paste which comes in contact with it, fixes it, so as to allow the remaining fluid portion to be poured off. A very thin coating of paste is thus left attached to the mould; when this is sufficiently dry, the mould is again filled for a short time with the creamy mixture, when a second thin deposit is formed upon the first. The mould having been dried in a warm room, the cast is taken out, examined, and touched upon by the modeler. Busts and statuettes are also formed in this way; but as they shrink as much as one-fourth during the firing, considerable dexterity is required to preserve their shape. The lace which is sometimes seen on these figures is real lace, dipped into slip, when the heat of the kiln destroys the thread, and solidifies the paste, which takes its place.

Encaustic tiles are made by what may be called a fourth process, namely, *veneering*. They consist of a body of red clay, faced with a finer clay for the pattern, and strengthened at the bottom with another clay, the junction of these layers apparently preventing warping. After the usual preparatory processes, the red clay is slapped into the form of a quadrangular block, from which the tile-maker cuts off a slab with a wire, and upon this the facing of finer clay, colored to the required tint, is batted out and slapped down. The bottom facing is added in a similar manner. The tile is then put into a box-press, when a plaster of Paris slab, with the pattern in relief, is brought down on the face of the tile, and impresses in the soft tinted clay the design, the hollow being afterward filled up with clay of another color. At the same time, the maker's name is stamped at the back, together with a few holes to make the mortar adhere. The colored clay, in a creamy state, is next poured over the face of the tile, so as completely to conceal it, and when, in the course of twenty-four hours, this colored slip has become hard, the superfluous clay is scraped away, the colored clay being left only in the hollows formed by the pattern-mould. The tile having been finished off with a knife, and defects corrected, is kept during a week in a warm room, called the *green-house*, and the drying is finished in a warmer room, called the *hot-house*, preparatory to firing.

The various articles of pottery, stone-ware, or porcelain having, by one or other of the processes named, been perfected as to form, and handles and other appendages, and solid ornaments added, are now in what is called the *green* state. The next process is to fix them, and deprive them of their plastic nature by the action of heat. The *potter's kiln* consists of a massive domed cylinder of brick-work, bound with iron, and protected from the weather by an outer conical hood or casing. The dome contains openings for the exit of the smoke, which escapes into the air through a chimney in the hood. Heat is supplied by means of six or eight fireplaces fixed round the cylinder, with proper circulating flues and dampers for regulating the draught. During the firing, the ware (unless of the commonest kind) is not exposed to the direct action of the fire, but is carefully packed in strong vessels, shaped very much like band-boxes; they are made of Staffordshire marl, and are called *seggars*. The pieces must be packed in the seggars in such a way as to economize space, and yet give them the full benefit of the heat; at the same time, they must be arranged according to their size and solidity, so that small and delicate articles may not vitrify under too strong a heat, and large ones have heat enough. Some articles admit of being placed in contact, so as to support each other and prevent distortion. When the pieces are large or complicated in shape, they may require special supports to prevent warping; these supports are of fire-clay, and nicely fit the parts supported. Articles in porcelain are sometimes separated during the firing by means of sand or powdered flint; but the contrivances of this kind are numerous. When the seggars are filled, they are conveyed to the furnace, and piled up so that the flat bottom of one seggar may form a cover to the open mouth of the seggar immediately beneath it, the surfaces being separated by a ring of soft clay, which forms a tight joint. As many as 30,000 pieces of ware may be included in one baking. When the seggars are properly arranged in piles, or *bungs*, as they are called, and steadied by means of short struts, the door of the kiln is closed with brick-work, the fires are lighted usually in the evening, and are urged during the whole of the night, so that flame may be seen issuing from the chimney. Early in the morning the man draws his first *watch*. Watches or trial-pieces are small rings of fire-clay, which vary in color with the temperature; a number of these are placed within the kiln in such positions that the man can withdraw them at pleasure by inserting a long iron rod through holes in the side of the kiln. The heat is regulated according to the aspect of these watches, and when, after thirty or forty hours, the firing appears to have been satisfactory, no more fuel is added, the fires are left to go out, and the kiln gradually cools during the next twenty or thirty hours. As much as fourteen tons of coal may be consumed in one firing. There can be no doubt that a very large proportion of this fuel is wastefully expended: our present abundant native store of coal leads to much extravagance in our various factories; and it has been suggested by M. Arnaux, a competent authority, to fire the ware by means of gas, which, he thinks, can be done with an ease and precision unattainable by the present system.

When the ware is removed from the kiln, its characters are found to have undergone a remarkable change. Instead of a soft, dull, friable or plastic material, we have a hard, brittle, resonant, light-colored, porous body. In this state it is called *biscuit*, from its resemblance to well-baked ship bread. Wine-coolers and similar porous articles, when brought to this state, are finished; but most articles, especially of earthenware, must be covered with some kind of vitreous glaze, to remove their porosity and liability to tarnish, and to render them fit for use. If colored ornaments have to be added, these are first put upon the biscuit, and the glaze, in the form of a white powder, is then made to cover the whole article, which, being passed a second time through the fire, the powder melts into a glass, which forms the ordinary surface of common wares. The firing is a costly process, from the great expenditure of time and fuel, and this second firing still further increases the cost of the ware. It thus became a great improvement when Wedgwood was able so to compound the ingredients of his ware that partial vitrification took place at the first firing, thereby depriving the ware of its porous character, and rendering a second firing unnecessary. So also, in the commonest kind of stone-ware, such as is made at the Lambeth potteries, the glazing is, by an ingenious device, effected simultaneously with the baking. When the ware has attained a very high temperature in the kiln, a quantity of moist salt (chloride of sodium) is thrown in; the salt is volatilized and decomposed in the presence of moisture, and by contact with the heated surfaces of the clay, hydrochloric acid is disengaged, and the ware becomes covered with silicate of soda, which, combining with the silicate of alumina of the ware forms a fusible double alkaline silicate or glaze on the surface.

The object of the glaze being to render the article impermeable by water, attempts have been made to accomplish that end in various ways. Certain rude nations render their wares impermeable by rubbing them while hot with tallow, which, becoming partially decomposed, fills up the pores, and imparts a black color. Even the vases of the artistic Etruscans and Greeks have not a vitreous but a carbonaceous glaze, which wears off in the handling. The wine and oil jars of Spain and Italy are made water-tight by the ancient method of rubbing them over with wax. The most common description of glaze is, as its name *glaze* or *glass* implies, vitreous. It is of two kinds, *transparent* and *opaque*. When the ware is of good color, and the ornaments are impressed upon it, the glaze may be transparent; but where the clay, otherwise good in quality, is bad in color, an opaque glaze, or *enamel*, as it is then called, is used. In some cases, articles made of a good clay, of a bad color, may, before firing, be dipped into a slip of white clay, and being thus veneered, admit of taking a transparent glaze. Glazes colored by means of a metallic oxide are also sometimes used. The glaze should not have too strong an affinity for the paste, or during the second firing it may be absorbed into the ware instead of remaining at the surface, to which it should adhere firmly, and expand and contract equally with the ware, so as not to be liable to craze or crack. Numerous substances are employed in the composition of glaze. For very hard ware, in which the point of fusion is high, the felspars and certain volcanic scoriæ are used; in other cases, common salt, potash, boracic acid, phosphate of lime, and sulphate of baryta, are the ingredients. Another class of glazes contains earthy and metallic substances, mixed or fritted into a glass; such are silica and lead, or enamels of silica, tin, and lead. Some glazes contain metallic oxides, such as those of manganese, lead, and copper. Metallic and earthy substances, if not previously fritted, form a glaze with the silica of the paste in the gloss oven. Such glazes, however, are commonly soft, and liable to be acted on by acid and fatty substances; so that lead glazes should be avoided for articles intended to receive food. In such cases, borax may be advantageously substituted for lead. A pure white paste is improved by a transparent glaze, but if of bad color, it may be dipped into opaque glazes even before the first firing. Glazes are made opaque by means of oxide of tin; color is given by the oxides of manganese, copper, and iron; while, by introducing these, together with the oxides of cobalt and of chromium, into opaque and transparent glazes, an agreeable variety is produced. Pegmatite forms a good glaze for hard porcelain; but for soft porcelain a glass is fritted and mixed with oxide of lead, or with earthy substances.

In applying the glaze to the biscuit, it is reduced to a fine powder, and mixed with water. When the biscuit is plunged into this mixture, the porous material immediately absorbs a quantity of the water, and leaves the powder equally distributed over its surface. When articles are glazed and fired at one operation, the ware in its green state is not absorbent, so that the glazing has to be put on with a brush. For articles which are glazed on the inside only, such as pipkins, the glaze is made creamy with water, and poured into the vessel and then out again, a sufficient quantity adhering to the surface by this means. Custom requires that jars shall have a portion of their surface of a deeper brown than the natural color of the material; they are therefore dipped to a certain height in a mixture of red ocher and clay slip. The glazing is completed during the firing by means of common salt, as already noticed.

The pieces having been covered with white powder, are arranged in seggars to protect them from the direct action of the fire in the gloss oven. They are separated from each other by means of supports, which present the smallest possible surface of contact. These supports, known as *cockspurs*, *triangles*, *stilts*, etc., have points projecting from them above and below, which serve to separate, while they support, the articles as they are piled up in the seggars. The seggars are piled up in the glaze-kiln in the same manner as in the biscuit-kiln, and the temperature is raised to a point sufficient to fuse the glaze into a transparent glass, and to unite it perfectly with the surfaces of the ware. To enable the workmen to determine when the proper temperature has been reached, watches, or rings of clay, covered with glaze, are placed in the oven, and drawn out from time to time.

THE ORNAMENTATION.

The love of ornament, which forms part of that higher sense of beauty common to our nature requires the addition of some kind of adornment to articles in common use. The rude pottery of savage nations is relieved in this way, and often with considerable taste. It may admit of question whether our own taste is equally correct in the elaborate decorations which we bestow upon articles intended for every-day use. Plates of Sèvres porcelain, richly decorated with landscapes, or portraits of distinguished individuals, may have a high artistic value, but are certainly not adapted to be placed before the company at a dinner-table. A dessert or dinner-plate is not in itself remarkable for beauty of form; but its effect is absolutely hideous when it is made to take a prominent part in decoration. In the palace at Fontainebleau we were introduced to a room, the walls of which were decorated with plates of Sèvres china, arranged in horizontal lines. In such an example, the costliness of the material and the skill of the artists were rendered simply rediculous. So also the rich blue and gold of a tea-service have too heavy an effect, when the feeling of grace and lightness ought to be inspired. The leading idea in ceramic ware should be that of purity. The white color would sufficiently suggest this if it were not concealed by ornament, just as that pure material glass, when not spoilt by the glass-cutter, reveals the unsullied transparency of the water or of the wine contained in it. The artist may exercise his taste in producing beauty of form, but the ornamentation of that form should be of the simplest character, only just calculated to relieve the beauty of the material. Our limited space will not allow us to enlarge on this subject, so that we at once proceed to a brief notice of the mechanical and chemical means by which ornaments are applied to pottery and porcelain.

When common ware is to be ornamented with a pattern, it is put on before the glazing. The blue pattern of an ordinary plate is printed on the biscuit with an ink composed of boiled linseed-oil, resin, tar, and oil of amber, colored by means of a mixture of oxide of cobalt, ground flint, and sulphate of baryta (fritted and ground), and blended with a flux of ground flint and thick glass powder, which serves to fix the color. The ink is made fluid by spreading it on a hot iron plate. It is taken up by means of a leathern dubber, and transferred to engraved copperplates, also heated, and the superfluous color is scraped off with a pallet-knife, and the surface of the plate is cleaned with a dossil. A sheet of yellow unsized paper is next dipped into soapy water, and placed on the copperplate, which is thus passed through a cylinder press. The pattern is thus transferred to the paper, which is taken by a girl called the *cutter*, who cuts away the unprinted portions, and leaves the pattern in separate parts. These are taken by a woman called the *transferrer*, who places each portion with its printed side next the biscuit, and rubs it with a flannel rubber, until the ink is properly absorbed. The pattern-papers are subsequently removed by placing the biscuit in water, and gently washing it with a brush. The biscuit is next dried in an oven, and is then ready for glazing; the heat of the gloss oven vitrifies the glaze, and allows the pattern to be seen through it. Instead of paper, a flexible sheet of glue, called a *paper* or *bat*, is in some cases used for transferring the design. The impression is taken in oil from the engraved plate, and after it has been transferred to the biscuit, the required color is dusted over it in a dry state. The sheet of glue can be cleaned with a sponge, and can be used over and over again.

When the pattern is required to produce high artistic effects of form and color, the work is performed by hand with a camel-hair pencil. The colors consist of metallic oxides ground up with such vitrifiable substances as glass, niter, and borax, oil of turpentine or of lavender being the usual vehicle. The greatest difficulty which the artist has to contend with arises from the fact, that the colors are for the most part dingy and unpleasing, and give no idea to an inexperienced eye of the intended effect.* It is not until the heat of the furnace has driven off the oil, and chemically combined the

---

* Attempts have been made to construct a pallet of enamel colors which do not change color in the firing, but only change from a dullness to a creaminess of texture. A case of this kind is mentioned by Brongniart, but the success attained by the inventor, M. Dihl, was only partial; since the rose tints, purple, and violet, produced by the precipitate of cassius, which cannot be prevented from changing under the action of heat, were omitted. Besides this, the action of the surface, and the different kind of glaze upon the colors were not taken into account.

ingredients of the colors, that the effect can be judged of. The artist has thus to work, as it were, in the dark: he is not cheered with the idea of progress, as in the ordinary oil-painting, where the work seems to grow into life, and to develop new details of beauty at every touch. Even after the first firing, it by no means follows that success has been attained. The work may have to be retouched, and again passed through the fire, or it may be injured by one or other of the numerous accidents to which a work is liable which has to pass through the fire.

The colors used are formed by the combination of certain metallic oxides and salts with certain fluxes, by means of heat, which enables them to fuse into colored glasses. The oxides are usually those of chromium, of iron, of uranium, of manganese, of zinc, of cobalt, of antimony, of copper, of tin, and of iridium. The salts and other bodies used for imparting color are the chromates of iron, of baryta, and of lead, the chloride of silver, the purple precipitate of cassius, burnt umber and burnt sienna, red and yellow ochers, etc. Some of these develop their colors under the influence of the highest temperature of the porcelain furnace, and are hence called by the French chemists *couleurs de grand feu;* others, and by far the larger number, are termed *muffle-colors*, inasmuch as they become developed under the more moderate heat of the muffle, which is a kind of seggar, in which the painted ware is inclosed, to protect it from the fuel. The first class of colors is limited to the blue produced by oxide of cobalt, the green of oxide of chromium, the brown produced by iron, manganese, and chromate of iron, the yellows from oxide of titanium, and the uranium blacks. Those colors form the grounds of hard porcelain, and as the heat employed in firing it is capable of fusing felspar, that substance is used as the flux. For an indigo blue, four parts oxide of cobalt and seven parts felspar, or for a pale blue, one part oxide of cobalt and thirty parts felspar, are well pounded, mixed by repeated siftings, and vitrified in a crucible in the porcelain furnace. The resulting glass is reduced to powder, ground up with a volatile oil, and applied to the surface of the biscuit, which, being again raised to the high temperature of the porcelain furnace, the color fuses, and becomes incorporated with the substance of the ware. The high temperature required for cobalt has, however, this inconvenience, that a portion of it becomes volatile, so as to affect objects placed near it. In this way a white vase in the same furnace may derive a blue tint from the vapor of the cobalt. This color is also uncertain in its results: it sometimes leaves white uncolored patches, or forms a dull granular surface. Oxide of chromium may be employed without a flux to give a green color to hard porcelain; but as it does not, under such circumstances, penetrate the ware, it is liable to scale off. A bluish-green is produced from three parts oxide of cobalt, one part oxide of chromium, and one-tenth of felspar, without fritting. Mixtures of the oxides of iron, manganese, and cobalt, produce a fine black, and by omitting the cobalt, various shades of brown.

The muffle colors are too numerous to be stated here; they are fired at a temperature equal to about the fusing point of silver. Many of them would become more brilliant and solid under a greater heat, but this would be injurious to those colors which are obtained from the purple precipitate of cassius,* on which the artist relies for some of his finest effects, such as fine purple, violet, and carmine tints.

In preparing metallic oxides and their fluxes sound chemical knowledge is required, otherwise the results cannot be depended on. The chemist relies on the stability of nature, as revealed to him by his science: he reduces his materials to a state of chemical purity, and compounds them according to the law of definite proportions. In order, for example, that the yellow color imparted by chromate of lead shall be identical at all times, the compound must obviously consist of nothing but equal equivalents of oxide of lead and chromic acid. In such case, if the pigment be applied at different times under the same circumstances, it will produce precisely similar results; but if either of the proximate elements of the salt be impure, no reliance can be placed on the compound. Different specimens will produce different results, although the same mode of applying them be always observed. In some cases, however, not even the chemical purity of the ingredients will insure harmonious results. The physical condition of one of the ingredients may be of importance, as in

---

* This pigment is formed by adding a solution of gold in *aqua regia* to one of chloride of tin.

the case of oxide of zinc, an ingredient in some of the enamel greens, yellows, yellow-browns, and blues. If the oxide be lumpy, granular, dense, and friable, it will produce a dull pigment, although chemically pure, while a light flocculent impalpable oxide, chemically identical with the former, will give satisfactory results. It is further necessary that solutions of a metal be made at the same temperature, that the acids which dissolve it be of the same strength, that the precipitate be neither more nor less rapid on one occasion than on another. Such conditions as these require to be carefully studied and noted, as, indeed, has been done in the laboratory at Sèvres, where minute records are kept of the processes required for compounding the colors.

But even when such conditions as the above are known and observed, there are others so slight as scarcely to be appreciable, but which, nevertheless, have an influence on the color. With certain delicate pigments, the porphyrization or grinding with water or oil a little more or less, the difference of touch of different artists in laying on the same pigment, will produce differences in tone, although all the other conditions be strictly observed.

Dumas defines the process of painting on hard porcelain to be the art of soldering by heat to a layer of the glaze a layer of fusible color, the dilatation of which shall be the same as that of the glaze and the body of the ware. The function of the flux is to envelop the color and attach it to the glaze. In most cases it has no action on the color, but is simply mechanically mixed with it: the flux, however, must mix with the glaze. That muffle colors do not penetrate the porcelain, may be proved by boiling in nitric acid a piece of painted ware after it has been fired, when the colors will disappear. As the flux is only a mechanical vehicle for the color, it must vary with the color; but the necessity for mixing or blending colors greatly limits the range of fluxes. A common flux is the silicate of lead, or a mixture of this with borax. Now the borax cannot be replaced by the fixed alkalies, on account of the readiness with which soda or potash becomes displaced in order to form other compounds. They have also a tendency to make the colors scale off. The mode of using the fluxes varies with the color: in some cases it may be ground up in proper proportions with the color; in others it must be previously fritted with the color. The first mode is adopted when the coloring oxide is readily altered by heat; but when the oxide requires a high temperature to bring out its characteristic color, the second method is adopted.

Not the least among the difficulties of enamel-painting is the high temperature required for the vitrification of the colors. The lowest heat of the muffle is about 1100° Fahrenheit; while some oxides do not develop their color below 1850°. In the regulation of the furnace, the most successful method is to begin with a low heat, and urge it rapidly up to its maximum, and as rapidly to lower the heat. A moderate heat, long continued, may produce devitrification—that is, the elements of the flux may separate, and combine again in a different manner, so as to produce an opaque substance known as *Reaumur's porcelain*. There is danger in the opposite extreme; for if the temperature be carried too high, some of the more delicate tints, such as the roses and the grays, become faint or vanish altogether, while the hardier greens, blues, and blacks remain. On the other hand, if the maximum temperature be not quite reached, the colors do not present that peculiar creaminess and glossiness which is characteristic of the art. The temperature is regulated by means of watches, consisting of small slabs of porcelain smeared with some trial color, usually the carmine produced by the purple precipitate of cassius. This forms a useful exponent of all the other pigments: it varies greatly in tint according to the temperature; so that, by arranging a scale of temperatures corresponding with a scale of tints, a tolerably accurate thermoscope may be formed. Brongniart invented a pyrometer for estimating the temperature of the interior of the muffle; its action depends on the expansion of a bar of fine silver, nearly eight inches long, introduced into the muffle, and connected with a graduated scale on the outside.

The kind of fuel used for heating the muffle has an influence on the colors; for although the muffle may consist of an iron box heated only on the outside, it is almost impossible to prevent some of the products of combustion from entering it. The smoke of ordinary coal is especially injurious to the colors, from the presence of sulphurous acid, which is also given off from coke. Wood has its pyroligneous acids, and even charcoal gives off carbonic acid. The presence of an acid is so injurious in the muffle, that the reds produced from green vitriol, before being used, must be thoroughly washed, to get rid of the last traces of acid. It is also stated that a muffle in which cuperose has been calcined, cannot be used for the firing of colors.

There is also a difficulty connected with the use of oxide of lead, which is required in the preparation of certain colors, but is injurious to the development of some others. Again, the fixed alkalies used in the composition of the glaze may react on the coloring oxides, especially at the maximum temperature. In this way, the oxide of chrome will produce yellow instead of green. The oxide of lead, the potash, and the soda, may not only act injuriously by contact, but by becoming volatilized, they may injure every color in the muffle. Moreover, the oxide of tin, used in certain glazes, may impart its own opacity to the colors. It is also a curious fact, that different kinds of kaolin are not all equally favorable to the devolopment of the colors put upon the ware; and we are informed, that the kaolin of Ebreuil will not allow of the development of any color derived from gold.

The metals used for imparting color, hitherto referred to, have been in the form of oxides, etc., and used with a flux. The gold used in gilding porcelain is first dissolved in *aqua regia;* the acid is driven off by heat, when the gold remains in a state of minute division. It may also be precipitated by means of sulphate of iron. In this minutely divided state it is mixed with one-twelfth of its weight of oxide of bismuth, together with a small quantity of borax and gum-water, and applied to the ware by means of a hair pencil. If the article is to have only a circular line of gold, it is placed upon a small table or whirler, and the artist, steadying his hand on a rest, applies the pencil to the article, while with the other hand he causes the table to revolve. The gold ornaments come out of the fire with a wretched dingy hue; but the luster of the gold is brought out by burnishing with agate and blood-stone, and the gilding is cleaned with vinegar or white-lead.

Other metals which, like gold, do not become readily oxidized, are applied to stone-ware, and form what are called *metallic lusters.* The silver-white hue known as *silver luster* is obtained from platinum by dissolving the metal in *aqua regia*, and pouring the saturated solution into boiling water. This is poured into a warm solution of sal ammoniac, when the metal forms a yellow precipitate, which, after having been washed and dried, is applied to the ware by means of a flat brush, and the article is then passed through the muffle-kiln. A sufficient body of luster may be obtained by repeating the operation; and should the articles come out of the muffle black, friction with cotton will give the required luster. A platinum luster resembling that of polished steel is obtained by dropping a solution of equal parts of tar and sulphur in hot linseed-oil, known as *spirit of tar*, into the acid solution of platinum. The mixture is spread over the ware, and passed through the muffle as before. *Gold luster* is obtained by precipitating a solution of gold in *aqua regia* by means of ammonia: it has fulminating properties, and must therefore be mixed with the essential oil of turpentine while moist, and in this state applied to the ware. After the firing, the luster will be brought out by friction with linen. The *lustre cantharide* of the French, which is remarkable for its iridescence, is obtained from chloride of silver, partly decomposed by means of combustible vapors. For this purpose a mixture of a lead glass, oxide of bismuth, and chloride of silver, is applied to the ware. This is then raised to a red heat in the muffle, when a fuliginous smoke is introduced, which effects the partial decomposition required. An *iron luster* is obtained by mixing a solution of iron or steel in hydrochloric acid with spirit of tar, and applying it to the ware. Silver and platinum lusters are usually laid upon a white ground; gold and copper lusters have the best effect on colored grounds. The paste body for lustrous ware is usually made for the purpose, of four parts clay, four of flint, four of kaolin, and six of felspar. Its color is brown, but it is coated with a lead glaze composed of sixty parts litharge, thirty-six of felspar, and fifteen of flint.

GLASS.

ITS HISTORY AND MANUFACTURE.

GLASS.

HISTORY AND MANUFACTURE.

THE general term *glass* is employed by chemists to denote all mineral substances which, on the application of heat, pass through a state of fusion into hard and brittle masses, and which, though not always transparent, exhibit a lustrous fracture when broken. The glass of commerce, however, to which our remarks are restricted, or the transparent and artificial substance which is usually distinguished by the generic name, is produced by the igneous fusion of siliceous earth with certain alkaline earths or salts, or with metallic oxides.

The etymology of the word has been much disputed. It is derived by some from the Latin *glacies*, ice, its resemblance to which is thought to have suggested the title. Others have remarked, that the common Latin designation of this substance is *vitrum;* and as the Romans applied this term, in common with the word *glastum*, to the plant which we call woad, they have deduced it from the latter of these, either because the ashes of this plant were used in the manufacture of glass, or because it exhibited something of the bluish color which is procured from woad. *Glassum*, the name given to amber by the ancient Gauls and Britons, has also been assigned as the origin of the word. But none of these etymons appear very satisfactory. The most plausible theory is that which derives the term from the Saxon verb *glis-nian*, or the German *gleissen, splendere,* which are probably contractions of the Anglo-Saxon *ge-lixan*, to shine, to be bright. This view is in a great degree confirmed by the sense in which the term glass and its derivatives are employed by our older writers, who frequently apply it to shining or glittering substances, without reference to color or transparency.

In the most remote ages the art of blowing glass into bottles, making it into vases, coloring it to imitate precious stones, melting it in enormous masses to make pillars, rolling and polishing it into mirrors, and tinting it in parts, were all perfectly well known. For its origin we must look to Egypt, the parent of so many collateral arts. The story of the Israelites having set fire to a forest, and the heat becoming so intense that it made the niter and sand melt and flow along the mountain side, and that they afterward did artificially what had been the result of accident, may be set down as equally fabulous with the story of the pirates, who are said to have landed on the sea-beach, and wishing to make their cauldron boil, piled up some vitreous stones and placed them on a quantity of sea-weed and blocks of wood, causing so strong a heat that the stones were softened and ran down on the sand, which melting and mixing with the alkali became a diaphanous and glassy mass. The fictitious character of both these stories is proved by the simple fact that it requires the most intense furnace heat to insure the combination of the sand with the niter.

Under these circumstances we are justified in believing that glass-making had its origin at the same time with the baking of bricks and pottery. The smelting of ores, too, required a furnace sufficiently intense to fuse the silicates analogous to glass, and hence it may be safely inferred, that in the age when melting and working metal was known the art of making glass was also practiced. In the book of Job the most precious things are compared to wisdom, but still more precious are gold and glass. The Hebrews must have become acquainted with glass while in Egypt, and in consequence of their proximity to the Phœnicians; and it is now generally believed that these two nations had the merit of originating and establishing its manufacture. The Athenian embassadors, in order to give an idea of the magnificence displayed at the court of the great King of Persia, said that they drank in cups of glass and gold. Some writers affirm that the Egyptians in some instances sealed up their dead in a coating of glass, and glass-houses are said not to have been uncommon in that wonderful country. Some authors ascribe, with very plausible reason, the discovery of glass-making to the priests of Vulcan at Thebes and Memphis, the greatest chemists in the ancient world. The Egyptians are also known to have made enamels of divers colors which they applied on pottery, magnificent specimens of which are still extant, and are called Egyptian porcelain. These are chiefly covered with beautiful blue or green, and groups of flowers or designs are traced in black. Glass beads and other ornaments made of that substance, skillfully manufactured and beautifully colored, have been found adorning mummies, which are known to be upward of three thousand years old. It is certain that Tyre, Sidon, and Alexandria, were long celebrated for their glass, and furnished the greater proportion of that used at Rome. Under the Roman Empire the Egyptians still preserved their superiority in the art of glass-making, and it is said that Aurelian caused them to pay their tribute in that manufacture. Adrian mentions that he had received drinking-glasses of various colors from a priest of a famous temple in Egypt, and gives instructions that they are not to be used but on the greatest occasions, and on the most solemn feast days. To these places the art was exclusively confined for some centuries, and was an article of luxury, being chiefly in the form of urns or drinking-cups of the most elaborate workmanship, and exquisitely embellished with raised, chased, or ornamented figures. The Barberini or Portland Vase, composed of deep blue glass, with figures of a delicate white opaque substance raised in relief, is a splendid specimen, and was found in the tomb of Alexander Severus, who died A.D. 285.

The art of glass-making seems to have been introduced into Italy by the Romans after their conquests in Asia in the time of Cicero, and the first glass works there were said to have been near the Flaminian Circus. It is highly probable that these workmen were imported from Egypt. The use of glass seems rapidly to have increased, and to have become very common, for we find an emperor in the third century of the Christian era saying, that he was disgusted with so low and vulgar an object as glass, and that he would only drink from vessels of gold. By this time the manufacture of glass was so considerable that an impost was laid on it, and it was extensively employed in the decorations of buildings, while in glass mosaics were combined the most brilliant colors.

From the circumstance of colored glass beads and amulets having been found among Druidical remains in England, it has been argued by Pennant and others, that the art of making glass was known in Britain before its invasion by the Romans. It can hardly, however, be believed that a people who had made very trifling advances in civilization, and who, it is known, were entirely unacquainted with any other art, should be found not only conversant with the manufacture of glass, a complicated and highly ingenious process, but should excel in it; for the beads and amulets spoken of are of exquisite workmanship, and beautifully colored in imitation of the rarest and most precious stones. There seems little doubt, therefore, that the ancient Britons procured these in the course of traffic with the Syrians, who visited the island, as we do those in the South Seas, to drive a trade with their savage inhabitants in toys and trinkets, giving them these in exchange for skins or other natural productions. By whatever means, however, these ornaments came into Britain, it is certain that they were in extensive use, though principally for religious purposes, long prior to the Roman invasion, as they are found in barrows or tumuli of a much older date. One at Stonehenge, in particular, on being opened, was found to be filled with them.

Glain Neidyr, or Druidical glass rings, generally about half as wide as our finger rings, but much thicker, have frequently been found. The vulgar superstition regarding these was, that they were produced by snakes joining their heads together and hissing, when a kind of bubble like a ring was formed round the head of them, which the others, continuing to hiss, blew on till it came off at the tail, when it immediately hardened into a glass ring. Success was thought to attend any one who was fortunate enough to find one of those snake-stones. They were evidently beads of glass employed by the Druids, under the name of charms, to deceive the vulgar. They are usually of a green color, but some of them are blue, and others variegated with wavy streaks of blue, red, and white.

Glass utensils have been found in Herculaneum, which city was destroyed by an eruption of Mount Vesuvius in the reign of Titus (A.D. 79). A plate of glass also found there has occasioned much speculation as to its uses. Similar plates, to which Pliny gives the name of *vitreæ cameræ*, seem to have been employed, in a manner not very well understood by us, as paneling for their rooms. It is disputed whether or not glass was used in Herculaneum for windows.

Dion Cassius and Petronius Arbiter concur in their account of the discovery of malleable or ductile glass by a celebrated Roman architect, whose success in the restoration to its position of a portico which leaned to one side had roused the envy and jealousy of Tiberius, and occasioned his banishment from Rome. Thinking that his discovery would disarm the emperor's wrath, the artist appeared before him bearing a glass vessel, which he dashed upon the ground. Notwithstanding the violence of the blow, it was merely dimpled as if it had been brass. Taking a hammer from his breast, he then beat it out into its original shape; but instead of giving him the reward which he had expected, the emperor ordered the unfortunate artisan to be beheaded, remarking, that if his discovery were known, gold would soon be held of as little value as common clay. This is probably another version of the story told by Pliny, of an artificer who made the same discovery, and whose workshop was demolished by those who had an interest in preventing the introduction of an article which would lower the value of gold, silver, and brass. Although it might not be justifiable to give unqualified disbelief to these stories, yet the knowledge we at present possess would restrict the possibility of such a discovery within the narrowest limits. The union of the properties of malleability and vitrification seems to be incompatible. Some metallic substances, by the application of intense heat, are reduced to the state of glass, but at the same time lose their malleability; which fact would seem to imply that it is impossible to communicate the latter property to glass. The extraordinary stories above mentioned have, however, been rationally enough explained by modern chemists. It has been observed by Kunckel, that a composition having a glossy appearance, and sufficiently pliant to be wrought by the hammer, may be formed: and by Neumann, that, in the fusion of muriate of silver, a kind of glass is formed, which may be shaped or beaten into different figures, and may be pronounced in some degree ductile. Blancourt, in his *L'Art de la Verrerie*, mentions an artist who presented a bust of ductile glass to Cardinal Richelieu, minister of Louis XIII. But he does not seem to have been more fortunate than his predecessors; for he was doomed to imprisonment for life, for " the politic reasons," as Blancourt with much simplicity observes (we quote from the translation published in 1699), " which, it is believed, the cardinal entertained from the consideration of the consequences of that secret," which no doubt led him to fear lest the established interests of French glass manufacturers might be injured by the discovery. From expressions used by Blancourt in other parts of his work, we think, that by malleable glass, such as was produced by this artist, he understood some composition similar to those which Kunckel and Neumann discovered, and was not very exact in limiting the term to that vitreous substance which we now generally understand when we speak of glass.

The precise period at which the making of window-glass came into practice is not now certainly known. The Roman windows were filled with a semi-transparent substance called *lapis specularis*, a fossil of the class of mica which readily splits into thin smooth laminæ or plates. This substance is found in masses of ten or twelve inches in breadth, and three in thickness; and, when sliced, very much resembles horn, instead of which it is to this day often employed by lantern-makers. The Romans were chiefly supplied with this article from the island of Cyprus, where it abounds. So good a substitute for glass is it said to have been, that, besides being employed for the admission of light into the Roman houses, it was also used in the construction of hot-houses, for raising and protecting delicate plants; so that, by using it, the Emperor Tiberius had cucumbers at his table throughout the whole year. It is still much employed in Russia instead of glass for windows.

There is no positive mention of the use of glass for windows before the time of Lactantius, at the close of the third century. But the passage in that writer which records the fact (*De Opif. Dei*, cap. 8), also shows that the *lapis specularis* still retained its place. Glass windows are distinctly mentioned by St. Jerome, as being in use in his time (A.D. 422). After this period we meet with frequent mention of them. Joannes Phillipinus (A.D. 630) states that glass was fastened into the windows with plaster.

The Venerable Bede asserts that glass windows were first introduced into England in the year 674, by the Abbot Benedict, who brought over artificers skilled in the art of making window-glass, to glaze the church and monastery of Wearmouth. The use of window-glass, however, was then, and for many centuries afterward, confined entirely to buildings appropriated to religious purposes; but in the fourteenth century it was so much in demand, though still confined to sacred edifices and ornamental purposes, that glazing had become a regular trade. This appears from a contract entered into by the church authorities of York Cathedral in 1338 with a glazier, to glaze the west windows of that structure; a piece of work which he undertook to perform at the rate of sixpence per foot for white glass, and one shilling per foot for colored. Glass windows, however, did not become common in England till the close of the twelfth century. Until this period they were rarely to be found in private houses, and were deemed a great luxury, and a token of great magnificence. The windows of the houses were till then filled with oiled paper, or wooden lattices. In cathedrals, these and sheets of linen supplied the place of glass till the eighth century; in meaner edifices lattices continued in use till the eighteenth.

The glass of the Venetians was superior to any made elsewhere, and for many years commanded the market of nearly all Europe. Their most extensive glass-works were established at Murano, a small village in the neighborhood of Venice; but the produce was always recognized by the name of Venetian glass. Baron von Lowhen, in his *Analysis of Nobility in its Origin*, states that, " so useful were the glass-makers at one period in Venice, and so great the revenue accruing to the republic from their manufacture, that, to encourage the men engaged in it to remain in Murano, the senate made them all burgesses of Venice, and allowed nobles to marry their daughters; whereas, if a nobleman marries the daughter of any other tradesman, the issue were not reputed noble."

The skill of the Venetians in glass-making was especially remarkable in the excellence of their mirrors. Beckman, who has minutely investigated the subject, is of opinion that the manufacture of glass mirrors certainly was attempted, but not with complete success, in Sidon, at a very early period; but that they fell into disuse, and were almost forgotten until the thirteenth century. Previously to this period, plates of polished metal were used at the toilet; and in the rudeness of the first ideas which suggested the substitution of glass, the plates were made of a deep black color to imitate them. Black foil even was laid behind them to increase their opacity. The metal mirrors, however, remained in use long after the introduction of their fragile rivals, but at length they wholly disappeared; a result effected chiefly by the skill of the Venetians, who improved their manufacture to such a degree that they speedily acquired a celebrity which secured an immense sale for them throughout all Europe.

From Italy the art of glass-making found its way into France, where an attempt was made, in the year 1634, to rival the Venetians in the manufacture of mirrors. The first essay was unsuccessful; but another, made in 1665, under the patronage of the celebrated Colbert, in which French workmen who had acquired a knowledge of the art at Murano were employed, had better fortune. But a few years afterward, this establishment, which was situated in the village of Tourlaville, near Cherbourg in Lower Normandy, was also threatened with ruin by a discovery or rather improvement in the art of glass-making, effected by one Abraham Thevart. This improvement consisted in casting plates of much larger dimensions than it had hitherto been deemed possible to do. Thevart's first plates were cast at Paris, and astonished every artist by their magnitude. They were eighty-four inches long and fifty inches wide, whereas none previously made exceeded forty-five or fifty inches in length. Thevart was bound by his patent to make all his plates at least sixty inches in length and forty in breadth. In 1695 the two companies, Thevart's and that at Tourlaville, united their interest, but were so unsuccessful, that, in 1701, they were unable to pay their debts, and were in consequence compelled to discharge most of the workmen, and abandon several of their furnaces. Next year, however, a company was formed under the management of Antoine d'Agincourt, who re-engaged the discharged workmen; and the works realized considerable profits to the proprietors, a circumstance which is attributed wholly to the prudent management of D'Agincourt.

Early in the fourteenth century the French government made a concession in favor of glass-making, by decreeing that not only should no derogation from nobility follow the practice of the art, but that none save gentlemen, or the sons of noblemen, should venture to engage in any of its branches, even as working artisans. This limitation was accompanied by a grant of a royal charter of incorporation, conveying important privileges, under which the occupation became eventually a source of great wealth to several families of distinction.

It has been said that the manufacturing of window-glass was first introduced into England in the year 1557. But a contract, quoted by Horace Walpole in his *Anecdotes of Painting*, proves that this article was made in England upward of a century before that period. This curious document is dated in 1439, and appears to be a contract between the Countess of Warwick and John Prudde of Westminster, glazier, whom she employed, with other tradesmen, to erect and embellish a magnificent tomb for the earl, her husband. John Prudde is thereby bound to use " no glass of England, but glass from beyond seas;" a stipulation which, besides showing that the art of making window-glass was known and practiced in England in the fifteenth century, seems also to indicate that it was inferior to what could be obtained from abroad. The finer sort of window-glass was made at Crutched Friars, London, in 1557. In the year 1635, Sir Robert Maxwell introduced the use of coal fuel instead of wood, and procured workmen from Venice; but many years elapsed before the English manufactories equaled the Venetians and French in the quality of these articles. The first flint-glass made in England was manufactured at the Savoy House, in the Strand; and the first plate-glass for looking-glasses, coach windows, and similar purposes, was made at Lambeth, by Venetian workmen, brought over in 1670, by the Duke of Buckingham. From that period the English glass manufactories, aided by the liberal bounties granted them in cash upon glass sold for export, became powerful and successful rivals of the Venetians and French manufactories. The bounty on glass exported, which the government paid to the manufacturer, was not derived from any tax by impost, or excise, previously laid; for all such were returned to the manufacturer together with the bounty, thereby lessening the actual cost of the article from 25 to 50 per cent., and enabling the English exporter to compete successfully in foreign markets. This bounty provision was annulled during the premiership of Sir Robert Peel, together with all the excise duty on home consumption.

The art of glass-making was introduced into Scotland in the reign of James VI. An exclusive right to manufacture it within the kingdom, for the space of thirty-one years, was granted by that monarch to Lord George Hay, in the year 1610. This right his lordship transferred in 1627, for a considerable sum, to Thomas Robinson, merchant-tailor in London, who again disposed of it for 250*l*., to Sir Robert Mansell, vice-admiral of England. The first manufactory of glass in Scotland, an extremely rude one, was established at Wemyss in Fife. Regular works were afterward commenced at Prestonpans, Leith, and Dumbarton. Crown-glass is now manufactured at Warrington, St. Helens, Eccleston, Old Swan, and Newton, Lancashire; at Birmingham, Hunslet near Leeds, and Bristol. It is also manufactured of excellent quality on the Tyne and Wear. Great improvements have recently been made in the manufacture of crown-glass; and we believe this article, as made in England, is superior in quality to that of any other nation.

The manufacture of glass was introduced into the American States in 1780 by Robert Hewes, a citizen of Boston, who erected a factory in the then forest of New Hampshire. The chief aim of Mr. Hewes was to supply window-glass, but he did not succeed. Another attempt was made in 1800, when a factory was built in Boston for making crown window-glass; but this was unsuccessful, till a German named Link, in 1803, took charge of the works, and the State of Massachusetts agreed to pay the proprietors a bounty on every table of window-glass they made; after which the manufacture was carried on successfully, the glass steadily improving in quality, and becoming famed through all the States as Boston window-glass. The same company, in the year 1822, erected new and more extensive works at Boston. The mystery attached to the art of glass-making followed it into America. The glass-blower was considered a magician, and myriads visited the newly-erected works, looking on the man who could transmute earthy and opaque matter into a transparent and brilliant substance, as an alchemist who could transmute base metal into gold.

Since the manufacture of flint-glass was introduced into the Eastern States there have been above forty companies formed from time to time, nearly thirty of which have proved failures. There are now ten in operation, two of which are at East Cambridge, three at South Boston, one at Sandwich, three near New York City, and one at Philadelphia. 48,000 tons of coal, 6500 tons of silex, 2500 tons ash, niter, etc., and 3800 tons of lead are annually consumed in the manufacture of flint-glass.

In the vicinity of Pittsburg, in the Western States, are nine manufactories of flint-glass and ten of window-glass, and in the river towns are fifteen window-glass factories.

There is good reason for supposing that the art of coloring glass is coeval with the art of glass-making itself. It is certain that the art was known in Egypt at least 3000 years ago. We have already mentioned the beautiful imitations of precious stones, found adorning mummies which are known to have existed for that time. We meet with frequent mention of specimens of Eastern workmanship of comsummate beauty, upon which great value was placed. The works of Caylus and Winkelmann furnish some striking instances of ancient skill in the formation of pictures by means of delicate glass fibers of various hues, which, after being fitted together with the utmost nicety, were conglutinated by fusion into a solid mass. The art of combining the various colors so as to produce pictures, such as is now practiced, is comparatively of recent date. The earliest specimens of this kind of work discover a fictitious joining of different pieces of glass, differently tinged, and so arranged as, by a species of mosaic work, to produce the figure or figures wanted. The various pieces are held together generally by a vein of lead, run upon the back of the picture, precisely at their junction.

It has already been stated that the Romans combined the most brilliant colors in their mosaics; and there can be little doubt that the mosaics gave the first idea of painted or stained glass for windows in the early Christain churches. In all the early specimens of Norman glass, similar coloring and design are to be traced. Starting from the fourth century, there is frequent mention of colored glass windows by Greek and Latin authors. St. John Chrysostom and St.

Jerome talk of "windows of divers colors;" and Lactantius says "that the soul perceives objects through our bodily eyes as through windows garnished with transparent glass." The early basilicas were all adorned with colored glass, and the early Christain poets sung in ecstasies of the effect produced by the windows at sunrise. In the sixth century, Prudetia, speaking of one of these structures, says:—"The magnificence of this temple is truly regal. The pious prince who consecrated it has caused the vaults to be painted at great expense, and has clothed it with golden walls, so that the light of day may repeat the fire of the morning. In the windows is placed glass of various colors, which shine like meadows decked in the flowers of spring." An inscription on St^a Agnese states, that that basilica, rebuilt by the Emperor Honorius, was decorated with glass, which produced the most magnificent effect. In the sixth century, Sancta Sophia, at Constantinople, also received painted windows, which Paul the Silent praises highly. Procopius says, that day seemed to be born under the vaults of the temple; and after such glowing description it cannot be doubted that the glass was stained, not colorless.

The use of colored glass, however, was not confined to Greece and Italy. It rapidly appeared in Gaul. Gregory of Tours, in the sixth century, also tells us that the church of St. Julien de Brionde, in that town, had colored glass windows; and the Bishop of Poictiers, describing Nôtre-Dame of Paris, admired the effect produced by the light falling upon the vaults and walls after passing through the painted glass, and compares it to the first tints of the morning sun.

In England, St. Wilfred, who lived early in the eighth century, is said to have been the first to introduce painted glass windows, and for that purpose had workmen brought from France or Italy.

The first painted glass executed in England was in the time of King John; previously to this, all stained or painted glass was imported from Italy. The next notice of it occurs in the reign of Henry III. The treasurer of that monarch orders that there be painted, on three glass windows in the chapel of St. John, a little Virgin Mary holding the child, and the Trinity, and St. John the Apostle. Some time after, he issues another mandate for two painted windows in the hall.

Even at this early period, however, England boasted of eminent native artists in glass painting, amongst the first of whom was John Thornton, glazier of Coventry. This person was employed, in the time of Henry IV., by the dean and chapter of York cathedral, to paint the eastern window of that splendid edifice; and for the beautiful and masterly workmanship which he exhibited in this specimen of his skill, he received four shillings per week of regular wages. He was bound to finish the work in less than three years, and to receive, over and above the weekly allowance, one hundred shillings for each year; and if the work was done to the satisfaction of his employers, he was to receive, on its completion, a further sum of 10*l*.

From this period downward there have been many skillful native artists, although the Reformation greatly impeded the progress of the art, by banishing the ungodly ostentation of ornamented windows from churches; indeed, so serious was this interruption, that the art had nearly altogether disappeared in the time of Elizabeth. Amongst the most eminent glass painters who first appeared upon the revival of the art, were Isaac Oliver, born in 1616, and William Price, who lived about the close of the seventeenth century. This artist was succeeded by a person at Birmingham, who, in 1757, fitted up a window for Lord Lyttleton, in the church of Hagley. To him succeeded one Pecket of York, who attained considerable notoriety, but who was entirely ignorant of the true principles of the art.

During all this time, however, and indeed until a comparatively recent date, painted glass was regarded as too costly and too magnificent an article to be otherwise employed than in decorating religious edifices or the palaces of nobles; and even in the the latter case it was but sparingly used. Modern improvement has placed this beautiful ornament within the reach of very ordinary circumstances; and the art of staining glass is now practiced with great success, and is extensively used in decorating our domestic as well as our palatial and ecclesiastical architecture.

The colors of modern artists, we venture to allege, notwithstanding what is often urged to the contrary, equal in variety and richness those of the ancients, and, with the superior knowledge which we now possess of the principles of drawing, and of bringing several colors together on a single sheet, encouragement alone is wanting to attract artists of talent and inventive genius to the pursuit of the art, and to carry it to a greater height of excellence than it has ever reached in the hands of their predecessors.

MANUFACTURE OF CROWN-GLASS.

In order to secure success to his operations, the glass manufacturer must bestow the utmost care upon the erection of his furnaces. They must be well and substantially built, of the best materials, of the most approved construction, and under the direction of a builder of tried skill and extensive experience. A false economy in these respects cannot fail of leading to the most ruinous results.

Crown-glass is the best kind of glass now employed in the glazing of windows, and is so called to distinguish it from the common, broad, or spread glass, which was in use before the introduction of crown-glass, but which, on account of its inferior quality, is now rarely used. In the manufacture of crown-glass the following furnaces and arches are required, viz., calcar arch, main furnace, bottoming hole, flashing furnace, nose hole, and annealing kiln.

A Calcar Arch for burning frit is a common reverberatory furnace, and is about ten feet long, seven feet wide and two feet high. The crown and sides are built of fire-brick, and the other parts of common brick. The bottom should be carefully joined and cemented, as the salt is apt to ooze through it.

The Main or *Glass-making Furnace* is an oblong, built in the center of a brick cone, large enough to contain within it two or three pots at each side of the grate-room, which is either divided, or runs the whole length of the furnace, as the manufacturer likes.

The arch is of an elliptic form. A barrel arch, that is, an arch shaped like the half of a barrel cut longwise through the center, is sometimes used. But this soon gives way when used in the manufacture of crown-glass, although it does very well in the clay furnace for bottle-houses.

The best stone in England for building furnaces is fire-stone from Coxgreen, in the neighborhood of Newcastle. Its quality is a close *grain*, and it contains a greater quantity of talc than the common fire-stone, which seems to be the chief reason of its resisting the fire better. The great danger in building furnaces is, lest the cement at the top should give way with the excessive heat, and by dropping into the pots, spoil the metal. The top should therefore be built with stones only, as loose as they can hold together after the centers are removed, and without any cement whatever. The stones expand and come quite close together when annealing; an operation which takes from eight to fourteen days at most. There is thus less risk of any thing dropping from the roof of the furnace.

The inside of the furnace is built either of Stourbridge fire-clay annealed, or the Newcastle fire-stone, to the thickness of sixteen inches. The outside is built of common brick about nine inches in thickness.

The furnace is thrown over an ash-pit, or cave as it is called, which admits the atmospheric air, and promotes the combustion of the furnace. This cave is built of stone until it comes beneath the grate-room, when it is formed of fire-brick. The abutments are useful for binding and keeping the furnace together, and are built of masonry. The furnaces are stoutly clasped with iron all round, to keep them tight. In four-pot furnaces this is unnecessary, provided there be four good abutments.

Bottoming Hole. The interior is of common fire-brick, the mouth either of common fire-brick or Stourbridge clay, and the outside entirely of common brick.

Flushing Furnace. The outside is built of common brick, the inside of fire-brick, and the mouth or nose of Stourbridge fire-clay.

Nose Hole. This is a small aperture off the flashing furnace, and of the same materials.

Annealing Kiln. It is built of common brick, except around the grate-room, where fire-brick is used.

The materials of which crown-glass is usually composed are kelp and fine white sand. Pearl ashes, or certain other alkalies, sometimes supply the place of the former of these substances. The quality of kelp is extremely various. That from Orkney is superior to what is made in Ireland, the Hebriles, or the lower parts of Scotland. It is found to contain less alkali, and to produce glass of a better color.* For the glass-maker's purposes the kelp of the Orkneys is decidedly the best. It is freer from sulphur than the others, the presence of which makes the glass green, crude, and fretful. The following is the course pursued in the preparation of kelp. The fuci are cut from the rocks in the months of May, June, and July. They are then brought to the shore, and, after being spread out and dried, are thrown into a pit lined with stones, in which a large fire has been previously kindled. On this fire the weed is heaped from time to time, until a large mass is accumulated, and the whole is reduced to a state of fusion. It is then well mixed and leveled, and allowed to cool. When sufficiently cold, it is taken from the pit, and broken into portable masses, for the convenience of transportation. To prevent the dissipation of the alkali, a thing very apt to occur, the greatest care is necessary in every part of this process; in the gathering and drying, as well as in the burning of the fuci; in the treatment of the mass whilst in a state of fusion; and in its exposure to the atmosphere during these operations. Kelp burners are but too frequently guilty of carelessness in this respect. In some places they burn the fuci in pits which are not lined with stones, and, of consequence, sand and earthy substances mingle with the fused mass.† It is no uncommon thing for the makers to increase the weight of kelp intentionally, foolishly thinking to procure a higher price for it by so doing. Such adulteration is, however, at once detected by the kelp merchant, and the article, which might otherwise have brought a good price, is reduced to less than a third of its value. The inferiority of the Lowland kelp to

---

* Some eminent chemists assert that, although the usual quantity of kelp be added in the manufacture of glass, the weight of the glass produced is nothing more than the original weight of the sand. But this is not the case with the Orkney kelp, for though it has less alkali, it contains more insoluble matter than the West Highland kelp, and of course produces a larger quantity of glass. The West Highland yields glass of very inferior color to that procured by the use of Orkney kelp.

† The best mode of preparing kelp, as invented by Colonel Fullarton, is by burning it in a reverberatory furnace, and throwing it down in the form of cakes, in the same manner as frit, which we shall afterward have occasion to describe. When so prepared, it is more fit for the glass manufacturer, being free of extraneous matter. This method is now employed by extensive makers of kelp in Ireland.

that of Orkney and the Hebrides, may with safety be attributed as much to to this practice as to the inferiority of the fuci. Some idea, but at best a very uncertain one, of the quality of kelp may be formed by the examination of its external appearance. A chemical analysis of its properties can alone give security to the manufacturer. In preparing it for the manufacture of glass, it is first broken into small pieces, either by the hand or by a machine called a stamper. It is then put into a mill and ground into a fine powder, stones and all other extraneous matter being picked out. The powder is afterward passed through brass wire sieves.

With regard to the other component part of window-glass, namely, sand, that of the best description is procured from Lynn Regis, in Norfolk. That procured from Alum Bay, on the Western coast of the Isle of Wight, is also of excellent quality. The superiority of this sand arises from the circumstance of its containing a greater quantity of minute transparent crystals than is found in the sand of any other place in the country. In preparing the sand, it is usually washed in a large vat with boiling or cold water, until the water runs off quite clear. The sand is then put into a calcining arch, where it is subjected to a strong heat for twenty-four hours. During this time it is kept red hot, and immediately on being taken out is plunged into pure cold water. This has the effect of dividing the particles of sand, and making it unite more readily with the alkali during the process of calcining. Some use niter during this process, which consumes any sulphureous matter that may be present, or extraneous substances of an animal or vegetable nature, and reduces them to an earth not injurious to glass. When this operation is completed, it is removed into the mixing room, where the proportions of material are adjusted and mingled together, previously to their being *fritted* or calcined. Here the materials, the sand and the kelp powder, are carefully proportioned, generally in the degree of eleven of kelp to seven of sand, some manufacturers using eleven to eight, which are mixed up according to the judgment of the mixer. The majority of glass manufacturers are now giving up the use of kelp. Within the last few years the improvements in the manufacture of carbonate of soda have been very great, while it has also fallen considerably in price. Instead, therefore, of using such an impure alkali as kelp with sand, carbonate of soda with sand and lime is employed, which gives glass of as good a color as plate, and is attended with many other advantages which the other materials do not possess. Manufacturers, instead of kelp, purchase sulphur, and with it make sulphuric acid. With sulphuric acid and muriate of soda they make sulphate of soda, to which lime, coal, etc. are added, and thus produce carbonate of soda, which, with sand and lime, is made into glass. The operations for preparing these materials are carried on within their own premises by several extensive glass manufacturers. The following mixture has been found to produce an excellent quality of glass: 3 cwts. Lynn sand; $2\frac{1}{2}$ ditto carbonate of soda; 14 lbs. niter; 14 ditto lime; 7 ditto charcoal; one-fourth of the above weight of cullet.

This mixture will make a very excellent glass when the furnace is kept at a proper heat. The proportions must, of course, be regulated in some degree by the heat which the furnace attains. The addition of any other ingredient will injure the quality and color of the glass. It may be either fritted or not before being put into the pots. The use of this mixture saves coals, time, and wages, as the founding occupies from sixteen to twenty hours only, whilst in other cases the time occupied by this process is from twenty to twenty-four hours. It can also be blown into a thinner and finer substance, and is thus liable to a less duty. When the sand and kelp are thoroughly mixed, the compost is put into a calcining arch or reverberatory furnace, where it is subjected to a heat so strong as to reduce it to a semi-fluid state. Whilst in this state, it is stirred without intermission, to prevent the formation of knots containing more sand than the rest of the batch, an effect resulting from the dissipation of the alkali by excess of heat. The process of calcining requires more or less time according to the varying properties of the ingredients composing the batch. From three to four hours is the time usually occupied by each batch. The frit, as the substance is now called, is taken from the furnace, spread upon a plate of iron whilst yet hot, and, before it becomes quite cold, divided into large cakes. In the opinion of many, it cannot be too old for use; as when new the glass made from it is full of what are called seeds. It is commonly kept about six months by opulent manufacturers. The last operation consists in throwing the frit into the melting-pots.

To prevent stones or clay from the furnace falling into the pot, those used in making flint-glass are always covered in on the top; and the same thing has been tried in crown pots, made with two openings, one in the front and one in the back, the back one to be plugged up when beginning to work from the front of the pot. This method succeeded very well, but was abandoned from the length of time it required; a circumstance which more than counterbalanced its advantages.

These pots or crucibles are made of the finest clay. Great care is requisite in the selection, and in cleansing it from extraneous particles, the presence of which, even in the smallest degree, will injure the pot. A fine powder procured by grinding old crucibles is generally mixed, in a proportion seldom larger than a fourth, with what is termed the virgin clay. This mixture dries more rapidly, contracts less while drying, and presents a firmer resistance to the action of the fire and alkali used in the composition of glass than the mere unmixed clay. These ingredients having been mixed, they are wrought into a paste in a large trough, and carried to the pot loft, covered in such a way as to exclude dust and other minute particles. Here a workman kneads this paste by tramping it with his naked feet, turning it from time to time until it becomes as tough as putty. It is then made into rolls, and wrought, layer upon layer, into a solid and compact body, every care being taken to keep it free of vacuities, as latent air would, by its expansion in the furnace, cause an immediate rupture of the pots.

After pots are made, very great care is necessary to bring them to the proper state of dryness before taking them to the annealing or pot arch. In drying they commonly shrink about two inches in the circumference. When pots are made during summer, the natural temperature is sufficient. In winter they are kept in a temperature of from fifty to fifty-five degrees Fahrenheit. They remain in the room where they are made for a period varying from nine to twelve months. Being afterward removed to another apartment, where the heat is from eighty to ninety degrees Fahrenheit, they are kept there for about four weeks. They are then removed for four or five days, more or less, according to their previous state of dryness, to the annealing arch, which is gradually and cautiously heated up till it reaches the temperature of the working furnace, whither, after being sufficiently annealed, they are carried as quickly as possible. Pots last upon an average from eight to ten weeks Their value is usually estimated at 8*l.* or 10*l.* each.

To the frit thrown into these pots there is added a proportion, about an eighth, of cullet or broken crown-glass. After this has been done, the furnace is raised to the highest possible degree of heat. The pots are filled every third hour or so, according as the frit melts, till they are completely full. The intensity of the heat is then increased, if possible, till the metal, as it is now called, is reduced to fine liquid glass, which is then ready for the operations of the workman.* From twenty-four to thirty hours in all are required for this process, which is called founding.†

The furnace is slackened for about two hours, and the metal being now in a workable state, the first operator who approaches the furnace is called the *skimmer*, who skims off all extraneous substances from the metal. Next follows the *gatherer*, who is provided with an iron pipe or tube, six or seven feet in length. (See Fig. 1.)

Having previously heated that end of the tube which takes up the glass, he dips it into the pot of metal; and by turning it gently round, gathers about one and a half pound of liquid glass on the end of it. Having allowed this to cool for a little, he again dips it into the pot, and gathers an additional quantity, of from two and a half to three pounds. This is also permitted to cool as before, when the operation of dipping is again repeated, and a sufficient quantity of metal, from nine to ten pounds weight, is gathered, to form what is technically termed a table or sheet of glass. The rod, thus loaded, is held for a few seconds in a perpendicular position, that the metal may distribute itself equally on all sides, and that it may, by its own weight, be lengthened out beyond the rod. The operator then moulds the metal into a regular form, by rolling it on a smooth iron plate, called

---

* The *sandiver* or glass gall is removed while the furnace is at its extreme degree of heat.

† A piece of wood about eleven inches long by seven broad, with a hole three inches by one inch, forms an excellent protection to the eyes from the heat to which they are exposed when examining the metal in the pots.

GLASS.]   [Plate 1.

Fig 1.

Fig 2.

Fig 3.

the "marver," a term corrupted from the French word *marbre*.

He then blows strongly through the tube, when his breath penetrating the red-hot mass of glass, causes it to swell out into a hollow pear-shaped vessel. (See Fig. 2.)

The tube with the elongated sphere of glass at the end of it is then handed to the *blower*, who heats it a second and third time at the furnace, pressing the end, between each blowing, against the bullion bar, so called from the part thus pressed forming the center of the sheet or bull's eye. (See Fig. 3.)

By the dexterous management of this operation, the glass is brought into a somewhat spherical form.

The blower heats a third time at the bottoming hole, and blows the metal into a full-sized globe. (See Fig. 4.)

When this part of the process has been completed, and the glass has been allowed to cool a little, it is rested on the casher box, and an iron rod, called a "*pontil*" or punty rod, on which a little hot metal has been previously gathered, is applied to the flattened side, exactly opposite the tube, which is detached by touching it with a piece of iron, dipped beforehand in cold water, leaving a circular hole in the glass of about two inches diameter. The operation of attaching the punty is shown by Fig. 5.

Taking hold of the punty rod, the workman presents the glass to another part of the furnace called the "nose hole," where the aperture made by its separation from the tube is now presented and kept until it has become sufficiently ductile to fit it for the operation of the flashing furnace. Whilst here, it is turned dexterously round, slowly at first, and afterward with increasing rapidity; and the glass yielding to the centrifugal force, the aperture just mentioned becomes enlarged. (See Fig. 6.)

The workman, taking great care to preserve, by a regular motion, the circular figure of the glass, proceeds to whirl it round with increasing velocity, until the aperture suddenly flies open with a loud ruffling noise, which has been aptly compared to the unfurling of a flag in a strong breeze; and the glass becomes a circular plane or sheet, of from four to four and a half feet diameter, of equal thickness throughout, except at the point called bullion or bull's eye, where it is attached to the iron rod. Figure 7 will give some idea of this very beautiful part of the process.

The sheet of glass, now fully expanded, is moved round with a moderate velocity until it is sufficiently cool to retain its form. It is carried to the mouth of the kiln or annealing arch, where it is rested on a bed of sand, and detached from the punty rod by shears. The sheet or table is then lifted on a wide pronged fork, called a faucet (see Fig. 8), and put into the arch to be tempered, where it is ranged with many others set up edgewise, and supported by iron frames to prevent their bending. From four to six hundred tables are placed in one kiln.

The kiln having been clayed up, the fire is permitted to die out, and the heat diminished as gradually as possible. When the glass is properly annealed, and sufficiently cold to admit of its being handled, it is withdrawn from the oven, after the removal of the wall built into the front of the arch, and is then quite ready for the glazier's use. It is first, however, removed to the manufacturer's warehouse, where the circular sheets are cut into halves, and assorted into the different qualities, known to the tradesmen by the names of seconds, thirds, and fourths.

We conclude our remarks on the manufacture by observing, that the quality of glass does not depend upon the mixtures alone, but also upon the treatment it receives after it has been made, the quality of the coals, and management of the furnaces. Cleanliness in every department of the manufacture, a general knowledge of chemistry, and of the art in all its details, with the most unremitting industry, and skill in the direction and government of the operatives, are all essentially necessary for the production of good glass.

MANUFACTURE OF BRITISH SHEET-GLASS.

This article is manufactured by Messrs. Chances of West Bromwich, near Birmingham, and Messrs. Hartley and Company, Sunderland, who, after having visited the glass manufactories of France, Belgium, and Germany, commenced, in 1832, the making of British sheet. The principle upon which it is manufactured is the same as that acted upon in the making of common or green glass. The metal is formed into cylinders, and then flattened into sheets. The French, Belgians, and Germans, having pursued this system for the last fifty years in making their window-glass, have much improved the old mode of making it; and as the parties who are now manufacturing this article in England are crown-glass makers, and have imported all the improvements adopted in the making of sheet-glass in France, Belgium, and Germany, and combined with these the improvements which their experience as crown-glass makers had taught them during the same period, they have surpassed the French, Belgians, and Germans in sheet-glass, and can now compete with them in all parts of the world.

There is no crown-glass made in France; and their window-glass, though superior to our broad or common glass, is not equal to the British sheet. In Germany there is little crown-glass made, and that of a very inferior quality. The greater part of the glass made in that country is sheet, and it is of much better quality than the French or Belgian. In Germany a common sort of glass is made, in the following manner: Three or four workmen form a partnership, and, having fixed upon a place in the woods where clay and sand are easily met with, they proceed to build a glass-house with wood and clay. They then make the pots, and, from the ashes of the wood which they burn, obtain potash, which, after it has been mixed with sand, they melt into glass. They blow the metal into cylinders, flatten it into sheets, cut, pack, in short, perform the whole operations from first to last, themselves. A good deal of the best of the glass made in this manner is sent to Nuremburg, where it is polished and sent into Holland. Some of it is sometimes smuggled into this country, and is known by the designation of Dutch glass.

The expense of making British sheet is about the same as making crown-glass, excepting in the case of large squares, when it is much less; in crown-glass it is very difficult to get a square 34 x 22, but in sheet-glass the common size is 40 x 30; nay, sheets are sometimes made as large as 50 x 36. Its other advantage over crown is, that it has none of that wavy or curved appearance, by which the vision is so much distorted in crown-glass; but, at the same time, sheet-glass has rather an unpleasant appearance when viewed from the outside of a building, in consequence of an unevenness of surface, technically termed *cockled;* when viewed, however, from the inside, it is difficult to distinguish it from plate-glass.

The materials employed in the making of sheet-glass are the same as those used in the making of crown-glass. The large melting furnace is also very similar; in France and Belgium they usually contain six or eight pots, but at the British manufactories such furnaces contain ten pots, each containing seven cwts. of metal, which requires fourteen hours to melt.

In a line with each pot, and four feet from the ground, are erected ten stages, with an open space between each, of about two feet, through which the workman swings his glass when making cylinders. When the metal is ready for working, the ten workmen take their stations, each having his own pot and stage, and also an assistant, and commence making the cylinders, as follows: After gathering the quantity of metal required (which varies from three to twenty pounds), the workman places it in a horizontal position upon a wooden block, which has been hollowed, so that, when the workman turns the metal, it shall form it into a solid cylindrical mass. In the mean time, the assistant, with a sponge in his hand, and a bucket of water by his side, lets a fine stream of water run into the block, which keeps the wood from burning, and also gives a brilliancy to the surface of the glass. The water, the moment it comes in contact with the glass, is raised to the boiling point, and, in that state, does no injury to the metal; but it is only when the metal is at a high temperature that such is the case; for, whenever the glass is cooled to a certain degree, it immediately cracks upon coming in contact with water. When the workman perceives that the mass of metal is sufficiently formed and cooled, he raises the pipe to his mouth at an angle of about seventy-five degrees, and commences blowing it, at the same time continuing to turn it in the wood block, till he perceives the diameter to be of the requisite dimensions, which are usually about ten inches. He then reheats this cylindrical mass, and when it is sufficiently softened, commences swinging it over his head, continuing to reheat and swing till he has made it the desired length, which is commonly about forty inches. It is now in a cylindrical form, forty inches long and ten inches in diameter, one end being closed and the other having the pipe attached to it. The workman now begins to open the end which is closed, for which purpose he incloses the air in the cylinder, by stopping the aperture of the pipe with his finger; and then placing the closed end of the cylinder toward the fire, it becomes softened, while at the same time the air within is expanding, and, in about thirty seconds, the glass becomes too soft to retain it, and bursts, a small aperture being formed at the point of the cylinder. The workman then turns the cylinder round very quickly, and, by keeping it warm at the same time, flashes it out perfectly straight; the other end, which is attached to the pipe, has now to be cut off. This is done in the following manner: The workman having gathered a small quantity of metal on the pontil, draws it out into a thread of about one-eighth of an inch in diameter, laps it round the pipe end of the cylinder, and, after letting it remain there for about five seconds, withdraws it suddenly, and immediately applies a cold iron to the heated part, which occasions such a sudden contraction, that it cracks off where the hot string of glass has been placed round it. The workman having now formed a perfect cylinder of forty inches in length and ten in diameter, has, before it can be flattened, to split it on one side, so that it can be opened out; but before doing this, he is obliged to let it cool, and then, laying the cylinder horizontally upon a bench, draws a red hot iron two or three times along the inner surface. The cylinder, thus heated, splits along the heated part, owing to the expansion of the glass when heated, its cylindrical form preventing its breaking at the point of expansion.

The blower having now completed his cylinder, hands it over to the flattener to make it into a flat sheet; to accomplish which, two furnaces are built together, the one for flattening the cylinders, the other for annealing the sheets, the former being kept at a much higher temperature than the latter. The cylinder after being gradually reheated, is placed in the center of the flattening furnace, upon a smooth stone, with the cracked side upward. In a short time it becomes softened by the heat, and by its own weight falls out into a flat square sheet of forty inches by thirty. The flattener, with a piece of charred wood, rubs it quite smooth, and then places it in the annealing arch, where it remains about three days to be annealed. A workman will make sixty cylinders 40 x 50 in one day; and a flattener can flatten the same number in the same time. This glass can be made of any thickness from one-twentieth to half an inch.

The same enterprising companies also manufacture extensively every variety of colored glass used by the glass-stainer, which is gathered, rolled, blown, and flattened, in a similar way with the sheet-glass, the pot-metal being gathered from one pot, and the flashed glass from two, one containing colorless and the other colored metal, which being blown and distended together are combined, while each portion retains its individual character. These oval and square glass shades used for covering French clocks and other ornaments, as well as glass dishes for dairy purposes, are also made by these parties; and since the abolition of the glass duties are much in demand.

There is another species of glass called broad or common window-glass, which is formed of the coarsest materials. The ingredients usually employed are, six measures of soap-boilers' waste, three of kelp, and three or four of sand. After these have been fritted for from twenty to thirty hours, they are removed while red hot to the pots in the working furnace, where, in the space of from twelve to fifteen hours, they are reduced to a fluid state. The metal is taken out in the manner already described, and blown into globes about a foot in diameter. A piece of iron dipped in cold water is run along them, and produces a crack nearly rectilinear; and, while yet warm and ductile, these spheres are opened out and flattened on a smooth iron plate at the mouth of the furnace.

MANUFACTURE OF PLATE-GLASS.

This description of glass may be manufactured in the same manner as broad window-glass, or by casting the materials in a state of fusion upon a flat surface. Little correct information has been published relative to the manufacture of plate-glass, from the reluctance of proprietors to permit their works to be examined by individuals who are capable of giving an intelligible account of them. If such are permitted to scan the mysteries, they are generally restricted to keep secret the information which they have acquired. The late Mr. Parks the chemist, however, seems to have been exempted from this condition, and after having visited the premises of the British Plate-Glass Company, at Ravenhead, in Lancashire, he published a short account of the process as there carried on. Besides the above manufactory in Britain, may be mentioned that of Messrs. Swinburne and Company, South Shields, the Thames Plate-Glass Company, the Union Plate-Glass Company, St. Helen's; and W. A. A. West's, Eccleston. Plate-glass is also made at St. Gobain in France, besides other places upon the Continent.

The following is Loysel's account of the relative proportions of the materials used at St. Gobain, in the manufacture of plate-glass: White sand, 100 parts; carbonate of lime, 12 ditto; soda, 45 to 48 ditto; fragments of glass of like quality, 100 ditto; oxide of manganese, $\frac{1}{4}$. The following proportions of ingredients are said to produce the best description of this article: Lynn sand which has been well washed and dried, 720 parts; alkaline salt containing 40 per cent. of soda, 450 ditto; lime slacked and sifted, 80 ditto; niter, 25 ditto; broken plate-glass, 425 ditto. These quantities produce one pot of metal, which yields 1200 pounds of glass.

Great nicety must be observed in conducting the processes of this manufacture. The materials must be selected with the utmost care. The sand should be of the whitest and finest description, and well washed and passed through a sieve, previously to being mixed with the other ingredients. Soda is always preferred to potash, because it imparts a higher degree of fluidity to the glass, and also because the impurities which it contains are more easily dissipated by the heat. Lime acts as a flux, and manganese has the effect of giving a slightly reddish hue to the mixture by which the colors of the other materials are neutralized, so that scarcely any appreciable tint remains. Cobalt is likewise used in some manufactories, much for the same purpose as manganese. The broken glass or *cullet* as it is technically called, is those fragmentary portions which are cut from the plates when they are squared, or that which may flow over in the process of casting. The sand, lime, soda, and manganese, being properly mingled in the proportions above given, are fritted in small furnaces, where the temperature is gradually raised to a red or white heat, and there maintained until no more vapor is evolved, nor change undergone by the mixture. This process occupies six hours, and after its completion the other ingredients are added, consisting of cullet and cobalt. At St. Gobain there are two kinds of crucibles employed; the one in which the glass is melted is called a *pot*, and has the shape of an inverted truncated cone; the other is entitled a *cuvette;* it is kept empty in the furnace, and exposed to the full degree of heat. Forty hours are requisite to vitrify the materials properly, and bring the glass to a fit state for casting. The pots are skimmed in the manner already described. When the liquid mass has been properly refined, the cuvette is filled by a copper ladle, and after sufficient time is allowed for the bubbles created by this disturbance to escape, it is removed to the table where the plates are cast. Copper was the metal of which tables were formerly constructed; but cast iron has now been found to answer the purpose completely, and it is greatly superior to copper in this respect, that it remains uninjured during all the sudden transitions of temperature to which it must be subjected. The British Plate-Glass Company were the first to introduce this improvement. They procured a plate fifteen feet long, nine feet wide, and six inches thick. The sides are provided with metallic ribs, the depth being exactly the measure of thickness which it is desired the glass should be of. During the casting there is a similar rib temporararily attached to the lower end of the table. The cuvette being filled with melted glass, it is withdrawn from the furnace by means of a crane, taken to the upper end of the casting table, and after being properly scummed, and elevated to a sufficient height by means of a crane, it is emptied of its contents. The surface of the melted matter is then smoothed by means of a large hollow copper cylinder, which extends across the table, resting upon the side ribs. This is set in motion, and rolled over the glass, by which process it is spread out into a sheet of uniform breadth and thickness. When the plate has become completely hardened, it is carefully inspected, to see that no flaws or bubbles appear on the surface. Should any be found, the sheet is immediately divided by cutting through them. It is afterward removed to the annealing oven, where it is placed in a horizontal position, and remains for about fifteen days. When glass is in a high state of fluidity it is liable to be injured even by a draught of air, so that the apartment must be kept as free as possible from disturbance. The opening or shutting of a door, by

GLASS.] [PLATE 2.

Fig 4.

Fig 5.

setting the air in motion, might impair the value of the plate. After having been withdrawn from the annealing oven, they have to undergo the operations of squaring, grinding, polishing, and silvering, before they are fit for the market. These processes have thus been described by a late writer upon the subject:

"The first process—that of squaring and smoothing the edges—is performed by passing a rough diamond along the surface of the glass, guided by a square rule; the diamond cuts to a certain depth into the substance, when, by gently striking the glass with a small hammer underneath the part which is cut, the piece comes away; and the roughnesses of the edge then left are removed by pincers. The plate is then taken to the grinding apartment.

"The next step is to imbed each of the plates upon a table or frame adjusted horizontally, and made of either freestone or wood, cementing the glass securely thereto by plaster of Paris. One plate being then reversed and suspended over another, the material employed in grinding their surfaces is introduced between the two, and they are made to rub steadily and evenly upon each other by means of machinery set in motion by a steam-engine." River sand and water were formerly used for the purpose of abrading the surface, but ground flint is now substituted, as answering the purpose better. When one side of each plate has been sufficiently ground, it is loosened from the frame, and turned over, so as to present the other surface to be ground in the same manner. Some degree of pressure is employed, by loading the upper plate with weights, as the grinding of each side approaches completion. The process thus described used formerly to last during three entire days, but this time is now much abridged. The greatest attention is required in order to finish with the surfaces perfectly level and parallel, for which end a rule and plumb-line are employed.

By means of this grinding, the plates will have been made level; but they are too rough to receive a polish. To fit them for this, they must again be ground with emery powder of increasing degrees of fineness. The preparation and sorting of this powder are effected in the following simple and ingenious manner: "A considerable quantity of emery is put into a vessel containing water, and is stirred about violently until the whole is mechanically mixed with the water. Emery is absolutely insoluble by such means; and if the mixture were left at rest during a sufficient time, the whole would subside in layers; the coarsest and heaviest particles sinking first, and so on successively, until the very finest particles would range themselves as the upper stratum. Previously to this, however, and while these finest grains are still suspended in the water, it is poured off into a separate vessel, and the emery is there allowed to settle. A fresh supply of water is poured into the first vessel, the contents of which are again violently agitated, and allowed partially to subside as before. A shorter interval is allowed for this than in the first case; and then the liquor is poured off into a third vessel, by which means emery of the second degree of fineness is separated. This operation is repeated in order to obtain powders having five different degrees of fineness. The deposits are then separately dried upon a stove to a consistence proper for making them up into small balls, in which form they are delivered to the workmen.

"In this further rubbing together, or, as it is called, *smoothing* of the glass plates, it must be understood that the coarsest emery is first used, and so on, substituting powders having increasing degrees of fineness as the work proceeds."[*]

These processes finished, the glass, although perfectly even, appears opaque or deadened on the surface, and requires polishing. This is effected in the following manner: A piece of wood is covered with numerous folds of woolen cloth, the layers being divided by some carded wool interposed between each, the whole forming a tolerably hard but elastic cushion, which is furnished with a handle. The plate is laid upon a bed of plaster, as already described, and the cushion being wetted, is covered with the red oxide of iron (the *colcothar* of commerce), and moved backward and forward upon the surface of the plate. Lastly, if the glass be intended for mirrors, it is silvered, that is, covered on one side with a thin coating of amalgam of tin and mercury.

The process of *blowing* plate-glass differs so slightly from the methods used in producing broad glass, that they need not be here repeated. Any difference that does exist, arises

---

[*] Glass Manufacture, Lardner's *Cabinet Cyclopædia*, No. 26.

from the great bulk and weight of the mass of glass operated upon.

STAINED OR PAINTED GLASS.

In an age like the present, when a high state of civilization and refinement demands the most careful and diligent cultivation of those arts which minister to the gratification of refined taste, Glass Painting, as an art now acknowledged indispensable in the decoration of our churches, palaces, etc., has assumed an importance not attained at any former period; and from its progress in connection with our present advances in artistic knowledge, we may safely infer, that if the noble elements in its nature and capabilities be fairly and legitimately applied, it will become the most potent agent in advancing the standard of architectural decoration. Within these few years in our own country, its progress has been altogether wonderful. Since the abolition of the duties on glass, Britain has produced the rarest and the richest colored glass in endless profusion and variety. Light has been admitted into gorgeous apartments through domes filled with colored glass of colossal dimensions, the decorations of which, worked out from the designs of the architects, enhance and give power to the other ornamentation of the interiors, and produce a *coup-d'œil* not previously known nor dreamt of. Stained glass windows for churches have been, and are being executed in Britain, which, for appropriate design, brilliant color, subdued tone, symmetrical proportion, minute manipulation, and variety of carefully considered detail, stand comparison with the best existing specimens of mediæval times. Windows for palatial and baronial structures have also been recently produced in this country, whereon are traced in imperishable lines, and blazoned in unfading colors, correct effigies of historical personages, representations of historical events, genealogical arrangements of heraldic bearings, and other legends and memorials, which will convey to distant times a favorable idea of the state of British art in the nineteenth century. The earliest record which we possess concerning the existence of this beautiful art is of the age of Pope Leo III., that is, about the year 800, a period in which many of the most magnificent ecclesiastical edifices on the Continent were erected. It is not known with certainty when stained glass was made use of for pictorial or figure subjects, but the historian of a monastery at Dijon, writing in the eleventh century, says, that there existed in his time in the church of his monastery some very ancient glass representing the mystery of the Holy Eucharist, and that this glass picture had been taken from the old church previous to its restoration. The earliest specimens of stained glass are composed of small pieces of glass, imbued throughout with color, united by grooved leaden joinings. It has been suggested that this arose from the glass-makers of that period not being able to make it in larger pieces. If so, in so far as sparkling brilliancy in glass decorations is a desideratum, it might almost have been as well for the art that the manufacturers of the colored metals had still been in the same position. Brilliancy is always increased in the same ratio with the number of pieces of glass in the composition.

Nothing could be more instructive or interesting than an investigation of the relative merits of the existing specimens of the art during the six centuries it was so diligently and effectively cultivated in connection with ecclesiastical architecture; but this inquiry would be too extensive to be opened up here. The following remarks, therefore, are confined to a few of the leading points in the glass of the various styles which prevailed in succession from the eleventh to the seventeenth century, in which may be traced very clearly the progress of the architects, under whom the glass painters worked, from clumsy and servile imitators to bold and original designers.

The Norman style, in its early period, was a direct though imperfect imitation of Roman architecture; but when pointed architecture had attained its greatest perfection, its chief feature was originality. During the gradual development of this, its peculiar characteristic, the openings in the walls by degrees were enlarged, until they ultimately became the principal points, and it was requisite that they should be judiciously decorated; and in no branch of art connected with pointed architcture can its onward movement be more clearly traced than in the painted glass windows.

The painted glass of the eleventh and twelfth centuries, like the Norman architecture of which it formed a part, was stately and of a magnificent character. The colors were of the most vivid and most positive description. There was no spot left for the eye to repose on; no neutral tints nor secondary colors were introduced. The whole of the ground and foliage were filled with intense color, ruby and blue invariably preponderating. The same love of violent and striking contrast as is peculiar to man in a state of semi-barbarism was manifested in the coloring of the windows of that period, and the general effect must have been congenial to the romantic and martial spirit of the age of chivalry. The leading forms were massive and simple, consisting chiefly of the circle and square, filled up chiefly with clumsy imitations of the foliated ornament to be found in Roman architecture. When figures were introduced, they were like those in the Bayeux tapestry, marvelously correct in costume, though disproportionate in drawing, and filled up with strong positive colors, flat, and the outline defined chiefly by the strong thick lines of the lead, resembling those highly titled personages represented on packs of cards, or those in Chinese processions, as delineated by the native artists of the Celestial empire.

In the thirteenth century, the painted glass, like the primary pointed or early English architecture of which it formed a part, was of a light and elegant character. The glass painter had then acquired a more correct idea of what constituted beauty, both in form and color. The positive colors were now used more sparingly, and indeed were almost confined to geometric bands, central points, and borders continued round each entire light. The general grounds were of a beautiful tint of neutral gray, produced by lines intersected at right angles, from which were relieved by bold lines scrolls of foliated ornament in clear colorless glass. The glass in the Sister windows of York Minster may be named as one of the finest existing specimens of this description. Figures and subjects also, when introduced, were better drawn than formerly. The faces were kept colorless, slightly shaded with brown of a rough gritty texture. The secondary colors were used in the draperies with a most delicious effect, softening and harmonizing the whole composition, and giving a lightness and variety previously unknown. In the leading forms of the ornamental portions also were seen repeated the geometric features of the building; and in the glass of the period we can recognize repetitions of the ground plans of the shafts, with the enrichments on the laps and on the mouldings of the windows and doorways. In the foliaged backgrounds, amid repetitions of the ancient Roman foliage, we now and then get fragments of simple foliage, such as trefoils, evidently taken from nature; and we are able to trace in progress an art which was shortly to become as original as beautiful, and dependent entirely on the artist's knowledge and appreciation of nature and geometry.

During the fourteenth and fifteenth centuries, when the secondary pointed or decorated style of architecture gradually developed its immense resources, and advanced steadily toward perfection, we find that the glass advanced in the same ratio as the art with which it was associated. In accordance with certain fixed rules of proportion, the glass artists elongated, intersected, diversified, and arranged rectangular, triangular, and curvilinear figures, and made these harmonious combinations their leading points for color. They were thus enabled with certainty to produce a pleasing effect, and to fill up the detail according to their own fancy, with an imitation of the common weeds, flowers, and plants that they found growing around them. This principle was carried out in every portion of the detail in the remarkable structures then erected; and the exquisite imitations of vegetables and plants on the carving of the caps, friezes, and mouldings, show the extraordinary love of nature which must have animated these fraternities of artistic minds by whom these details were worked out. The monks of Melrose made "gude kail," says the old song, and from the exquisite manner in which that vegetable has been carved on some of the portions of that fine old abbey, one would conclude that the carvers must have shared largely and appreciated highly the "gude kail" of the holy brotherhood, a feeling no doubt also entertained by the glass painters of the structure whose works would doubtless exhibit similar genius with the glass of that period, which was well characterized by a rich juicy natural freshness, as well as an easy play of elegant outline and graceful proportion. In many instances also, the gray background produced by intersected lines was abandoned, and a tint of rough gray obscure subtituted, which imparted to the whole a softer effect, and gave a better relief to the outlined foliage of which the diapering was composed. At this period also, glass painting had attracted artists of high genius, and the figures and subjects in the glass of the period are perfect specimens of what the art ought to be. These artists tested its capabilities, and how well and thoroughly they did so may be seen in Cologne Cathedral, where the flowing, bold, and elegant outline, the rough semi-transparent texture, the calm expressive countenances and attitudes of the figures of Durer and his cotemporaries, fairly set aside and overpower the glowing, brilliant, but frowsy specimens there of modern German art, as practiced in Munich, under the auspices of the Bavarian government. Perhaps, also, the ornamental glass of the period in that structure will be found equal to any in the world for geometrical symmetry and natural foliage, the latter imitated from the common weeds and plants indigenous to the locality. So fascinating and far-famed were the painted glass windows of that period, so novel were the effects produced by the rich semi-transparent shadows and reflected lights, that Mr. Eastlake conjectures that the increase of color in shade which is so remarkable in the Venetian and early Flemish pictures may have been suggested by the slight shading on the stained glass through which it was transmitted. Over the Lady Chapel in the north aisle of York Minster are two windows of this period remarkable for the brilliancy and quiet feeling formerly alluded to as indispensable in glass painting of a high character.

After the decorated period painted glass degenerated first into the flat, tame insipidity of the perpendicular style, and then ran riot in the extravagant tortuosities and monstrosities of the capped, jeweled and double gilt details of Elizabethan architecture, which it seems a fallacy to suppose was imported from Italy. At the time of its introduction a strong tide of feeling had set in against every thing that pertained to the Roman Catholic religion, and it seems unlikely that after having diverged from the style with which that religion had so long been identified, the nation should have imported any thing from Italy, its headquarters and chief stronghold. It rather seems probable that at a time when attempts were made to get quit of every existing form and style of architectural decoration, there would be awakened a strong desire for novelty; and when it is remembered that the newly discovered continent of America was visited by crowds of adventurers, it will not appear unlikely that many of these adventurers must have been delighted and dazzled with the magnificent and unique architectural structures which adorned the ancient cities of Central America, and may have imported home and introduced into English architecture many of the features which we are accustomed to believe were originated in the Elizabethan era. The painted glass of that period partakes of the same character; and in Du Paix's great work on New Spain will be found something very like the origin of many of these peculiarities, eccentricities, and heterogeneous conglomerations which characterize the wood and stone carving, as well as the wall and window decorations of that extraordinary style of architecture.

From this rapid sketch of the history of the rise and decadence of painted glass, it appears that there is no limit to its capabilities; and that forming, as it does, a leading architectural decoration, it is as well adapted to one style as another. If the principles of harmonious coloring and symmetrical proportion be carefully attended to, as was the case in the best specimens of the art in the mediæval period, painted glass must ever be regarded as one of the most attractive decorations for church or mansion. It is no doubt a species of mosaic, and the artist must generally depend on harmonious combinations of color and continuity and firmness of outline for the effect he intends to produce, as the brilliant coloring and mosaic character are lost in the same ratio as shading is introduced.

It is also true that windows are generally intended for the transmission of light, and that in some cases the sacrifice of light required for pictorial effect cannot be made. Yet who can resist the attractions of such pictorial glass as is to be found in the windows of St. Gudule at Brussels, or St. Jans Kirk at Gouda, where the principles of chiaro-oscuro and perspective are fully developed, where foreground and distance hold their proper places, and where the lights and brilliant colors are arranged in a manner to rival the best specimens of the ancient masters of painting in oil. This mode of treatment is not to be advocated for general use, but where there is light enough and to spare, and where men of high artistic powers apply themselves to glass painting, they may safely be left to their own genius, and allowed to render their conceptions as vividly and perfectly in glass as others do on can-

GLASS.] [PLATE 3.

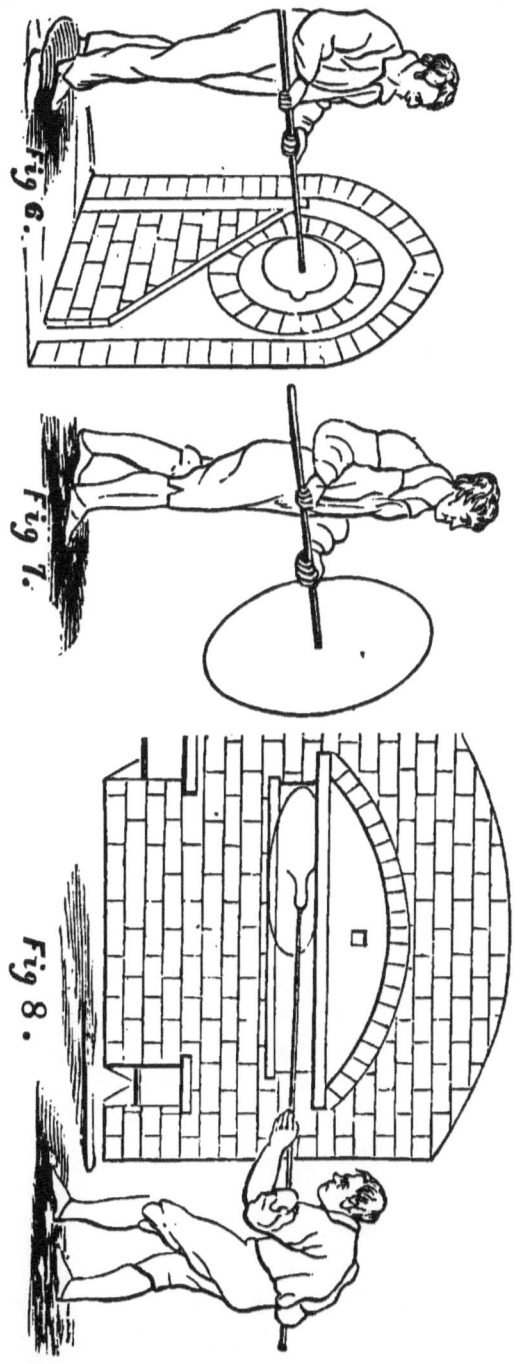

vas. The dull, heavy, uniform opacity which pervaded the glass of the last century, when it was made up in squares, the colors fused, and the whole work looked like cloth transparencies, is not to be tolerated. The brightest colors that can be produced in pot metals joined in the ancient way by leading ought always to be used; and although in general the effect of mosaics in low relief may be preferable, yet the shading and toning requisite to give full effect to a good pictorial design may be given without detracting greatly from the light.

Glass is the most enduring species of artistic medium, and it is to be hoped that this quality will yet cause eminent artists to leave the impress of their genius on painted windows. Had the art of painting on glass been known in the age of Phidias, we might have had preserved, in colors as vivid as when the works were executed, the Jupiter of Homer by Apelles, the pictorial embodiment of the Athenian character by Parrhasius. A singular fact illustrative of the durability of painted glass may here be stated in connection with York Minster. When the nave of that fine structure was destroyed by fire, the heat was so intense that many of the stones were calcined. When the leaden framing of the windows melted, the glass made of many small pieces fell down undamaged, and was afterward carefully rebuilt in new leading and fixed in its original place, where it now remains the most fragile yet the most enduring portion of the ancient structure.

In ornamental painted glass the positive colors ought generally to be used sparingly, and confined to the chief points in the composition. When overloaded with color, the sparkling brilliancy so desirable in painted glass is entirely lost. The general ground of the window should be of a neutral tint suitable in tone to its character and situation. In a southern aspect this tone should be of a cool gray, and the positive colors, blue, green, and purple, ought to predominate over ruby, yellow, and orange. In a northern aspect the general ground should be of a warm sunny tint, and the warm ought to predominate over the cold colors. An eastern window ought to approximate in color to a northern, a western to a south window.

It is always desirable to have a combination of straight

and curved lines in the leading forms of painted glass. As in the human body, the effect of the elliptic curvature òf the muscles is enhanced by the angular position of the straight lines on which they are placed, or by the sharp square indications of the bony extremities, in like manner the curvilineal lines in ornamental decorations appear to more advantage when balanced by a harmonious proportion of straight lines. A very important feature in glass is diapered work in the backgrounds, a great variety of designs for which may be obtained from plants and flowers by the wayside, in the field, or the garden; and the more homely these are, they are often the more suggestive and pleasing. Borderings are almost indispensable in all ornamental painted-glass windows. They bind together what might otherwise be disjointed and scattered, and afford scope for endless variety of design, both in form and color. Heraldic symbols and emblazonments have always been amongst the most attractive features in stained-glass windows. The points which most shields form for a balance of positive color; the crests, mantling, supporters, and mottoes, twisting or twining either quaintly or gracefully through the composition, not to speak of the interesting nature of heraldry as a guide through the intricate mazes of family connection, wending through the depths of ages—all tend to render it the most admirable field on which the glass painter can be engaged. For hall or library windows such devices are very appropriate, and indeed so highly are they appreciated, and so much is painted glass now coming into repute, that there is scarcely a new house of any pretensions without its library or staircase stained-glass window. In towns where back drawing-room windows look into mean or filthy lanes, what a delightful remedy is found in light sparkling stained glass. Either heraldic blazon, family monograms, or ornamental devices, may be used; and if the inner window be fitted up flush with the inner wall, and the room lighted at night mainly from lights placed between the outer and inner windows, the effect is chaste and beautiful.

Restorations of windows connected with ancient edifices afford fine media for embodying local legends or historical local incidents. In new structures for public purposes what place so fitting or so striking as the windows for representations of men eminent in connection with such institutions.

Monumental windows have recently been introduced into churches with excellent effect, and they afford scope for invention as various as the characters of those whose virtues they are designed to commemorate. In churches even in Scotland stained glass is rapidly assuming its ancient importance, and there can be little doubt that it will ultimately be so much encouraged and cultivated that the windows of our public edifices will be the honored medium of transmitting to remote posterity the works of the master-minds of British art.

Except in the name, painting on glass has no resemblance to any other department of the pictorial art but that of porcelain. Both the colors, and the process of their application throughout, are entirely different. Where animal and vegetable substances are freely used as coloring matter in every other department of pictorial art, they are wholly excluded in that of glass painting, where all the pigments used are subjected, after being laid on, to the operation of fire, to make them penetrate the body of the glass, or become fused on its surface—a process which would wholly destroy the coloring properties of such substances. All the colors employed in glass painting and staining are oxides of metals or minerals, as gold, silver, cobalt, which not only stand the fire, but require the powerful interference of that agent to bring out their brilliancy and transparency. Some colors, on the application of heat, penetrate the body of the glass, and from this circumstance, are called stains; while others, being mixed with a vitreous substance called flux, become fused or vitrified on the surface. The former produces a variety of colors, and all of them are perfectly transparent. The produce of the latter are only semi-transparent, but they may be made to yield any color or tint required.

In preparing these colors, the most important point to be attended to is, to have all those that are to be used at the same time of an equal degree of softness. To attain this, those that are hard, and require a great degree of heat to make them effective, must be fixed first; leaving the soft colors, for which a slight heat only is necessary, to the last. If used promiscuously, and without regard to this precaution, some of the colors would be rendered too fluid, while others would be insufficiently fused, and the work in consequence spoiled.

GLAZING OF WINDOWS.

Putty, an important and indispensable article in the glazier's trade, is composed of whiting and linseed oil. Chalk is sometimes used instead of the former, but the expense and labor incurred in preparing it is much greater, and besides it generally contains sand, so that it is no object to the glazier to employ it. Whiting is in every way to be preferred; it must be thoroughly dried before the oil is added to it, otherwise the union will not be effected, or at least it will be very imperfect.

After the whiting has been thoroughly dried and prepared, it ought to be passed through a very fine sieve, and all the remaining lumps and knots pulverized, and then also passed through the sieve. Great care must be taken to keep the whiting free of sand and other extraneous substances.

When putty is to be made, put the proper quantity of oil into a tub or other open vessel, and gradually add the whiting whilst yet in a hot state, at the same time keeping the whole in motion with a stick, until it becomes of a sufficient consistency to admit of being wrought by the hand on a board or table. Having been removed thither from the tub, it must be wrought up with dry whiting, until it is converted into a compact mass. When brought to this state, it ought to be put into a hollowed stone or mortar, and beat with a wooden mallet till it becomes soft and tenacious, when more whiting must be added, until it has attained a proper consistency.

When putty is required of a superior degree of fineness, and which will also dry quickly, add a little sugar of lead or litharge; and if an increase of strength be wanted, a little white lead.

When the panes have been fitted into the checks of the sashes, they must be removed, and the checks well bedded with beat putty. This done, the panes are again returned to their places, and gently pressed or lodged in the bedding, the workman, as it were, humoring the glass should it be bent or twisted, and taking care that there is no hard extraneous substance mingled with the putty, which might endanger, if not actually break the glass. When a pane is perfectly bedded it lies quite firm, and does not spring from the putty; but when, either from a perverse bend or twist in the glass, or any other accidental cause, it happens that it cannot be made to go quite close to the check, the vacant space must be carefully and neatly filled upon the back puttying, otherwise the window will not be impervious to the weather, and will be very apt to fall into decay by the admission of moisture.

The convex or round side of the pane, where such a shape occurs, should be presented to the outside, and the concave or hollow to the inside. When thus placed, they resist the weather better than if the hollow sides were exposed to it.

After the pane has been bedded, the next process is the outside puttying. This putty should be kept in the fore check, about the thirty-second part of an inch below the level of the inside check, so as to allow the thin layer of paint which binds these two substances together to join the putty and glass; and that it may not offend the eye by being seen from the inside; and that, when it is painted, the brush may not encroach on any visible part of the pane, leaving those ragged lines or marks which are so often seen from the inside on ill-finished windows, and which are so displeasing to the eye. This operation, and finishing the corners, are two nice points in the art, and therefore, when properly done, discover the neat-handed and skillful workman.

All frames or sashes of windows ought, before being glazed, to receive one or two coats of white paint, to which a small portion of red lead has been added to facilitate its drying, and to give increased strength and durability to the paint.

Lattice or Lead Windows.—This antique and singularly beautiful style of glazing has unaccountably fallen much into disuse, although of late years it has certainly undergone something like a resuscitation, in consequence of a revival of the public taste for stained glass, and a growing predilection for Mediæval architecture in churches, cottages, and the like. For these, and for staircase windows, and indeed all windows similarly situated, as in halls, lobbies, or the like, it is particularly adapted.

It may be proper to premise, that lead windows require

stained or colored glass for producing their fullest and best effects, and it was with stained glass only that they were originally constructed; but very neat and elegant windows are executed in this style with plain glass, where variety and beauty of figure are made to compensate in some measure for the absence of color.

Lead windows may be made to any pattern, and in this there is great scope for the display of a correct taste. In the time of Elizabeth, this branch of the glazier's art was carried to great excellence, especially by one Walter Geddes, who was employed in glazing most of the royal and public buildings of that period. Geddes executed in this style some windows of transcendent beauty, displaying an endless variety of the most elegant and elaborate figures. The most useful and most common description of plain glass lead windows, however, are those of the diamond or lozenge shape; but, as already said, they may be made to any pattern desired.

The lead work can be adapted with ease to any pattern that may be chosen for the glass; and it can likewise be made to any breadth, from one-eighth to five-eighths of an inch.

The apparatus and tools necessary for producing this are, a glazier's vice or lead mill, moulds for casting the lead into slender bars or rods of about eighteen inches in length, which is the first process; and a three-fourth inch chisel; a hardwood fillet for forcing the glass into the grooves in the lead frame-work; and an opener or wedge tool, made also of hard wood, or ebony, for laying open the grooves for the reception of the glass; two copper bolts for soldering, the end formed like an egg.

The lead intended to be employed in window-making must be soft, and of the very best quality; and great care must be taken to have the moulds properly tempered, otherwise the lead will not be equally diffused in them, and the castings consequently not perfectly solid throughout, as they ought to be.

The castings are, as already noticed, usually about eighteen inches in length, and are afterward extended by the mill to the length of five or six feet.

It may not be unnecessary to add, that the mill not only extends the lead, and reduces it at the pleasure of the operator to the dimensions required, but at the same time forms the grooves into which the edge of the glass is afterward introduced in forming the window.

When the lead has been prepared in the manner described, the glazier ought to proceed to cut out the panes wanted. For this operation he must prepare by first outlining the full dimensions of the window, and then lining it off to the pattern required, shaping the panes accordingly. If the window is of a large size, this may be done by compartments, to be afterward united, and thus be more conveniently wrought.

When all the glass has been cut for the window, the next thing to be done is to open the grooves in the lead with the opener or wedge tool. The panes are then, in order that they may be water-tight, fastened very firmly into the grooves with the wooden fillet already spoken of (which may be fixed on the handle of the chisel or cutting-tool), the parallel lines of lead being secured in their proper places on the board, when the window is of the diamond shape, by a small nail at either end, until the course is finished, when the work is permanently fastened by running a small quantity of solder gently over the two connecting pieces of lead at each joint, or angular point. When the window has been completed, it should be removed from the working board to a flat table, and there covered with a thick layer of cement, composed of white lead, lamp-black, red lead, litharge, and boiled linseed oil, with a half-worn paint-brush, and the composition carefully rubbed into every joint. This will render the window completely impervious to the weather, as the cement, if properly laid on, will fill every chink, where it will soon become as hard and durable as any other of the materials of which the window is composed.

The window, on being fitted into the frame, that is, on being set in its place in the building for which it is intended, ought to be supported with iron rods, extending three-eighths of an inch beyond the breadth of the frame on each side, running across it at the distance of from twelve to fourteen inches from each other, and secured to the lead frame-work at intervals with copper wire.

THE CUTTING DIAMOND.

Before the introduction of the diamond as an agent in cutting glass, that operation was performed by means of emery, sharp pointed instruments of the hardest steel, and sometimes red-hot iron. These were the only contrivances known and practiced by the ancient glaziers.

In considering the diamond in its relations to the purposes of the window-glass cutter, there occur some circumstances not unworthy of remark. Amongst these, it may be noticed, that the cutting point of the diamond must be a natural one; an artificial point, however perfectly formed, will only scratch the glass, not cut it. The diamond of a ring, for instance, will not cut a pane, but merely mark it with rough superficial lines, which penetrate but a very little way inward. Artificial points, corners, or angles, therefore, produced by cutting the diamond, are adapted only for writing or for drawing figures on glass, and such were those used by Schwanhard, Rost, and the other old artists who were celebrated for ornamenting glass vessels. The cutting diamond does not write so well on glass, from the circumstance of its being apt to enter too deeply, and take too firm a hold of the surface, and thus become intractable. It may be further noticed, that an accidental point produced by fracturing the diamond, is as unfit for cutting as an artificial one. Such a point will also merely scratch the glass. No point, in short, that is not given by the natural formation of the mineral, will answer the purposes of the window-glass cutter.

The large sparks, as the diamonds used for cutting glass are called, are generally preferred to the small ones, from the circumstance of their being likely to possess (although this is by no means invariably the case) a number of cutting points; while the very small sparks are not always found to possess more than one. Thus, if the point of the latter is worn or broken off, although the spark be turned, and reset in its socket, it will still be without the power of cutting, and consequently useless; while the former, on undergoing the same operation, will present a new and effective point.

The large sparks are called *mother sparks*, and are sometimes cut down into as many smaller fragments, bearing the same name, as there are natural points in them. Each of these, therefore, can have only one cutting point, and are consequently only proportionately valuable to the glazier, since they cannot be restored by resetting.

The Setting of Diamonds is a process with which every

glazier ought to be acquainted; nor is it an art of difficult acquirement; some practice, and a little patience, are all that is necessary.

After having selected a stone, as clear and pellucid as possible, and of an octahedral shape, or as near to that form as it can be procured, the workman proceeds to ascertain which is its cutting point, or, if it has more than one, which is the best. This will be found to be that point which has the cutting edges of the crystal placed exactly at right angles to each other, and passing precisely through a point of intersection made by the crossing of the edges.

He then provides a piece of copper or brass wire, a quarter of an inch in diameter, having a hole drilled in one of its ends large enough to contain three-fourths of the diamond to be set. Having temporarily secured the diamond in this hole, the setter ascertains the cutting point by trying it on a piece of glass; and when he has discovered it, he marks its position by making a slight notch in the wire with a file or otherwise, exactly opposite to the cutting point, as a guide to him in his operations when he comes to fix it permanently in the socket head of the handle. When doing this, care is taken to keep it exactly parallel with the inclined plane of the socket head.

The cutting point having been ascertained, and the diamond fixed into its place, the wire is then cut off about a quarter of an inch below the diamond, and filed down to fit exactly into the aperture in the socket head, into which it must be soldered. The rough or superfluous metal around the stone is removed with a file; and, lastly, the setting is polished with emery or sand-paper. Such is the most approved method of setting new diamonds, and it applies equally to the resetting of old ones. But in the latter case, the first process, that of detaching the stone from its bed, is accomplished either by means of a knife, or by applying the blow-pipe.

The art of managing the diamonds in glass-cutting, so as to produce effective results, can only be attained by considerable experience. The diamond must be held in a particular position, and with a particular inclination, otherwise it will not cut, and the slightest deviation from either renders an attempt to do so abortive. In the hands of an inexperienced person it merely scratches the glass, leaving a long rough furrow, but no fissure. The glazier judges by his ear of the cut made. When the cut is a clean and effective one, the diamond produces, in the act of being drawn along, a sharp, keen, and equal sound. When the cut is not a good one, this sound is harsh, grating, and irregular. On perceiving this, the operator alters the inclination and position of his diamond, until the proper sound is emitted, when he proceeds with his cut.

The diamonds employed in glass-cutting are of the description known by the technical name of *bort*, a classification which includes all such pieces as are too small to be cut, or are of a bad color, and consequently unfit for ornamental purposes. These are accordingly selected from the better sort, and sold separately, at an inferior price.

Though there are many substances that will scratch glass, the diamond was thought to be the only one that would cut it; but some experiments of Dr. Wollaston have shown that this is not strictly correct. That eminent philosopher gave to pieces of sapphire, ruby, spinel ruby, rock crystal, and some other substances, that peculiar curvilinear edge which forms the cutting point in the diamond, and in which, and in its hardness, its singular property of cutting entirely lies, and with these succeeded in cutting glass with a perfectly clear fissure. They lasted, however, but for a very short time, soon losing their edge, although prepared at a great expense of labor and care; while the diamond comes ready formed from the hand of nature, and will last for many years.

MANUFACTURE OF FLINT-GLASS OR CRYSTAL.

This branch may be defined the art of forming useful and ornamental articles of glass, and is the most ancient department of glass manufacture. The manipulatory processes have scarcely been varied and only slightly extended since the earliest times. The progress of chemistry has supplied purer materials but introduced few new ones. Thus we find that baryta has replaced the lead, and soda the potash in ancient glass, while in the production of colored glasses purer and additional metallic oxides are used. Yet this art has shown less tractability in the hands of the improver than perhaps any other industrial art.

The best flint-glass or crystal is composed of silica, potash, and lead, the average proportion being one-half sand, one-third red lead or litharge, one-sixth carbonate of potash, and a little saltpeter, manganese, and white arsenic to correct and improve the color or accidental impurities of the other materials. For inferior glass, or "tail metal" as it is technically called, soda is substituted for potash, and baryta for lead or litharge. In still cheaper "metal" for common small phials or bottles, a mixture approximating that for window-glass is used. For optical purposes the proportion of lead is increased to improve the refractive properties, which increase in proportion to the density of the medium. The specific gravity of the metals varies from about 3·6 to 2·5.

The furnaces employed are generally circular, and contain eight or ten pots. The "found," as the period of melting the materials is termed, commences generally on a Friday evening. The materials or " batch," and a portion of broken glass or " cullet" being mixed together are gradually introduced into the heated pot. The grate is in the center of the furnace, and there are flues at the back of the piers between the arches. As the batch melts there is a considerable evolution of gases, which at length subsides, when the metal begins to "fine" and reaches the " crisis." It is then cooled until about the consistency of thick honey. The evolution of the gases disperses air-bubbles through it; and the glass-maker endeavors so to regulate the heat of the furnace that the bubbles may rise to the surface, burst, and leave the metal plain and fine, but if the heat be continued beyond the crisis, the quality of the metal is deteriorated. For some time after the greater part of the gas has escaped, little bells or beads, technically called "seeds," rise and are extricated more freely by agitation or alteration of temperature. If the metal becomes solid while these bubbles are rising, it retains them, and if the " crisis " is not quickly passed, although the seed may be overcome by long-continued fusion, yet bad color and other defects arise. Strings and striæ, which upon close examination may be found in nearly all glass, are very common and troublesome. They may be caused by improper mixing of the materials, separation in the pot of metal of different densities, large grains of sand or pieces of refractory clay. But as strings and striæ in clear ice give pure water when melted, so in glass, mechanical rather than chemical means must be used for their prevention and cure. For optical glass Bontemps has carried out the recommendation of Faraday, and by systematically stirring the fluid glass has nearly reduced the manufacture of optical glass for large lenses to a certainty.

Crystal glass is popularly called colorless, but a practiced eye quickly detects color, which is more readily perceptible in the mass. It is probable that even pure silica, oxide of lead, and carbonate of potash will not produce colorless glass, but that there is a color proper to glass as there is to air and water. But the main causes of color in crystal are slight impurities, consisting of the oxides of iron or compounds of sulphur or carbon. A large excess of lead gives a yellow color —the oxides of iron, orange or olive-green tints, and compounds of sulphur or carbon, orange or blue. The peroxide of iron gives orange of a light tint, compared with the olive-green produced by the same quantity of the protoxide of iron. The addition of the black oxide of manganese or of saltpeter, produces purple, peroxidizes the oxide of iron, and, combined, forms what is called white, but practically an approach to black, and by a large dose of these materials glass of opaque blackness may be produced. Saltpeter also peroxidizes the iron, and heightens the color due to manganese. Purity of materials is essential to success, and oxide of manganese was formerly called glass-makers' soap; but although it reduces the color arising from iron, it does not annihilate it. Glass rendered colorless by manganese becomes pink by exposure to the direct rays of the sun, and if too much is used in the "batch" the metal is rendered pink, and is called high-colored. Glass with too little manganese has a "low color." The high color may be reduced by the deoxidizing agency of a pole of wood, with which, in such case, the metal is stirred. Some of the high color is lost in the annealing, and thick vessels remaining long in the "leir" or oven lose more than the light articles which are passed quickly through; therefore to obtain equality of color, the metal for thick goods must be highest colored. Arsenious acid is also employed as a corrective of color. Sulphur is a powerful agent in coloring glass. Sometimes a pot of metal foams while melting and is of a dark amber or orange color, which occasionally it retains when cool, or at other times changes to the light blue tint of the common soda-water bottle. Both tints are caused by the presence of sulphur, the orange by the larger quantity. One part of sulphur to two hundred of glass produces a dark color; hence, by adding a sulphuret to the melted metal the tints can be deepened at will. Splitgerber shows that glass containing one of sulphur in three hundred of glass becomes at a moderate low red heat nearly black and opaque, but becomes more transparent at a higher temperature. Similar changes are produced by heat on sulphur in its pure state. At its melting-point it is lemon yellow; at higher temperatures it becomes orange, and gradually deepens nearly black, and at a still stronger heat is volatilized in yellow vapors. Similar results are obtained with glass colored by gold, silver, and copper; glass colored by sulphur takes a deeper stain from silver than other glass, but if overheated becomes a light greenish-yellow on the reverse, and dark chestnut on the obverse, and is rendered useless. In Bohemia, glass consisting only of potash, silica, and lime is stained of a bright scarlet color by copper: the process is not followed in Britain, probably in consequence of British glass always containing lead or soda.

The metallic colors used for flint glass are cobalt for blue; chromium or a mixture of iron and copper for green; manganese for purple; copper for deep scarlet or light blue; gold for crimson; antimony or iron for yellow; uranium for topaz. Glass colored by the oxide of uranium exposed in a dark room to the dim light of the electric Aurora becomes translucent and illuminated throughout, and is partially so when exposed to the hydrogen flame. White enamel is obtained by the addition of the oxides of arsenic, tin, fluor spar, or phosphate of lime, and colored enamels are produced by adding the appropriate metallic oxides.

In the manipulation of the glass the men are arranged in sets of four, called chairs, and there are generally four chairs to a furnace. The principal workman of each chair is called the *gaffer*, the second the *servitor*, the third the *foot-maker*, and a boy completes the set. The wages of these men vary (in Great Britain) from 60s. to 20s. per week. The work is heavy, and requires such skill and dexterity that few first-class workmen are found. The men work in six-hour shifts, there being a complete double set. The first operation of the glass-blower is to skim the metal, as most impurities float on its surface, and this is done with an iron rod heated at its extremity and dipped into the metal, a little of which adheres. This is flattened on an iron plate and repeatedly introduced, gradually growing larger until it gathers and removes all the floating matter from the surface of the metal. The operation of making crystal articles then goes on as follows.

An iron tube is heated at the end and dipped into the semi-fluid metal, a portion of which is collected, withdrawn from the pot, and then rolled on an iron plate called the *marver*, until it has acquired a circular shape. The *marver* also equalizes the heat of the gathering, which the iron tube cools and stiffens, and which requires to be equally ductile in all its parts. The servitor now prepares a *post*, as a flattened round hot lump of metal on a punty or iron rod is called, and applies it to the end of the globe. The two masses of glass are thus united together, and that attached to the hollow tube is separated by touching it, near to where the tube enters the globe, with a small piece of iron wetted with water. By this means the glass cracks, and a smart blow on the iron tube completes the disunion. The workman now takes the punty from his assistant, and laying it on his chair arm, rolls it backward and forward with his left arm, while with his right he moulds it into the various shapes required, by means of a very few simple instruments. By one of these called a procello, the blades of which are attached by an elastic bow like a pair of sugar tongs, the dimensions of the vessel can be enlarged or contracted at pleasure. Any superfluous material is cut away by a pair of scissors. For smoothing and equalizing the sides of the vessel a piece of wood is used. After the article is finished it is detached from the punty and carried on a pronged stick to the annealing oven or leir.

For a fluted or ribbed *cane*, as a solid glass rod is technically called, the metal is forced into a mould of the requisite shape and then withdrawn; after which, if attached to another *post* and the two punties be twisted and drawn in opposite directions, the ribs become spiral lines, which become more acute as the drawing is extended. Venetian filigree work is produced in this way; and if in the hollow flutes of the mould colored glass or enamels are inserted, and the gathering introduced, the colored glass or enamels are welded to and withdrawn with it. When again heated, and twisted or drawn, these streaks of color or enamel become spiral, and ornament the surface. If before being drawn the mass be redipped into the pot of crystal glass and then twisted, the spiral lines of color or enamel become internal. By the repetition of this process spirals can be formed within spirals, and by placing these filigree *canes* side by side and welding them together, very curious and intricate patterns are obtained. By the ordinary process of blowing, vessels are formed with smooth and concentric interior and exterior surfaces, and do not exhibit the brilliancy of the crystal so much as when it has numerous inequalities. The most brilliant effect is produced by cutting, but moulding is much cheaper, and this branch of the art has now reached a high state of excellence. The moulds are generally of iron highly polished, and kept a little below a red heat. The surface of the metal is injured by contact with the mould, but its transparency is restored by being reheated. A very exact regulation of the temperature is necessary in reheating fine mouldings; too little heat does not give the "fire polish," too much softens the metal and obliterates the mouldings. The moulds for pressed goods are made in pieces so hinged or connected as to close and leave a vacuity, the form of the article required, the hollow in which is not however produced by blowing but by the plunger of the press under which the mould is placed. The required quantity of metal is then dropt in, when the plunger descends and forces it into all parts of the cavity, completing the formation of the article, which is then stuck to a punty, and fire-polished and annealed.

What is called cased glass is crystal covered with coats of colored glass. It is thus obtained. The gathering of crystal is thrust into a colored or enameled shell, which is previously prepared. The welding is completed by reheating; and two or more coats of different colors or enamels may thus be employed. When cut through to the crystal in various figures, the edges of the different colors on enamel are seen.

The Venetian frosted glass is obtained by immersing the hot metal gathering in cold water, quickly withdrawing it, reheating and expanding it by blowing, before it becomes so hot as to weld together the numerous cracks on the surface caused by the cold water. These cracks only penetrate where the metal has been cooled by the water, and remain as depressions until the article is finished.

Venetian vitro-di-trono consists of spiral lines of enamels or colors, crossing each other diamond-wise, in the body of the glass, and inclosing an air-bubble in the center of each diamond. It is thus formed: a gathering is blown in the mould with the necessary canes twisted and blown out as formerly described for spiral filigree, the canes being left projecting from the outside like ribs or flutes. A similar piece is made and turned inside out. The projecting canes on this piece are on the inside, and the spiral lines reversed. The one piece is then placed under the other, and both are welded together. The ribs or flutes projecting from the two surfaces in contact inclose air in the diamonds, which gradually assumes the bubble shape. The vessel is then formed in the ordinary manner. The most beautiful regularity of lines is thus obtained; and when the ends are closed by the procellos, the lines are drawn to a center as regularly arranged as if they had been turned in an engine.

Incrustations are formed by placing the substance to be incased on the surface of the article and dropping melted metal on it, or by preparing an open tube of glass, inserting the object, and welding the open end. By suction instead of blowing, the metal is collapsed on the object, and the air withdrawn. From the unequal contraction between the object and the crystal by which it is surrounded there is much difficulty in the annealing, and to avoid the risk of breakage the object should be made of materials expanding and contracting like the glass itself.

The round, heavy paper weights containing various ornaments apparently in the body of the metal are made as follows: Canes are made to the required pattern—say, for example, a star within a tube. A gathering of white enamel is formed in a star-shaped mould, and coated with crystal. After this is marvered, it is dipped into a colored enamel, and drawn out into a cane; and if this is covered with crystal, the eye cannot detect the junction of the external crystal with that of the cane, but the enamel casing will appear as a tube with the star standing in the center. Devices of numerous kinds are thus made in canes, and then welded together. The end is then ground, and after being heated and incased in crystal, the lens-like shape of the paper weights adds to the effect by magnifying the incrusted canes.

The light-refracting properties of crystal are best shown by cutting and polishing. Stones of various textures, or wood, sand, or emery, in water, are used with the metal mills, water only with the stones, and pumice-stone and putty-powder with the wood for smoothing and polishing. The articles are held in the hand, and applied to the mill while rotating. The punty marks are ground off tumblers, wine-glasses, and such like, by boys holding them on small stone mills. Ground or frosted glass is made by rubbing the surface with sand and water. Iron tools fixed on a lathe and moistened with sand and water are used to rough out the stoppers and necks of bottles, which are completed by hand with emery and water. The neighborhood of the coal-fields is of course the chief seat of the manufacture, and probably the best crystal of Great Britain is now made in Manchester.

BOTTLE-GLASS.

The common green or bottle-glass is made of the coarsest materials; sand, lime, sometimes clay, any kind of alkali or alkaline ashes, whose cheapness may recommend it to the manufacturer, and sometimes the vitreous slag produced from the fusion of iron ore. The mixture most commonly used is soap maker's waste, in the proportion of three measures to one measure of sand. The green color of this glass is occasioned by the existence of a portion of iron in the sand, and, it may be, also in the vegetable ashes of which it is composed.

When castor-oil or champagne bottles are wanted, a portion of crown-glass cullet is added, to improve the color. The impurity of the alkali, and the abundance of fluxing materials of an earthy nature, combined with the intense heat to which they are subjected, occasion the existence of but a very small proportion of real saline matter in the glass, and thereby render it better than flint-glass for holding fluids possessing corrosive properties.

The soap-maker's waste is generally calcined in two coarse arches, which are kept at a strong red heat from twenty-four to thirty hours, the time required to melt the materials and work them into glass, which is termed a journey. After the soap-maker's waste is taken out of the arch, it is ground and mixed with sand in the proportions already mentioned. This mixture is put into the fine arches, and again calcined during the working journey, which occupies ten or twelve hours more. When the journey is over, the pots are again filled with the red-hot materials out of the fine calcining arch. Six hours are required to melt this additional quantity of materials. The pots are again filled up, and in about four hours this filling is also melted. The furnace is then kept at the highest possible degree of heat, and in the course of from twelve to sixteen hours, according as the experience of the founder may determine, the materials in the pots are formed into a liquid glass fit for making bottles. The furnace is now checked by closing the doors of the cave, and the metal cooling, it becomes more dense, and all the extraneous matter not formed into glass floats upon the top. Before beginning to work, this is skimmed off in the way already described in our account of crown-glass making. A sufficient quantity of coals is added at intervals, to keep the furnace at a working heat till the journey is finished.

After the pots have been skimmed the person who begins the work is the gatherer, who, after heating the pipe, gathers on it a small quantity of metal. After allowing this to cool a little, he again gathers such a quantity as he conceives to be sufficient to make a bottle. This is then handed to the blower, who, while blowing through the tube, rolls the metal upon a stone, at the same time turning the neck of the bottle. He then puts the metal into a brass or cast-iron mould of the shape of the bottle wanted, and, continuing to blow through the tube, brings it to the desired form. The patent mould now in use is made of brass, the inside finely polished, divided into two pieces, which the workman, by pressing a spring with his foot, opens and shuts at pleasure. The blower then hands it to the finisher, who touches the neck of the bottle with a small piece of iron dipped in water, which cuts it completely off from the pipe. He next attaches the punty, which is a little metal gathering from the pot, to the bottom of the bottle, and thereby gives it the shape which it usually presents. This punty may be used for from eighteen to twenty-four dozen of bottles. It is occasionally dipped into sand to prevent its adhering to the bottle. The finisher then warms the bottle at the furnace, and taking a small quantity of metal on what is termed a ring iron, he turns it once round the mouth, forming the ring seen at the mouth of bottles. He then employs the shears to give shape to the neck. One of the blades of the shears has a piece of brass in the center, tapered like a common cork, which forms the inside mouth; to the other is attached a piece of brass, used to form the ring. The bottle is then lifted by the neck on a fork by a little fellow about ten years of age, and carried to the annealing arch, where the bottles are placed in bins, above one another. This arch is kept a little below melting heat, till the whole quantity, which amounts to ten or twelve gross in each arch, is deposited, when the fire is allowed to die out.

HISTORY

AND

PROCESS OF MAKING GAS-LIGHT.

GAS-LIGHT.

Light, whether obtained from natural or artificial sources, is so necessary for the correct and successful execution of almost every operation of human industry, that whatever is calculated to simplify the means of procuring it, or to increase its intensity, cannot fail to be attended with the most beneficial consequences to civilized society. For every purpose to which it is applied, it must be admitted that the light of day, when it can be enjoyed freely and without interruption, is by far the most suitable; but in large and crowded cities, as well as in situations less favorable in point of climate, where the sun is sometimes shrouded for days together in dense and impenetrable clouds, it becomes expedient to compensate for the absence of his rays by artificial substitutes, which, however inferior in brilliancy and general usefulness, may nevertheless answer sufficiently well in those cases where a less ample supply of light is requisite.

Some substances, denominated *phosphorescent*, have the property of absorbing the solar rays, on being exposed for a short time to their influence, and of emitting the light which they thus imbibe when they are afterward placed in the dark; but the feeble and transient illumination which they shed, though sufficient to indicate their luminous condition, is totally unfit to afford such a supply of light as is necessary for conducting any of the operations of art, which require care and precision for their performance.

There are, however, a variety of inflammable substances, both of animal and vegetable origin, which, during the process of combustion, give out light as well as heat; and hence, from the earliest periods of human society, it has been customary to burn substances of that description for the purpose of obtaining artificial light. These substances, which were generally of a fatty or oleaginous nature, are composed chiefly of carbon and hydrogen. When they are exposed to a certain high temperature, they are resolved into some of the compound gases which result from the union of these elements, particularly carburetted, and bi-carburetted hydrogen, or olefiant gas, both of which are highly inflammable, and yield, during their combustion, a fine white light. In order to facilitate the decomposition, and to carry on the combustion with due economy, a quantity of some fibrous substance, in the form of a wick, is connected with the oleaginous matter, for the purpose of causing it to burn slowly and effectually. Accordingly, if the flame be suddenly extinguished, the inflammable gas which is formed by the decomposition of the matter in immediate contact with the wick is observed to escape from it, and may be again set on fire by the application of a lighted taper.

When it is required to convey from place to place the light obtained from these substances, no arrangement is found to be more convenient for their decomposition than that which is effected by means of the wick; but if the light is to remain in a permanent position, it will frequently be more advantageous to resolve the oleaginous matter into gas, and then to transmit it, in that state, through pipes, to the various points where it is to be consumed.

Although the different substances which have been used from the earliest times for yielding artificial light have always been actually resolved into gas before they underwent the process of combustion, that fact was entirely unknown until pneumatic chemistry unfolded the properties of the aërial bodies, which perform so many important functions in the economy of nature, as well as in the processes of the arts. It was then discovered that hydrogen, one of the component parts of water, was a highly inflammable gas, capable of being produced under a great variety of circumstances: from vegetable matter decaying in stagnant water, forming what is called light carburetted hydrogen, a stream of which, when ignited, produces the natural phenomenon known as "ignis-fatuus, or Will-o'-the-Wisp:" from coal, oils, and fatty substances, when, in combination with larger proportions of carbon, it forms gases of high illuminating powers. The use of gas for the purpose of illumination is therefore of recent date; but although late in its origin, the successive improvements which the invention has received, and continues to receive, from the joint labors of chemists and practical engineers, have tended greatly to simplify the processes for producing the gas, and for improving its quality and means of distribution.

In many parts of the world there are certain deposits of petroleum or naphtha which furnish gaseous matter; and this issuing from some fissure in the earth, becomes ignited by lightning or some other cause, and continues to burn for a long period. Such a flame is regarded by an ignorant people with superstitious reverence, and has been sufficient to found a religious sect of fire-worshipers. Deposits of coal, or of bituminous schist, sometimes furnish the gaseous matter for such flames. The practical Chinese, about thirty miles from Pekin, are said to make use of this gas in the boiling and evaporating of salt brine, and for lighting their streets and houses.* "Burning fountains," as they are sometimes called, are not uncommon, and their origin is the same. In 1851, in boring for water on Chat Moss, on the line of railway between Manchester and Liverpool, a stream of gas suddenly issued up the bore, floated along the surface of the ground, and caught fire on the application of flame. A pipe was inserted into the bore, and a flame eight or nine feet long was thus produced.

In 1667, Mr. Shirley describes in the Philosophical Transactions of the Royal Society a burning spring in the coal district of Wigan in Lancashire: he traced its origin to the underlying beds of coal. In 1726, Dr. Hales, in his work on *Vegetable Statics*, gives an experiment on the distillation of coal, by which it appears that 158 grains of Newcastle coal yielded 180 cubic inches of inflammable air. In 1733, Sir James Lowther sent to the Royal Society specimens of inflammable air from a coal-mine near Whitehaven. The gas was collected in bladders, and a number of experiments were tried on it.

It appears, however, that the Rev. John Clayton had performed some experiments on the distillation of coal some years previous to the publication of Dr. Hales's book; but

---

* So do the practical Americans.

he did not publish an account of them until 1739, and this account consists of an extract from a letter written by Clayton to the Hon. Robt. Boyle, who died in 1691, and was probably written sometime before this year. It is inserted in the Transactions of the Royal Society for the year 1739; and is probably the earliest evidence of the possibility of extracting from coal, by means of heat, a permanently elastic fluid of an inflammable nature. We shall therefore give the account of the discovery in his own words: Having introduced a quantity of coal into a retort, and placed it over an open fire, he states that " at first there came over only phlegm, afterward a black oil, and then likewise a spirit arose which I could noways condense; but it forced my lute, or broke my glasses. Once when it had forced my lute, coming close thereto in order to try to repair it, I observed that the spirit which issued out caught fire at the flame of the candle, and continued burning with violence as it issued out in a stream, which I blew out and lighted again alternately for several times. I then had a mind to try if I could save any of this spirit; in order to which I took a turbinated receiver, and putting a candle to the pipe of the receiver whilst the spirit arose, I observed that it catched flame, and continued burning at the end of the pipe, though you could not discern what fed the flame. I then blew it out and lighted it again several times; after which I fixed a bladder, squeezed and void of air, to the pipe of the receiver. The oil and phlegm descended into the receiver, but the spirit still ascending, blew up the bladder. I then filled a good many bladders therewith, and might have filled an inconceivable number more; for the spirit continued to rise for several hours, and filled the bladders almost as fast as a man could have blown them with his mouth; and yet the quantity of coals I distilled were inconsiderable.

" I kept this spirit in the bladders a considerable time, and endeavored several ways to condense it, but in vain. And when I had a mind to divert strangers or friends, I have frequently taken one of these bladders, and pricking a hole therein with a pin, and compressing gently the bladder near the flame of a candle till it once took fire, it would then continue flaming till all the spirit was compressed out of the bladder; which was the more surprising because no one could descern any difference in the appearance between these bladders and those which are filled with common air." *

It is evident from this narrative, related with so much simplicity, that an accident which happened to Mr. Clayton's apparatus was the means of leading to the discovery of coal-gas; but it does not appear that he or any other individual thought of applying the discovery to any practical purpose until the year 1792, when Mr. Murdock, who then resided at Redruth, in Cornwall, commenced a series of experiments upon the properties of the gases contained in different substances. In the course of his researches he found that the gas obtained by the distillation from coal, peat, wood, and other inflammable substances, yielded a fine bright light during its combustion; and it occured to him, that by confining it in proper vessels and afterward expelling it through pipes, it might be employed as a convenient and economical substiute for lamps and candles.

Mr. Murdoch's attention to the subject having been interrupted for some time by his professional avocations, he resumed the consideration of it in 1797, when he exhibited publicly the results of his more mature plans for the preparation of coal-gas. The following year (being then connected with Messrs. Boulton and Watt's engineering workshop), he constructed an apparatus at the Soho foundery for lighting that establishment, with suitable arrangements for the purification of the gas; and these experiments, Dr. Henry states, " were continued with occasional interruptions until the epoch of the peace in 1802, when the illumination of the Soho manufactory afforded an opportunity of making a public display of the new lights; and they were made to constitute a principal feature in that exhibition."

In this brief sketch of the progress of gas-lighting, it may be noticed that the Lyceum theater in London was lighted with gas in the course of the years 1803-4, under the direction of Mr. Winsor, who is entitled to no small commendation for the warm interest which he took in drawing the public attention to the subject; and in 1804-5 Mr. Murdoch had an opportunity of carrying his plans into effect on a still larger scale, by means of the apparatus erected under his superintendence, in the ex-

---

* Mr. Clayton also alludes to the discovery of the gas which he obtained from coal, in a letter to the Royal Society, dated May 12, 1688.

tensive cotton mills of Messrs. Philips and Son of Manchester.

It has been alleged that gas-lights were used in France before they were known in this country; but as the earliest exhibition of these lights, on which the claim of priority of discovery is founded, took place at Paris in 1802, it is evident, from the foregoing statements, that the exhibition alluded to was ten years subsequent to the first experiments of Mr. Murdoch on the subject.

The practicability of lighting by means of coal-gas having been demonstrated by Mr. Murdoch, a number of scientific men applied their talents to the further development of the art. Dr. Henry, the celebrated chemist, lectured on the subject in 1804 and 1805, and furnished many hints for the improvement of the manufacture. Mr. Clegg, an engineer in the employment of Boulton and Watt, was a worthy successor of Murdoch, and for many years was the most eminent gas-engineer of this country. A good deal of the machinery of the gas-house in its present form was contrived by Mr. Clegg, and to him also we are indebted for the ingenious wet gas-meter. In 1813 Westminster bridge was first lighted with gas, and in the following year the streets of Westminster were thus lighted, and in 1816 gas became common in London. So rapid was the progress of this new mode of illumination, that in the course of a few years after it was first introduced, it was adopted by all the principal towns in the kingdom, for lighting streets as well as shops and public edifices. In private houses it found its way more slowly, partly from an apprehension, not entirely groundless, of the danger attending the use of it; and partly, from the annoyance which was experienced in many cases, through the careless and imperfect manner in which the service-pipes were at first fitted up. These inconveniences have been in a great measure, if not wholly, removed by a more enlarged knowledge of the management of gas; and at present there are few private houses in large towns which are not either partially or entirely lighted up by it. As the demand for gas increased, various improvements were from time to time introduced both in the mechanical arrangements, and in the chemical operations of the manufacture. The rapid increase in the population of the metropolis, and of all large towns, has naturally led to an increased consumption of gas, and the application of gas to the purposes of warming and cooking has also further increased the demand for it. Hence it has been not only necessary that new gas-works should be erected, for the supply of new districts, but that the resources of old works should be enlarged. It is only a few years ago that a gas-holder capable of storing 250,000 cubic feet of gas was regarded as of enormous size; at the present time, gas-holders are made of double that capacity, and we occasionally hear of them of the capacity of upward of a million cubic feet. There is one such at Philadelphia; it is 140 feet in diameter and 70 feet in height. Nor will such dimensions as these be regarded as superfluous when it is stated that some of the large metropolitan works send out each from a million to a million and a half cubic feet of gas in one night in mid-winter.* The Westminster gas-works alone are accustomed to supply as much as five millions cubic feet of gas in one night from their three stations.

Of the Site and general Arrangement of the Apparatus for the Production and Purification of Coal-Gas.

In describing the site and general arrangements of a gas establishment, it is not easy to give directions respecting points which must be regulated in every case by circumstances of a local nature; but when a choice of ground is in our power, a spot ought to be selected having a central situation

---

* A few years ago Mr. Hume, in the House of Commons, moved for a return, which has been published under the following title:—"Return or statement from every gas company established by act of parliament in the United Kingdom, stating the several acts of parliament under which established, the rates per 1000 cubic feet at which each company or corporation have supplied gas in each of the three years since 1846 to 1849, and the average prices of the coals used by the company in each year for the same period; also stating the amount of fixed capital invested by each gas company, and the rate per cent. of dividend to the shareholders or proprietors on their shares in each year since that date (in continuation of Parliamentary Paper No. 734 of Session 1847)." It appears from this document that the fifteen companies in London charged at the rate of 6s. per 1000 cubic feet of gas, with the exception of the City of London Company, which charged only 4s. with coals at 15s. 9d. per ton. The highest rate is 10s. per 1000 feet, charged at Inverness, with coals at 24s. 4d. per ton. Bury St. Edmunds charged 8s. 4d., with coals at 10s. 6d. Birmingham has rates of 6s. and 3s. 4½d., with coals at 15s. per ton.

with regard to the buildings, streets, etc., which are to be supplied with light, and standing as nearly as possible on a medium level with them. When the manufactory is placed considerably below that level, the gas is apt to be propelled with too much velocity through the burners; and when above it, an opposite inconvenience is experienced, the gas being in that case necessarily subjected to an extra pressure, by which the chance of its escape through any imperfection of the pipes is proportionally increased. Of the two evils, therefore, the least objectionable is that in which the situation of a gas-work is below the mean level of the streets.

But besides the conditions favorable to an equable and uniform distribution of the gas at the different points to which it may be conducted, there are other considerations scarcely less important, which in selecting a proper site for the erection of the establishment ought not to be disregarded. Among these may be reckoned a regular supply of water for the various manipulations of the work; and facility of access for the delivery of coal and the removal of the coke, tar and other products of the distillation. Railways are now so common that they are often as valuable to a gas-work as the vicinity of navigable water. In the Central Gas-consumer's works at Bow Common, which were laid out under the skillful scientific direction of Mr. Croll, a branch railway is connected with the lines which supply the coal, and is actually continued into the retort-house, so that the coal wagons only arrive at their final destination at the mouths of the retorts which are to be fed. But in fixing the situation of an establishment which is professedly erected for the public benefit, the comfort or the interest of individuals ought not to be entirely overlooked; for although a gas-work may not prove, under proper management, a nuisance, it can never be considered to be any advantage to the neighborhood in which it is placed.

The apparatus for the production and purification of coal-gas consists, in the first place, of suitable vessels for decomposing by heat the coal from which the gas is to be procured; secondly, of a series of pipes for conveying off the gas, and conducting it into proper receptacles, where it may be separated from the grosser products, which tend to impair the brilliancy of the light; thirdly, of the condensing apparatus, for removing more effectually the tar and other condensable substances that come over with the gas; fourthly, of the purifying apparatus, for abstracting the sulphuretted hydrogen, carbonic acid, etc., by which the gas is contaminated, and which if allowed to remain, would be injurious to the gas-fittings, to the books and furniture of rooms, or to the health of the consumer; and, fifthly, of the gasometer or gas-holder, with its tank, into which the gas is finally received in a purified state.

Of the Retorts, or Vessels for decomposing the Coal.

The vessels employed for the decomposition of coal and other substances capable of yielding carburetted hydrogen, by their destructive distillation, are formed of cast-iron, of clay, of brick, or of wrought iron, and are termed *retorts*. Various shapes have been adopted in the construction of these vessels; nor have their forms been more varied than the modes in which they have been disposed in the furnaces erected for their reception. In many instances they have been constructed of a cylindrical shape, varying in length and diameter. Those first employed were placed with their axis in a vertical direction; but experience soon showed that this position was extremely inconvenient, on account of the difficulty which it occasioned in removing the coke, and other residuary matters, after the coal had been carbonized. Attempts were made to remedy this inconvenience, by enlarging the size of the retort, and introducing the coal inclosed in a proper grating of iron, having the form of a cage. The increased dimensions of the retort, from which the principal advantage to be derived from this arrangement was expected, were found, however, to present great obstacles to the complete carbonization of the coal; for although the disengagement of gas during the first stages of the process was sufficiently copious, it diminished rapidly the longer the distillation was continued, in consequence of a crust of coke being formed next to the heated metal, which not only opposed the transmission of the heat to the internal mass of coal, but gradually prevented, by its accumulation, the extrication of the gas from the undecomposed portion of it.

The retorts were, therefore, next placed in a horizontal position, as being not only more favorable to the most economical distribution of the heat, but better adapted to the introduction of the coal, and the subsequent removal of the coke, after the carbonization was carried to a due extent. At first the heat was applied directly to the lower part of the retort, but it was soon observed that the high temperature to which it was necessary to expose it, for the perfect decomposition of the coal, proved destructive to it, and rendered it useless long before the upper part had sustained much injury. The next improvement was, accordingly, to interpose an arch of brickwork between it and the furnace, and to compensate for the diminished intensity of the heat by a more diffused distribution of it over the surface of the retort. This was effected by causing the flue of the furnace to return toward the mouth of the retort, and again conducting it in an opposite direction, till the heated air finally escaped into the chimney.

This arrangement continued for a long time in use, and seemed to admit of little improvement, unless with respect to the shape and dimensions of the retorts. The cylindrical form has the advantage of possessing great durability, but it is not so well fitted for rapid decomposition of the coal (on which depends much of the good qualities of the gas) as the elliptical shape. Flat-bottomed or D-shaped retorts have also been long in use: the small London D is about 12 inches wide by $12\frac{1}{2}$ inches deep, while the York D varies from 20 to 30 inches in width, and from 9 to 14 inches in height. Retorts are also made of a rectangular section, with the corners rounded and the roof arched. Elliptical retorts are varied into what are called *ear-shaped* or *kidney-shaped*, and it is not unusual to set retorts of different forms in the same bench, for the convenience of filling up the branches of the arch which incloses them. The length of retorts formerly varied from 6 to 9 feet; they are now in some cases made $19\frac{1}{2}$ feet in length and $12\frac{1}{2}$ inches in internal diameter, and are charged at both ends.

Iron retorts of from 6 to 9 feet in length carry a charge of from 120 to 200 lbs. of coal, which is usually renewed every six hours. Instead of the old method of charging with the shovel, which occupies at least half an hour, and entailed a great loss of gas, the whole charge is now deposited in an iron scoop, with a cross handle at the end, and it is lifted by three men, pushed into the retort, turned over, and the whole charge deposited at once, a contrivance which does not occupy more than 30 or 40 seconds. Indeed it is not uncommon for a bench of 7 retorts to be emptied and recharged in the brief space of 20 minutes. When square-backed retorts are used, the backs are apt to wear much more quickly than any other part, in consequence of the fierce heat which plays upon them; it is therefore sometimes usual to throw in a few shovelfuls of coal to the extreme end before depositing the charge with the scoop. This occupies more time in charging, but it has the effect of preserving the backs. The objection does not apply to retorts with circular ends.

Every retort is furnished with a separate mouth-piece, usually of cast-iron, with a socket for receiving the stand-pipe, and there is a movable lid attached to the mouth, together with an ear-box cast on each side of the retort for receiving the ears which support the lid. The ears hold a crossbar through which is passed a screw which presses on the lid, and secures it to the mouth-piece. That part of the lid which comes in contact with the edge of the mouth-piece has applied to it a lute of lime mortar and fire-clay, and when the lid is screwed up, a portion of this lute oozes out round the edges and forms a gas-tight joint.

In some cases the screw is got rid of by a more expeditious contrivance, in which the ears support an axis, which carries a lever formed at one end into a sort of cam, and bearing at the other end a ball of cast-iron about 4 inches in diameter. On lowering this ball the cam presses with great force against the back of the lid, and holds it securely; and if more force be required, a weight can be attached to the iron ball.

In attaching a mouth-piece to a clay retort, the end is notched with grooves for the purpose of holding the binding cement more securely. The mouth-piece is attached by means of bolts with T heads let into the body of the retort Iron cement is used, in which fire-clay takes the place of sulphur; this being spread over the joint, the mouth-piece is attached and screwed up.

The temperature best suited for the production of gas from coal, being what the workmen term a *bright red*, was found to be very destructive to the retorts when they were exposed to the direct action of the fuel; and accordingly means were employed to protect them from the rapid oxidation which they suffered under these circumstances, by interposing between them and the furnace a partition of fire tiles or arched bricks, with side flues for the admission of the heated air.

With the view of occupying less room, and saving the expense of fuel, several retorts are sometimes set together in an oven of brickwork, and heated by a smaller number of furnaces than there are retorts. By this arrangement the fuel is certainly economized, but the plan is liable to the objection, that when any one of the retorts is worn out, those connected with it cannot be used till the faulty one is replaced; and though various expedients have been proposed for obviating that inconvenience, none of them can be said to have effectually answered the purpose.

The fuel required for carbonizing a given quantity of coal may be stated to be, in general, from one-third to one-fourth of its weight for Newcastle coal. It is stated, that under Mr. Croll's method of setting, the carbonization is carried on by the combustion of only 12 per cent. of fuel, or that 100 tons of coal are carbonized by 12 tons of coke.

Various attempts have been made to render the retorts more independent of the laborers. In Mr. Brunton's retort, a hopper containing the charge of coal is attached to the mouth-piece. The charge is introduced by removing a slide, and a piston is then advanced for the purpose of pushing forward the coal, and ejecting the coke, the latter falling through a shoot at the further end of the retort, and thence into a cistern of water into which the lower end of the shoot dips. This retort is not of equal section throughout: it is 15 inches in diameter at the mouth, and 21 inches at the other end, the length being $4\frac{1}{2}$ feet. The advantages of this arrangement, independently of the saving of labor, are said to be an increased production of gas, and a consequent diminution in the amount of tar, naptha, and ammoniacal liquor, this diminution being stated at 50 per cent. less than the ordinary yield of those secondary products. Moreover, a good deal of bituminous vapor, and minutely divided carbon, which, under the usual arrangement, go to swell the increase of tar, become decomposed under the higher temperature of Mr. Brunton's retorts by passing over the red-hot coke, and forming illuminating gas. Indeed, it is now generally admitted as an axiom in gas-making, that the most productive yield of gas is under a high temperature; for it is possible under low heats to distil off the volatile parts of the coal as bituminous vapor only, without any production of carburetted hydrogen gas. By exposing the coal in a thin layer to a very high heat, the distillation is effected most rapidly and most profitably. Mr. Clegg describes a retort into which the coal is introduced by means of an endless web formed of iron plates, each 2 feet long, and 14 inches wide, and linked together by iron rods. The coal, broken small, is placed in a hopper, to which is attached a feeder with six radial projections. Each of the six partitions thus formed supplies sufficient coal to cover one plate of the web, with about 120 cubic inches of coal to the depth of $\frac{3}{8}$ths of an inch. The hopper, which contains 24 hours' charge of coal, is luted after each charge. The endless web is moved by passing over drums, one revolution of which every 15 minutes conveys the web through the retort, and effects the distillation of the coal. The coal is carried on the upper surface of the web, and as the web turns over the second drum the coke is discharged by a pipe into a vessel below, and the empty portion of the web returns to the hopper, and passing over the surface of the first drum receives another charge. The charge is so regulated, that about 100 square inches of heated surface in the retort is allowed for every pound of coal, which is said to yield 5·36 cubic feet of gas, or 12,000 cubic feet per ton of Wallsend coal. The charge for each retort is about 18 cwt. of coal for 24 hours, or about double the quantity under the old plan in retorts of similar dimensions. The coke is also said to be in much greater quantity. In the course of time the plates of the iron web become converted into steel, the value of which is sufficient for the purchase of a new web. Mr. Lowe has also introduced an arrangement for increasing the yield of gas by making the products of a new change pass over the portion of the retort which is already at a red heat. For this purpose the *reciprocating retort*, as it is called, is made of thrice the usual length, and is charged at both ends; but the dip pipe at one end is made to enter to a greater depth into the tar of the hydraulic main than at the other end; so that supposing both the dip pipes to be open, the products of distillation will of course be discharged into the main by the shorter pipe, where there is less pressure to be overcome. This pipe, however, is furnished with a cup-valve, which can be closed at pleasure; and when so closed, the products of distillation must escape by the longer dip-pipe. When the charge has been half worked off in one-half of the retort, a fresh charge is introduced into the other half, and the products of distillation of the new charge are made to pass over the incandescent coal, or that which has been about three or four hours under distillation. This is readily effected by closing or opening the shorter dip-pipe, according to the end of the retort last charged. The principle of the reciprocating retort has been adopted at different works, with variations in the practical details.

Of late years clay retorts have been largely introduced into gas works, and they are said to be more durable, and to stand a higher temperature than iron retorts, the latter working best at a cherry-red heat, and the former at a white heat, which is more favorable to the increased production of gas than the lower temperature. It is stated, that where a clay retort has yielded a million and a half cubic feet of gas, an iron one has furnished only 800,000 cubic feet. Clay retorts appear, from their greater porosity to leak more than iron ones; but after working some months, the pores become clogged with carbon, and the porosity is thus greatly diminished, and the leakage is even less than in iron retorts working under the same pressure.* As the demand for clay retorts increased, the manufacture of them improved, an example of which improvement is well illustrated in the case of the retorts of the Great Exhibition of 1851, exhibited by Messrs. Cowen of Blaydon Burn, near Newcastle-on-Tyne. When this firm first manufactured retorts about twenty years

---

* One of the greatest sources of loss in the manufacture of gas arises from the leakage, not only of the retorts and other apparatus within the works, but also of the mains, a loss amounting to from 10 to 30 and upward per cent. Mr. Croll estimates the loss at one-sixth of the gas sent out. The porosity of cast-iron pipes, not at their joints merely, but throughout their whole length, is evident from the saturation of the soil with gas in the immediate vicinity of the mains. Not only does the gas escape by exosmose into the air, but by the reverse process of endosmose, air enters the pipes in some cases, as Prof. Graham has found, to the extent of 25 per cent. Prof. Brande thinks that the fetid odor of the soil in contact with the gas mains is due to the exosmose of ammonia, rather than of tar and naptha, to which the ill odor is generally attributed.

ago, each retort was made in ten pieces, which number was reduced to four, then to three, and then to two; and in 1844 the retort was made complete in one piece of the dimensions of 10 feet in length, and 3 feet in internal width. The clay of which these retorts are manufactured is exposed to the weather for some years, and is frequently turned over, and the fragments of fossils picked out, by which means most of the iron is got rid of, which in other fire-clays is so injurious. Some of these retorts are stated to have continued in active use for 38 months, thus exhibiting four times the durability of iron ones.

Brick retorts, or rather *ovens*, have also been introduced, and are said to be very durable, and to work satisfactorily. In one case the charge is 5 or 6 cwt. of coal every twelve hours, and the yield 9000 cubic feet of gas for one ton of Welsh coal, and from 10,000 to 12,000 cubic feet from one ton of Newcastle coal. The fuel required for the carbonization of the coal is said to be unusually large. Wrought-iron retorts, made of thick boiler plates firmly riveted together, have also been tried to a limited extent.

When clay retorts came into general use, the circumstance that they required a much higher heat than iron retorts suggested the economical plan of heating the clay retorts by the direct action of the furnace, and arranging the iron retorts in a separate oven, heated by the same furnace, or within a system of return flues, where they would be submitted to a less intense heat. By this means Mr. Croll has found, that with two furnace grates of 252 square inches in each, he has been able to carbonize in 24 hours five tons of coal in the clay retorts of one bench, and three tons and a half in the iron retorts of the same bench, with such an economy of fuel, that only twelve per cent. of all the coke made is required for the furnaces; whereas, in most of the London works, nearly one-third of the coke made is consumed in heating the retorts.

The quantity of gas produced during the time the coal is undergoing decomposition is extremely variable. From a small retort, exposed for eighty-five minutes to a bright red heat, which was kept up with the utmost possible uniformity, the following results were obtained from eight pounds of the Wemyss coal:

	Cub. Ft.	Cub. In.
In 1st ten minutes	6	235
2d do	8	980
3d do	8	1254
4th do	5	784
5th do	4	1450
6th do	3	313
Last twenty-five minutes	6	1660
	43	1492

At the time the process was terminated the extraction of aëriform matter had nearly ceased, so that the quantity of gas yielded by a pound of the coal was about five and a half cubic feet. The same coal carbonized on the large scale yielded when the process was carried on for four hours, at the rate of four and one-third cubic feet of gas per pound. The weight of the coke in the above experiment was 32,050 grains; and as the weight of the gas, the specific gravity of which was ·65, must have been 15,026 grains, the tar and other residuary products, including the sulphuretted hydrogen abstracted by the process of purification, must have amounted to 8924 grains.

When the decomposition is effected on the large scale, the quantity of gas is found to vary with the quality of the coal, and the manner in which the operation is conducted. According to Mr. Peckston, a chaldron of Newcastle Wallsend coal yields 10,000 cubic feet, being at the rate of $370\frac{1}{3}$ cubic feet per cwt. The different kinds of Newcastle coal yield from 8000 to 12,500 cubic feet of gas per ton; the parrot or cannel coals furnish from 9000 to 15,000 feet per ton, the last named quantity being obtained from the Boghead cannel, in which case the specific gravity of the gas is ·752, and as much as 866 avoirdupois lbs. of gas are obtained from each ton of coal. The Wallsend Newcastle, known as Berwick and Craister's, only yields 449 lbs., and of the specific gravity ·470. Of the Derbyshire, Staffordshire, Welsh, and other varieties of coal, the yield varies from 6500 to about 11,000 cubic feet of gas per ton of coal. So that under the best methods of working it is of great importance to obtain a coal that is rich in bituminous matter.

It must not, however, be supposed that any thing like the above quantities of gas are obtained from coal in the practical working of it in the gas-house. The manufacturer is exposed to losses from a variety of causes, such as leakage, as already noticed, and also from the tendency of the carbon of the gas, or of the hydro-carburets distilled from the coal, to form deposits of charcoal which may attain an inch or more in thickness on the inner surface of the retorts, not only producing a loss of gas, but causing the retorts to burn out more quickly, and leading to expense and delay in removing the deposit. It was formerly supposed that this deposit was owing to the overheating of the retort, or to an excess of heating surface. It was found, however, by Mr. Grafton that the pressure to which the gas is subjected in the retort is the cause of the deposit. It is scarcely necessary to remark, that when an elastic body is generated in a close vessel, the pressure which it exerts upon such vessel depends greatly upon the resistance to which it is exposed in seeking to escape. In endeavoring to force its way by the dip-pipe through several inches of tar into the hydraulic main, the resistance thus offered produces a considerable pressure on the inner surface of the retorts. The passage of the gas through the washing vessels and lime purifiers increases this pressure, thereby promoting the deposit complained of, and causing an increased production of tar at the expense of the gas. Mr. Grafton found that by working the retorts under a pressure of 14 inches of water, a deposit of carbon one inch in thickness was formed within the retorts in one week, and in the course of two months it filled up nearly one-fourth of the retort. On working the retorts with no other pressure than that produced by the insertion of the dip-pipe half an inch into the fluid of the hydraulic main, little or no deposit took place in the retorts in four months with the same kind of coal. It is now common at many gas-works to introduce some kind of pumping apparatus, known under the name of the *exhauster* or *extractor*, between the hydraulic main and the condenser, or between this and the lime purifiers, by which means the pressure of the gas within the retorts can be reduced to any amount. It is, however, found desirable not to carry this reduction too far, lest atmospheric air should find its way into the retorts, and thus form an explosive mixture with the gas.

The quality of the gas yielded by coal varies greatly at different periods of the carbonizing process. The first products, when the coal has not been previously well dried, consist almost entirely of aqueous vapor and carbonic acid; these are succeeded by light carburetted hydrogen, olefiant gas, and sulphuretted hydrogen, which gradually diminish in quantity till toward the close of the process, when almost the only products are carbonic oxide and hydrogen. Hence, if the process be carried on too long, the gases obtained in the latter stages of it will not only be useless for the purpose of yielding light, but the fuel employed for their production will be expended in wasting the retorts to produce substances which are calculated to impair the illuminating power of the gases with which they are mixed. In the case of cannel coal, the interval between the charges of the retorts should not exceed three and a half or four hours; nor in the case of the Newcastle coal, which is not so easily decomposed, ought that interval to extend beyond six hours.

The Condensing Main and Dip Pipes.

From the retorts the gas, after its production, ascends by means of pipes, called *stand-pipes*, into what is termed the *condensing main*, which is a large cast-iron pipe, about twelve or fifteen inches in diameter, placed in a horizontal position, and supported by columns in front of the brickwork which contains the retorts. Wrought-iron hydraulic mains are now coming into use, and are preferable on account of their superior lightness and strength. This part of a gas apparatus is intended to serve a twofold purpose : First, to condense the tar and grosser products of distillation; and, secondly, to allow each of the retorts to be charged singly without permitting the gas produced from the rest, at the time that operation is going on, to make its escape. To accomplish these objects, one end of the condensing main is closed by a flanch; and the other, where it is connected with the pipes for conducting the gas toward the tar vessel and purifying apparatus, has, crossing it, in the inside, a semi-flanch or partition, occupying the lower half of the area of the section, by which the condensing vessel is always kept half full of liquid matter.

The stand-pipes are connected by a flanch with a branch-pipe rising from the upper side of the condensing main; and as the lower end of it dips about two inches below the level of the liquid matter, it is evident that no gas can return and escape, when the mouth-piece of the retort is removed, until it has forced the liquid matter over the bend, a result which is easily prevented by making it of a suitable length. The upper part of the branch of the dip-pipe is generally furnished with a ground plug to allow the removal of the tarry matter, which is apt to accumulate in a concrete state at the lower part of the pipe where it is nearest the furnace. The dip-pipes vary in diameter from $3\frac{1}{2}$ to 4 inches.

Of the Tar Apparatus.

After emerging from the lower end of the dip-pipe, the gas, now bereft of a considerable portion of the vapor of water, tar, and oleaginous matter, which ascends with it from the retort, is conveyed by pipes, for the purpose of being completely freed from these impurities, into contrivances where a more perfect condensation takes place. As the subsequent purification of the gas depends, in no small degree, upon the perfect separation of the tar and other condensable products, by which it is accompanied, the construction of the vessels best calculated for attaining that end is a matter of the utmost importance; and indeed it may be justly affirmed, that unless that separation be effectually accomplished, the action of the chemical agents to which the gas is afterward exposed, must be limited and imperfect.

The first contrivances employed for the purpose of condensation were all constructed on the supposition that the object would be best attained by causing the gas to travel through a great extent of pipes, surrounded by cold water, and winding through it like the worm of a still, or ascending upward and downward in a circuitous manner. An improvement on this form of condenser, and the one now in general use, consists of a series of upright pipes connected in pairs at the top by semi-circular pipes, and terminating at the bottom in a trough containing water, and divided by means of partitions in such a way that as the gas enters the trough from one pipe it passes up the next pipe and down into the next partition, and so on to the end of the condenser. The cooling power of this air-condenser, as it is called, is sometimes assisted by allowing cold water to trickle over the outer surface of the pipes. In passing through these pipes the gas is considerably reduced in temperature, and the tar and ammoniacal liquor condense, the tar subsiding to the bottom, and the ammoniacal liquor floating on the surface. In the course of time the water in the trough is entirely displaced by these two gaseous products, and as these accumulate they pass off into a tar-tank, from which either liquor can be re-removed by means of a pump adapted to the purpose.

Of the purifying Apparatus for separating the Gases unfit for the purposes of Illumination.

With the two compounds of hydrogen and carbon, viz., olefiant gas and light carburetted hydrogen, which are yielded by coal during its destructive distillation by heat, several other products are obtained, which are not only useless for the purpose of illumination, but are calculated to diminish the brilliancy of the light which is afforded by these gases, and even to prove a source of serious nuisance during their combustion. Among these products of a deleterious nature are carbonic acid and sulphuretted hydrogen; and in smaller quantity, carbonic oxide, nitrogen, and hydrogen. The first two are by far the most objectionable of these impurities; and fortunately their separation can be effected more easily than that of the others, the presence of which is of less importance.

Carbonic acid is readily absorbed by any of the alkalies or earthy bodies in a caustic state; and sulphuretted hydrogen, which possesses many of the properties of an acid, unites not only with the alkalies and alkaline earths, with which it forms a species of salts termed *hydrosulphurets*, but also with the metallic oxides, most of which it reduces.

The alkalies being too expensive to be used for separating carbonic acid and sulphuretted hydrogen from coal-gas, a more economical substitute, and which answers the purpose almost equally well, is found in quick-lime. This substance is accordingly used in every gas establishment on the large scale, in some form or other, in purifying the gas. It is employed in two states; either in the condition of a thin paste, which the workmen call the *cream* of lime, or of a moistened powder, such as lime assumes when it is slaked with a little more than the usual quantity of water. The apparatus must therefore be accomodated, in its construction and arrangement, to these different conditions of the purifying material.

When the lime is used in a liquid state, the gas is made to pass through it so as to be as much as possible exposed to its action; and it being highly conducive to the success of the purifying process that a succession of fresh portions of the liquid lime should be brought in contact with the gas as it passes through it, the material is kept in a state of constant agitation by means of machinery.

One of the objections against the method of purifying by the cream of lime, or lime in a liquid state, is, that unless the gas be previously freed entirely from tar, that substance enveloping it with a thin film of oleaginous matter, which has little tendency to unite with water, carries the gas along with it in rolling bubbles, so that the internal parts of it can thus scarcely ever come into contact with the purifying materials. In some arrangements mechanical contrivances are employed to agitate and disperse the gas, with the view of exposing every portion of it, more or less, to the action of the lime; but these modes of promoting the efficacy of the process cannot be resorted to without the aid of some moving power, which, in many cases, must necessarily be attended with considerable trouble, as well as additional expense. There is another objection to which this method of purification, even if it required not the assistance of machinery, must always be liable; namely, that the olefiant gas, upon which the illuminating power mainly depends, is largely absorbed by water, insomuch that either oil or coal gas, standing a few days over that fluid, suffers a great deterioration of its quality, and becomes in every respect less fit for the purposes of illumination. When lime is used in the dry state, or rather in the state of a moistened powder, for purifying coal gas, neither of these objections is applicable; and when the arrangements for that mode of purification are contrived with a due regard to the simplictiy and convenience of the manipulations, the separation of the useless and noxious gases is effected more easily, and not less effectually, than by the method of liquid lime. The abstraction of the sulphuretted hydrogen becomes more perfect by adding to the lime a small portion of the peroxide of manganese, which, being a cheap substance, adds very little to the expense of the process.

It is stated that a bushel of quick-lime is sufficient for the purification of 10,000 cubic feet of gas. By slaking and reducing it to powder its bulk is more than doubled; two bushels of hydrate of lime thus formed cover a surface of 25 square feet to a depth of $2\frac{1}{2}$ inches. At some works a bushel of slaked lime, or half a bushel of unslaked lime, is allowed for every ton of coals distilled. Some engineers estimate that 40 lbs. of lime are required for every 10,000 cubic feet of gas from average Newcastle coal. If more lime is required the coal must have been damp, or have contained more than the usual proportion of sulphur. Good Newcastle coal contains about one per cent. of sulphur; some kinds of *cannel* only one-half per cent. The capacity of dry lime purifiers is calculated on the assumption that 25 square feet of surface are required for 10,000 feet of gas. The purifiers are generally arranged in a set of four, three of which are usually at work while the fourth is being emptied. The spent lime contains hydrosulphuret of ammonia, and when exposed to the air it evolves sulphuretted hydrogen, carbonic acid taking its place. The poisonous liberated gas thus becomes a nuisance to the neighborhood, but it is sometimes got rid of before the purifier is emptied, by connecting each purifier with a large horizontal pipe which opens into the chimney-shaft of the retort-house, the powerful draught of which draws off all volatile matter from the lime, air instead of gas being let in at the bottom. The cover of the purifier can then be raised, and the lime be removed without annoyance to any one. The lime is burned in ovens, and is used a second time in the purifiers, after which it becomes refuse.

The quantity of lime necessary for purifying a given volume of coal-gas varies, as already stated, with the quantity of sulphur contained in the coal from which the gas is produced. It is proper, however, to examine at intervals, during the progress of the purification, the state of the gas by such chemical tests as are calculated to detect the presence of any of the deleterious substances with which it is usually contaminated. Thus carbonic acid is readily discovered by agitating a small portion of the coal-gas with lime water in a limpid state, the solution being quickly rendered turbid when the most minute quantity of that gas is present. Sulphuretted hydrogen is discovered with equal facility by causing a small current of coal-gas to play against a slip of paper moistened with a weak solution of acetate of lead, or nitrate of silver, both of which instantly become black when they are exposed to the action of sulphuretted hydyogen.

Of late years a variety of improvements have been introduced for purifying gas, which we now proceed briefly to notice. They are at present only in partial use, but are likely to lead to important results. Indeed the chemistry of the manufacture is just now in a transition state, and is receiving considerable attention from scientific men.

After the tar and ammonia have been for the most part extracted from the gas by the condenser, a further separation of ammonia is now frequently effected by passing the gas through layers of coke dust, cinder or breeze, or brick-dust, placed in trays or sieves, six or eight inches apart, in a vertical hollow shaft, and as the gas streams up through the porous column the ammonia is retained. This *scrubber*, as it is called, is sometimes used, in conjunction with a washing vessel, and sometimes the latter only is employed, with the advantage of separating a portion of sulphuretted hydrogen and carbonic acid as well as the ammonia; but the wash-vessel is said to remove much of the olefiant gas, the illuminating power of which is very high; an objection which does not apply to the scrubber. Mr. Croll has patented a method of separating ammonia by means of chloride of manganese, which has the effect also of removing much of the sulphide of carbon, of producing a saving of one-half or one-third of the lime required in the subsequent process, while a valuable product is formed by the chlorine of the manganese uniting with the ammonia, to form sal-ammoniac. Ammonia has also been separated by passing the gas through dilute sulphuric acid, the resulting sulphate of ammonia being also a valuable secondary product. The ammonia may also be separated by means of sulphate of manganese, chloride, or sulphate of zinc.

Formerly a good deal of ammonia passed off with the gas to the consumer, to the great injury of the gas meter, the gas fittings, and the furniture of houses. After the ammonia has been separated, the gas is passed into the dry-lime purifiers, which are preferable to the wet lime, not only for the reasons already stated, but on account of the less amount of pressure required to force the gas through them. The objection to dry lime is on account of the volatile nature of the offensive hydrosulphide of ammonia, which is only combined with it, so that when the purifiers are opened, and the spent lime taken out, the oxygen of the air combines with the hydrosulphide, evolving great heat, and filling the neighborhood with noxious odors. This serious objection is now obviated by getting rid of the ammonia between the condenser and the purifier: the salts separated by the dry lime are then no longer volatile, but, on the contrary, the spent lime becomes in some cases a valuable manure, consisting, as it does, of sulphate, carbonate, and cyanide of lime.

A method of purifying the gas, patented by Mr. Hills of Deptford, is now attracting considerable attention. It is based upon the property of the hydrated oxide of iron to decompose sulphuretted hydrogen, a portion of the sulphur forming a sulphide with the iron. Quick-lime is also used to separate carbonic acid, and the oxide of iron is mixed with sawdust or cinders (breeze) for the purpose of increasing the surfaces of contact, and this mixture is placed in the purifiers. When a sufficient quantity of gas has passed through it the purifiers are opened, and the mixture is exposed to the air, under which new condition it combines with oxygen, and again becomes fitted for use in the purifiers. The chemical changes which occur in these operations are the following:—The mixture of hydrated oxide of iron, etc., absorbs sulphuretted hydrogen $Fe_2O_3 + 3HS = Fe_2S_3 + 3HO$. The sulphide of iron, by exposure to the air, absorbs oxygen, and the sulphur is separated in an uncombined form $Fe_2S_3 + O_3 = Fe_2O_3 + S_3$. The mixed material can be again employed in the purification of the gas, and the process may be repeated until the accumulation of sulphur mechanically impairs the absorbent powers of the mixture. The sulpho-cyanogen which accompanies the gas is retained by the oxide of iron, and gradually accumulates in the mixture.

Chemists have also sought for substitutes for lime, or for means of diminishing the amount usually required. M. Penot recommends sulphate of lead for separating sulphide of hydrogen. Professor Graham proposes to add to the slaked lime one equivalent of crystallized sulphate of soda, which would absorb sulphide of hydrogen until two equivalents thereof were absorbed by one equivalent of lime; the lime is converted into sulphate, and the soda becomes bi-hydrosulphuret, which might be readily washed out of the lime, and again be converted into soda by roasting, and thus be used over and over again to mix with the lime. The secondary product formed in the manufacture of chloride of lime, viz., the mixture of chloride of manganese with sulphate of soda, has also been used as an efficient gas-purifier.

Gasometers for receiving and containing the Gas before it is consumed.

As many disadvantages would be experienced by attempting to adjust the production of the gas to the rate of its consumption, it is found to be more convenient, as well as more economical, to store up such a portion of it during the day as shall compensate for the deficiency of the supply that may be furnished during the time the gas is being consumed in the course of the evening. The capacity of the vessels used for this purpose, which are incorrectly called *gasometers* (for they do not *measure* the gas, but only act as *gas-holders*), must be regulated by a regard to that consideration.

The form of the gasometer is generally that of an inverted cylindrical cup, the diameter of which, when economy is studied, ought to be double of its depth, or at least not more than two or three inches less. Gasometers were formerly composed of sheet iron varying in weight from two to three lbs. to the square foot, well riveted at the joints, and kept in shape by means of stays and braces formed of cast or bar iron. The sheet iron was made to overlap at the joints—a slip of canvas, well besmeared with white-lead, being interposed to secure perfect tightness. The prismatic shape was also formerly adopted, but it was not found to be so convenient as the cylindrical, partly on account of the difficulty of making it retain its form, and partly on account of the greater quantity of material, compared with the capacity, that is necessary for its construction.

The gasometer on the old construction was furnished with a tank, of the same form with itself, but a little larger in dimensions, for containing the water, in which it was suspended at different altitudes, by means of a chain and counterpoise moving over pulleys. The tank was sometimes built of stone, but more frequently constructed of cast-iron plates bolted together by flanches, with an interval between them of about three-eighths of an inch, which was afterward filled up with iron cement.*

As the gasometer, when it is immersed in the water of the tank, suffers a loss of weight equal to that of the portion of the fluid it displaces, it is evident that unless some arrangement be made to counteract the varying pressure which must thus result from the different depths to which it may be immersed, the gas contained in the gasometer will be expelled, at different times, with a varying force. If, however, the weight of the chain of suspension, or rather the weight of that portion of it whose length is the same as the height to which the gasometer ascends, be equal to half the loss of weight which the gasometer sustains by immersion in water, a perfect compensation will be made, and an equilibrium will hold between the gasometer and its counterpoise at all altitudes. Thus, if the weight of the gasometer were five tons, or 11,200 lbs., and it lost by immersion a seventh part of its weight, or 1600 lbs., then the weight of that portion of the chain equal in length to the highest ascent of the gasometer would require to be 800 lbs., and the weight of the counterpoise 11,200 — 800, or 10,400 lbs.

	lbs.
For, the gasometer being immersed, its virtual weight is 11,200 — 1600, or	9,600
Weight of portion of chain now acting with the gasometer	800
Sum is the weight of counterpoise	10,400

Again:

The gasometer being elevated out of water, its weight is	11,200
Weight of chain now acting in opposition to it	800
Difference is the weight of counterpoise	10,400

* The following iron cement is recommended by Peckston : Take iron turnings or borings, and pound them in a mortar till they are small enough to pass through a fine sieve ; then, with one pound of these borings, so prepared, mix two ounces of sal-ammoniac in powder, and one ounce of flowers of sulphur, by rubbing them well together in a mortar ; and afterward keep the mixture dry till it may be wanted for use. When it is so, for every part thereof, by measure, take twenty parts of iron borings, prepared as above mentioned, and mix them well together in a mortar or other iron vessel. The compound is to be brought to a proper consistency by pouring water gently over it as it is mixing ; and when used it must be applied between the flanches by means of a blunted caulking iron.

Although the compensation, by this adjustment of the weight of the chain, answers the purpose in the most effectual manner, the following method is by some deemed preferable. Let the counterpoise consist of a long cylindrical or prismatic body, having the area of its horizontal section equal to the area of a similar section of the plates of the gasometer, and be allowed to descend into the water as the gasometer rises out of it. Also let the chain be of a weight equal (length for length) to a column of water of equal bulk with the counterpoise. Then, if the weight of the gasometer be, as already supposed, 11,200 lbs., the weight of the counterpoise must be the same; but the weight of that portion of the chain, which, by the above arrangement, was only equal to half the loss of weight sustained by the gasometer when immersed, must now be equal to the whole of that weight.

	lbs.
Then, the weight of the gasometer in the water is, as before	9,600
Weight of the chain now acting with the gasometer	1,600
Weight of counterpoise, now out of the water	11,200

Again:

	lbs.
The weight of the gasometer, out of the water, is	11,200
Weight of the chain, now acting in opposition to the gasometer	1,600
Weight of the counterpoise, in water	9,600

Though we have only shown, in both these modes of compensation, that an equilibrium between the gasometer and its counterpoise holds in the extreme cases, it would be easy to prove that the same thing must subsist at all the intermediate elevations of the gasometer. At the same time, it must be obvious that these contrivances, however well calculated they may be to secure the equilibrium alluded to, can have no effect in expelling the gas; and therefore, when it is wished that the contents of the gasometer shall issue from it under a certain pressure, the weight of the counterpoise must be diminished to a suitable extent. Thus, if it were required that the pressure employed for expelling the gas should be equal to that produced by a column of water three-fourths of an inch deep, then it would be necessary to diminish the weight of the counterpoise by the weight of a column of water having the same diameter with the gasometer, and an altitude of three-fourths of an inch. If the diameter of the gasometer above-mentioned were, for example, 35 feet or 420 inches, the weight of a cylindrical portion of water having that diameter, and a depth of three-fourths of an inch, would be, in grains,

$$420^2 \times 7854 \times \tfrac{3}{4} \times 252.5 = 26236876 \text{ grs. or } 3740 \text{ lbs.}$$

Hence it would be necessary to make the counterpoise 3740 lbs. lighter than it was supposed to be according to the above-mentioned arrangements, in order that the gas might issue from the gasometer under the pressure of three-fourths of an inch of water. If the calculation were conducted with extreme accuracy, the specific gravity of the gas ought also to be taken under consideration; but the object to be attained is not of so delicate a nature as to require an attention to such minute circumstances. Besides, we shall afterward find that the value of the arrangements we have described for obtaining a uniform and equable pressure is greatly diminished; and these are even entirely superseded by a contrivance called the governor.

Such is the old method of constructing gasometers. Of late years, however, a different system has prevailed. Instead of making them of heavy plate iron, strengthened by angle iron and stays, and of so great a weight as to require the above-described complex system of equilibrium chains and counterbalancing weights to relieve the gas from the great pressure to which it would otherwise be subjected, the gasholders are now made so light that they actually require to be loaded in order to supply the required pressure. The practice has even been introduced of not suspending the gas-holders at all, but regulating their rise and fall by means of guide-rods placed round the tank.

The pipes by which the gas is commonly introduced and conducted off being in many cases considerably below the level of the street pipes with which they communicate, are apt to be filled up in the course of time with the condensed water which passes off in a vaporous state with the gas. To remedy this inconvenience, it is necessary to place vessels for receiving that water in connection with the entrance and exit pipes, so contrived that the accumulated water may be easily removed from them when required.

Of the Main and Service Pipes.

The gas being duly purified and prepared for combustion, the next point to be considered is the transmission of it from the gasometer to the various places where it is to be consumed. As it must sometimes be conveyed, particularly in the case of large establishments, to the distance of several miles, it is evident that unless the diameters of the various pipes through which it is to be conducted have a due relation to the quantity of gas to be transmitted, there will be a danger either of incurring an unnecessary expense, by making the pipes too large; or, what is still worse, of being exposed to a deficiency of supply, by making them too small. The first object, therefore, to be ascertained by the engineer, is the probable number of lights that may be required in the various streets and lanes in which these pipes are to be laid; and these being known, the corresponding quantity of gas, according to the quality of it, may be afterward computed. With regard to the relative dimensions of the pipes at different distances from the gas-work, the only general rule to be observed is, that the sum of the areas of the sections of the main pipes proceeding immediately from the gasometer should be equal to the sum of the areas of the sections of the various branch-pipes which they supply with gas; and this rule, with some little modification, should be followed in the case of the subordinate ramifications.

In the case of good coal-gas, we may safely reckon that one-fourth of a cubic foot of it will furnish the light of a moulded candle for an hour, of which one pound will, when the candles are burnt in succession, last forty hours. On this supposition, and assuming that the pressure upon the gas in the gasometer is equal to three-fourths of an inch of water, the diameters of pipes necessary for conveying various quantities of gas may be stated as follows:

Diameter of Pipe in Inches.	Quantity of Gas in cubic feet per hour.	Equivalent Number of Candles.
¼	4	16
½	20	80
¾	50	200
1	90	360
2	380	1,520
3	880	3,520
4	1,580	6,320
5	2,480	9,920
6	3,580	14,320
7	4,880	19,520
8	6,380	25,520
9	8,090	32,360
10	10,000	40,000

This table has been deduced partly from theoretical considerations, and partly from the results of experiment. Peckston affirms, in his work on Gas-lighting, that a pipe ten inches in diameter, is capable of transmitting 50,000 cubic feet of gas per hour, under a pressure of one inch of water; while, according to the statement of Mr. Creighton, such a pipe would scarcely convey the tenth part of that quantity, under a pressure of from four-eighths to three-fourths of an inch of water. It is impossible to reconcile these discordant statements either by an allowance for the difference of pressure or the difference of the specific gravities of the gases; for it ought to be kept in view that the discharge of gas is directly proportioned to the square root of the height of the column of water by which it is pressed, and inversely as the square root of the specific gravity of the gas. Both of these propositions, however, must be greatly modified by friction, and consequently by the length of the pipes through which the gas is conveyed. In the supply stated to be furnished by pipes of different dimensions, we have deemed it safest rather to underrate the quantity than overrate it.

The main-pipes are usually made of cast-iron, joined together with socket-joints, in lengths of three yards. The depths of the sockets vary in pipes of different sizes from three to six inches, part of them being fitted with gasket to bring the centers of the pipes into line, and the remainder with lead after the gasket has been driven home with suitable chisels or caulking irons. The depth of lead to secure a good joint should not be less than an inch and a half; the interval between the spigot and the socket being from three-eighths to seven-eighths of an inch, according to the diameter of the pipe. Joints are now frequently made without lead. One plan is to caulk into the bottom of the socket, to the depth of two inches, white rope-yarn covered with putty, and to nearly fill up with tarred gaskets, leaving a *gate* into which is poured a composition of melted tallow and vegetable oil. Another plan is to bore the socket of the pipe with a slightly conical opening, the small end being similarly turned to fit the socket. The two ends of the pipe are coated with a mixture of white and red lead, and being brought together, are driven home by a mallet. Such a joint is said to be quite tight. Rings of vulcanized India-rubber have also been recommended for the joints of gas and water pipes.

As a considerable quantity of water is carried off by the gas in the state of vapor, which is afterward condensed in the pipes, some arrangement must be made for its collection and occasional removal; and accordingly, in laying the pipes, care must be taken to give them a regularity of declivity toward one or more points, where proper syphons, close vessels, and cocks must be placed, to receive and discharge the collected water. When these precautions are neglected, or when the levels are inaccurately taken, much annoyance is experienced; and as the evil can only be corrected by lifting and rejoining the pipes, the utmost attention should be paid to guard against it at first.

To convey the gas from the main-pipes, and distribute it through the various apartments of dwelling-houses, pipes made of block-tin are generally used; these being more durable and better adapted to the purpose than pipes composed of copper or any other metal. In arranging the interior fittings, the same precautions must be observed as were recommended in the case of the main pipes, viz., to give the various branches a due degree of inclination, so as to cause all the condensed water to flow to one or more points, where proper cocks must be placed for its removal. Unless this be done, the lights will be apt to flicker, or be extinguished at times altogether. Nor is it of trivial moment to enjoin the workmen, when they are soldering the service-pipes, to avoid with the utmost care allowing any of the melted metal to find its way into the inside of the pipes; it being in a great measure to this circumstance that the deficiency in the supply of gas, so frequently complained of, is owing.

Of the Governor or Regulator.

The quantity of gas consumed in large towns varying greatly at different times, it is evidently a matter of some importance to the public, as well as to the manufacturers of gas, that the supply of it should be duly adjusted to the consumption; so that when the lamps are once regulated to a proper height of flame, they may continue afterward to burn with the same steady light throughout the whole of the evening. Any contrivance that can accomplish so desirable an object must save a great deal of trouble to the consumer of gas, and much unnecessary waste of it to the manufacturer; and such is the design of the governor or regulator. Fig. 1 represents one of these contrivances, d being the pipe proceeding from the gasometer, by which the gas is admitted, and e the pipe by which it escapes; c is a valve of conical form, fitted to the seat i, and raised and depressed by means of the weight f attached to a cord passing over a pulley; bb is a cylindrical vessel formed of sheet-iron which ascends and descends in the exterior vessel aa, in which water is contained to the level represented. The gas, entering at d, passes through the valve, fills the upper part of the inverted vessel bb, which it thus partially raises, and escapes by e. If the pressure from the gasometer be unduly increased or diminished, the buoyancy of bb will be increased or diminished in like proportion, and the valve being by this means more or less closed, the quantity of gas escaping at e will be unaltered. And not only will the governor accommodate itself to the varying pressure of the gasometer, but also to the varying quantities of gas required to escape at e for the supply of the burners. Thus, if it were necessary that less gas should pass through *e*, in consequence of the extinction of a portion of the lights, the increased pressure which would thus be produced at the gasometer would raise the governor, and partially shut the valve, till the state of it was duly adapted to the requisite supply of gas.

When a large district is supplied by a single gas company, and different parts of the same district consume variable quantities of gas, variable pressures are required. One part of the district where there are numerous shops will consume more than another part which consists chiefly of private houses, so that the pressure for the former must be greater than that required for the latter. For example, the Westminster district has about 20 such divisions, comprising nearly 150 miles of main, and the varying pressures required for each division are managed as follows: In the superintendent's room there are a number of small gasometers, called *pressure indicators*, and over each is the name of the sub-district to be supplied. Each gasometer is about 12 inches in diameter. It is supported in a tank of water in such a manner that it can rise and fall with the varying pressure in the mains with which it is connected by a pipe. At the upper part of the gasometer is a rod, carrying a black lead pencil, which bears upon a cylinder which is covered with a sheet of paper, along the top of which are marked the twenty-four hours of the day. From these hours perpendicular lines are drawn to the bottom of the sheet, and there are also horizontal lines, and the bottom is divided into tenths. The cylinder is connected with a time-piece, so as to rotate on its axis, by which means the pencil draws a line opposite the hour when it is set going. If the pressure be constant for a number of hours, the pencil will of course describe a portion of the circle round the cylinder parallel with the top and bottom edges of the paper, or a straight line when the paper is unrolled; if the pressure vary, the line will be diagonal or zig-zag. At the end of twenty-four hours the paper is taken off the cylinder, and replaced by a new one. A collection of these papers for each district furnishes an index to the supply of gas at any hour of the day to the sub-district to which it refers.

It is often necessary to ascertain the pressure to which the gas is subjected in the various forms of apparatus used in the manufacture. For this purpose a simple gauge is attached thereto, consisting of a bent graduated glass tube containing a portion of water or of mercury. If one end of the tube be screwed into a vessel or an upright tube containing gas of the same pressure as that of the external air, the liquid will stand at the same height in the two limbs of the gauge. If the pressure be greater than that of the external air, the liquid will rise in the open limb, and the pressure of the gas will be 1, 2, or more inches, according to the height to which the liquid rises. But if the pressure of the gas be less than that of the atmosphere, the atmospheric pressure, which always acts at the open end of the tube, will prevail, and the liquid will be depressed in the open limb, and rise in the other.

The Gas-Meter.

The gas-meter is a simple but ingenious mechanical contrivance, the design of which is to measure and record the quantity of gas passing through a pipe in any given interval of time. Experience has proved it to be no less advantageous to the consumer than to the manufacturer of gas, by allowing the former to use gas without any unnecessary waste of it, and securing to the latter a fair and regular price for the quantity of it actually consumed.

There are two forms of meter in actual use, viz., the *wet* and the *dry*. The former, the invention of Mr. Clegg, is represented in the annexed figures. In the sections, figs. 2, 3, *ec* represent the outside case, having the form of a flat cylinder; *a* is a tube which enters at the center for admitting the gas, and *b*, fig. 2, is another for conveying it off to the burners; *gg* are two pivots, one supported by the tube *a*, and the other by an external water-tight cup, projecting from the outside casing, and in which is contained a toothed wheel *h*, fixed upon the pivot, and connected with a train of wheel-work (not shown in the figure) to register its revolutions. The pivots are fixed to and support a cylindrical drum-shaped vessel *ddd*, having openings *e, e, e, e*, internal partitions *ef, ef, ef, ef*, and a center piece *ffff*. The machine is filled with water, which is poured in at *h* up to the level of *i*, and gas being admitted under a small pressure at *a*, it enters into the upper part of the center piece, and forces its way through

[PLATE 1.

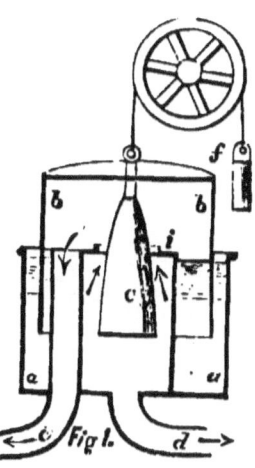

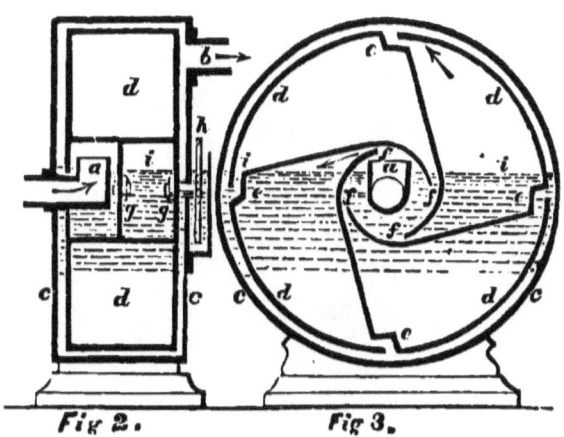

[PLATE 2.

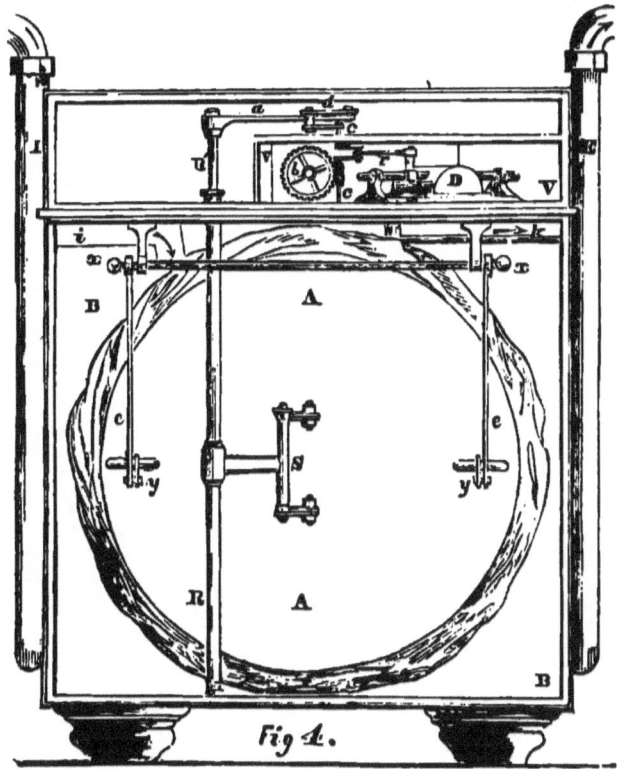

Fig 4.

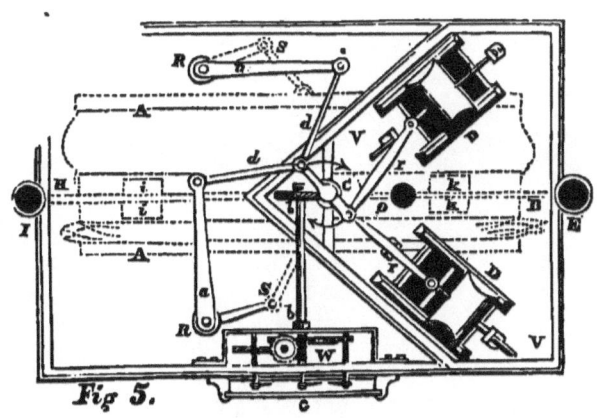

Fig 5.

such of the openings *f* as are from time to time above the surface of the water. By its action upon the partition which curves over the opening *a*, a rotatory motion is communicated to the cylinder; the gas from the opposite chamber being at the same time expelled by one of the openings *e*, and afterward escaping at *b*, as already mentioned.

As the quantity of gas which passes through the machine in any given time depends not only upon its internal dimensions and the number of revolutions which it performs, but also upon the level of the surface of the water in which the cylinder revolves, due care must be taken to maintain the water at the same level, for the regular action of the meter. This is easily accomplished, by pouring in water when necessary, till the superfluous quantity is discharged by an orifice properly placed for the purpose.

One great objection to the wet meter is, that the water is liable to freeze in winter, by which means the supply of gas is stopped; it has been proposed to use a solution of caustic potash or soda instead of water, as being less liable to freeze, and exerting a beneficial action on the gas by removing traces of carbonic acid or sulphide of hydrogen. A second objection is, that if the water level be lowered so that one compartment may at the same time communicate with the central and outer spaces *ff* and *de*, more gas will pass than can be registered, an effect sometimes produced by the dishonest consumer tilting forward the meter. In the dry meter, as its name implies, no liquid is used, and the gas is measured by the number of times that a certain bulk of it will fill a chamber constructed so as to contract and expand for the passage of the gas. These alternate contractions and expansions give motion to certain valves and arms, which, with the aid of a train of wheels, turn the hands of the dials as in the wet meter. The two forms of dry meter which have attracted most attention are Defrie's and Croll and Glover's. Defrie's meter consists of three measuring chambers separated by leathern partitions partially covered by metal plates, and as they expand by the pressure of the gas they assume the form of a cone on one side or other, the motion of which backward and forward drives the measuring machinery, and by an action somewhat similar to that of a three-throw pump, a continuous stream of gas is ejected. This incessant bending of the leather backward and forward causes it to wear rapidly, while the efficiency of the meter obviously depends on the soundness of the leather. In Messrs. Croll and Glover's meter the leather is applied in perhaps a less objectionable form. This meter consists of two short metal cylinders, each closed at one end; AA, fig. 4, representing one such end attached to a fixed central plate BB, by means of broad bands of leather, which act as hinges, allowing one side to swell out with gas, while the other parts with its gas by being pressed in toward the center plate. The to-and-fro motion of the discs which close the short cylinders affords means for measuring the gas. Each disc is kept in place by a hinge joint S attached to upright rods, RR'. There are also parallel motions exy attached to each disc, and to the top plate of the meter. As the gas passes into each cylinder and distends it, the rods RR', one on each side, are made to move each through the half of a circle by means of jointed levers S attached to them. At the top of each rod are two arms Rad, R'ad, fig. 5, each of which partaking of the motion of the rods RR describes alternately the arc of a circle, and a rotatory motion is obtained by means of connecting rods attached to these arms, and also to two other arms rr which work two D valves DD, each of which is made to slide backward and forward over three apertures, the two outer of which lead to the inside and outside of the cylinders respectively, and the middle aperture to the exit pipe E. It is the function of these valves to regulate the flow of gas into and out of the two chambers of each division of the meter. While the gas is flowing into one cylinder and distending it, the gas on the other side of this cylinder disc is expelled to the exit pipe E; as soon as this is done, the valve is reversed and gas enters on the side of the disc from whence it was last expelled. The process is then repeated by the other disc, and in this way a continuous flow of gas is obtained by means of the two valves DD, which being placed at right angles to the double-cranked shaft, and the two cranks on the shaft being at an angle of 45° to each other, it follows that as one valve closes the other opens, but the closed valve always begins to open before the other is quite shut. In fig. 5, the dotted portion represents one of the short cylinders A' distended with gas, and the other cylinder A collapsed.

We will now trace the course of the gas in its passage

through the meter. Suppose a continuous stream of gas under pressure to be passing down the inlet pipe I. On arriving at i it meets with a horizontal tube which conducts it by the aperture o, fig. 5, in the direction of the arrow into a triangular chamber VV. It then passes down an open slit of one of the valves, which we will call No. 1, and entering one of the cylinders, distends it and forces the gas which was on the outside of the disc to escape through slit No. 2, and so along a tube k leading to the exit pipe E. While this action is going on, that is, while the cylinder on one side is being distended, the cylinder on the other side is already full, the gas is shut off from it by the sliding valve D, and is made to pass on the outside, where exerting its pressure on the disc AA, it forces it inward, and the gas escapes along a short pipe attached to either side of the partition BB into slit No. 2, and so escapes to the exit pipe. The triangular chamber VV has no connection with the cylinders, etc. situated below it except through the tubes already indicated, and the train of wheels W, fig. 5; and the dials are also so boxed in as not to be exposed to the corrosive action of the gas. The rods R'R pass into this upper compartment through leather washers and a stuffing of wool. The cylinders are inclosed in an oblong box of iron plate or galvanized iron, so as to be completely concealed from view. The pressure to which the gas is subjected in order to force it along the mains is amply sufficient to work this meter. If the gas were subjected to the pressure of only half an inch of water, this quantity multiplied into the area of the disc, which in a ten-light meter is ten inches in diameter, amounts to many pounds.

The circular motion of the double crank is transmitted by means of an endless screw c, fig. 4, and a spur-wheel b along a wire bb, fig. 5, to a train of wheels W, which record their revolutions on the face of the dials G, also shown separately in fig. 6, registering the number of cubic feet of gas consumed, in units, tens, hundreds, thousands, etc. The top circle marks the units, the left-hand circle hundreds. The motion of the hand from 0 to 1 shows that 100 cubic feet of gas have passed through the meter, while a whole revolution of this hand registers ten times that quantity, or 1000 cubic feet. The motion of the hand of the center circle from 0 to 1 indicates 1000 feet, and a whole revolution 10,000 feet. The right-hand circle, in a similar manner, indicates in a whole revolution 100,000 feet. In reading off the numbers on the circles, we take the number at which the hand is pointing, or the lower of the two numbers that the hand is between. In fig. 6, beginning with hte right-hand dial, the hand is between 9 and 0, showing that nearly a whole revolution has been accomplished; we therefore write down:

> 90,000 for the right-hand dial,
> 8,000 for the middle dial,
> 700 for the left-hand dial.
> ——
> 98,700

If the collector, in taking the register three months before, had recorded the quantity as 73,200, this quantity, deducted from 98,700, gives 25,500 cubic feet as the consumption of gas for three months. The top or units dial is not used in registering, but it serves to indicate to the collector as well as to the consumer that the dial is acting properly, the more rapid motion of the hand facilitating this object.

Burners.

The most economical mode of consuming gas, so as to obtain from a given volume of it the greatest possible quantity of light, both in degree and duration, is a problem of no less importance than that of the most suitable arrangements for its production and purification. The presence of oxygen, in some form or another, being essentially necessary to produce ordinary combustion, it follows, that from whatever cause that principle may be deficient in quantity, the combustion must be imperfect; and when this is the case, the light yielded by the combustible body is also diminished in a proportional degree. On the other hand, if the quantity of oxygen brought into contact with the combustible body be more than sufficient for its entire combustion, the superfluous quantity of that gas, instead of augmenting the effect, can only lower the temperature, and diminish, it may be presumed, in a corresponding degree, the intensity of the light. This must be the consequence if the brilliancy of the light yielded by a combustible body depends at all upon the temperature to which it is exposed during its combustion; and that this is the case may be in-

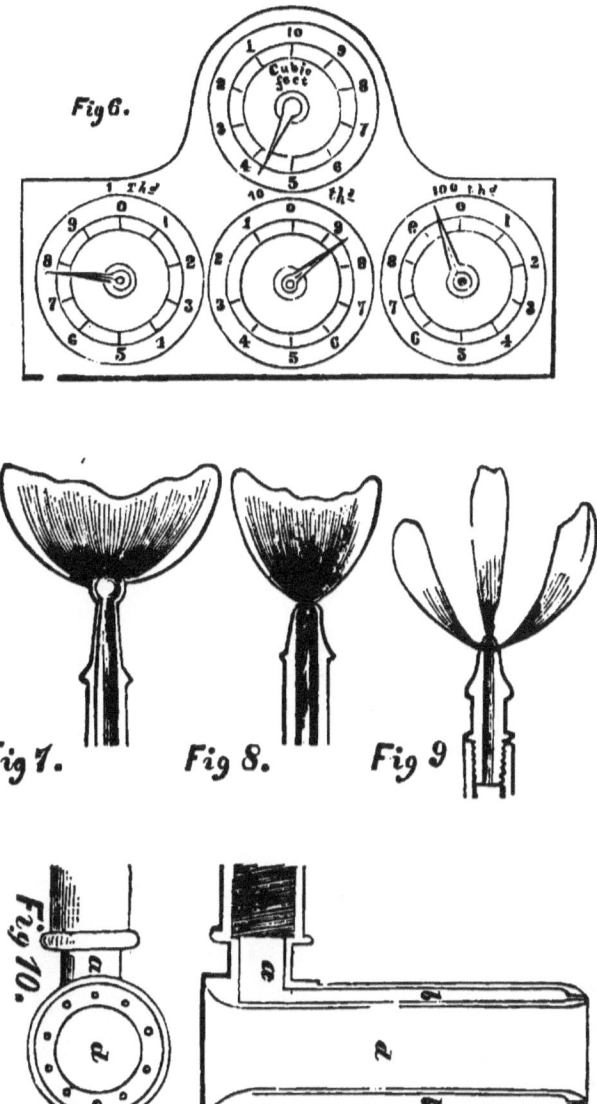

ferred from the simple fact of causing the flame of a jet of gas to play first against a sheet of ice, and then against a bar of red-hot iron, when the difference of the light will be such as to leave no doubt of the influence of temperature upon its intensity. A similar result is obtained by bringing the flames of two separate jets into contact, when an obvious increase of light is perceived. From these simple facts it may be inferred, that though a certain quantity of common air must be brought into contact with the inflamed gas to produce the greatest intensity of light, whatever exceeds that quantity will not only be useless, but by diminishing the temperature of the flame, must tend to impair the brilliancy of its light.

But although the immediate cause of the light is probably the high temperature to which the carbonaceous portion of the gas is exposed, the condition in which the carbon exists at the time it is so exposed is of the utmost importance to the effect. According to the opinion of Sir Humphry Davy, as adopted by Drs. Christison and Turner, " a white light is emitted only by those gases which contain an element of so fixed a nature as not to be volatilizable by the heat caused during the combustion of the gas; and that in coal-gas this fixed element is charcoal, formed by the gas undergoing decomposition before it is burnt. The white light is caused by the charcoal passing into a state, first of ignition, and then of combustion. Consequently no white light can be produced by coal or oil gas without previous decomposition of the gas."

" That the gas undergoes decomposition before it burns, and that the carbonaceous matter is burnt in the white part of the flame in the form of charcoal, is shown by placing a piece of wire-gauze horizontally across the white part of the flame, when a large quantity of charcoal will be seen to escape from it unburnt. And that this previous change is necessary to the production of a brilliant white light will appear, if we consider the kind of flame which is produced when decomposition does not previously take place. For example, if the gauze be brought down into the blue part, which always forms the base of the flame, no charcoal will be found to escape. Or, if the gauze be held at some distance above the burner, and the gas be kindled not below but above it, by which arrangement the air and the gas are well mixed previous to combustion, the flame is blue, and gives hardly any light. The reason is obviously, that in both cases the air is at once supplied in such quantity in proportion to the gas that the first effect of the heat is to burn the gas, not to decompose it." (*Edin. Phil. Journal*, No. xxv.)

To these statements it may be added, that if a jet of oil or coal gas, burning with a fine yellow flame in common air, be suddenly surrounded with an atmosphere of oxygen gas, the color instantly changes into a pale blue, yielding the most feeble light; nor does the flame recover its brilliancy until the oxygen is largely diluted with carbonic acid, when it burns for a short time with greater splendor than at first. For although the light is greatly enfeebled when the combustion of the gas takes place in pure oxygen, it becomes much more vivid when the combustion is carried on in air that is more largely charged with oxygen than common air. Hence the brilliancy of the light appears to depend upon two conditions: 1*st*, the perfect combustion of a portion of the gas in an undecomposed state; 2*dly*, the temperature produced by that combustion upon the residual part in a decomposed state. When a large portion of the gas is consumed in the first condition, the temperature is higher; but the undecomposed part is then too small in quantity to yield an intense light, in consequence of the attenuated state of the carbon; and, on the other hand, when a small portion of the gas is consumed in the undecomposed state, the temperature produced is too feeble to raise the temperature of the now partially decomposed part to a sufficient pitch for the full ignition of the carbon.

The conditions which thus seem to be necessary for obtaining the greatest portion of light from the combustion of a given quantity of gas, while they are perfectly consistent with the most anomalous facts presented by that process, so they appear to afford the only sure principles upon which we can proceed in the construction of gas-burners. One of the most obvious conclusions deducible from these principles is, that whatever be the form of the gas-burner its construction should be such that while it admits as much air as is necessary for the perfect combustion of the gas, it should never admit more than is barely sufficient for that purpose.

According to the experiments of Drs. Christison and Turner, the diameter best fitted for single-jet burners appears to be about one twenty-eighth of an inch for coal-gas, and one forty-fifth for oil-gas. As these dimensions, however, must vary with the quality of the gas, we consider one thirty-sixth of an inch to be more applicable to the gas obtained from cannel coal, if its specific gravity be not less than ·65. Every form of burner composed of separate jets, in which the gas is made to issue in a horizontal or oblique direction, gives a consumption which increases in a much faster ratio than the light which it yields; and consequently, however beautiful such burners may be in appearance, they are far from being economical.

One of the most useful forms of a burner with single jets, is where there are two holes, and their directions are so inclined as to cause the streams of issuing gas to cross, and exhibit during their combustion a broad continuous flame. This burner, which is termed a *swallow-tail*, is well adapted for street-lights, as it gives a powerful light and consumes a small quantity of gas. When the gas is emitted by a narrow slit at the top of the burner, the burner receives the name of a *bat-wing*. Specimens of common gas flames are represented in figs. 7, 8, 9.

But of all the forms of the burner, that upon the Argand principle, in which the holes are arranged in a circle, d, fig. 10, so as to allow the air to have access to the flame internally as well as externally, is the most economical, and the best calculated to secure the complete combustion of the gas. The diameter of the holes should, in this burner, be about the fortieth part of an inch for coal-gas of an ordinary good quality, and the distance between them should be such as to allow the separate flame of the different jets to unite together and form a continuous hollow cylinder of light. In fig 10, a is the pipe which supplies the gas, and bb the channel up which it passes to the holes shown in the lower figure.

The construction of burners, and the most economical mode of consuming gas, having been examined with much philosophical precision by Drs. Christison and Turner, we shall extract from their elaborate dissertation on the subject the most valuable and important conclusions which they have deduced from their experiments; and this we do with greater confidence, because the results they obtained coincide very exactly with those which the writer of this article procured when engaged in the same inquiry. The three leading points to which they directed their attention were, 1*st*, the length of flame most suitable for different burners; 2*dly*, the form, magnitude, and position of the orifices through which the gas is discharged, and 3*dly*, the modifications of the light produced by the glass chimney of the Argand burner.

With regard to the length of flame which afforded the greatest light compared with the expenditure of gas, they found that, in the case of the jet, the best length for coal-gas was about five inches, and for oil-gas about four inches. When the flame was kept shorter, the quantity of gas consumed was greater in comparison of the light which it yielded; but no advantage was gained by increasing the length beyond that mentioned as the most suitable for each gas; the combustion becoming less perfect and beginning to be accompanied with the escape of the carbon in the form of smoke. Thus they found that, in the case of coal-gas having the specific gravity ·602, while the lights emitted from a two-inch and a five-inch flame were as 556 to 1978, the corresponding expenditures were to each other as 605 to 1437. But the light, in an economical point of view, must be estimated inversely as the quantity of gas from which it is obtained; and hence the ratio of the lights, in reference to the expenditure, was as $\frac{556}{605}$ to $\frac{1978}{1437}$, being as 100 to 150.

In the case of Argand burners, the augmentation of the light in a ratio greater than the expenditure was exemplified in a still more remarkable degree. Thus the following results were obtained with coal-gas of the specific gravity ·605, by elevating the flame of a five-holed burner, successively from half an inch to five inches:

Length of Flame.	Half-Inch.	One-Inch.	Two-Inch.	Three-Inch.	Four-Inch.	Five-Inch.
Light	18·4	92·5	259·9	308·9	332·4	425·7
Expenditure	83·7	148	203·3	241·4	265·7	318·1
Ratio of light to expenditure	100	282	560	582	582	604

Hence the light is increased about six times for the same expenditure by raising the flame from half an inch to three or four inches; but very little is gained by any additional increase of the flame beyond that length, in the description of burners with which the experiments were made.

These facts receive a satisfactory explanation from the general principles which we have already laid down with respect to the combustion of the luminiferous gases. When the flame is short, the supply of oxygen for the combustion is too great; almost the whole of the gas is thus consumed before any portion of it can undergo the decomposition which is necessary for the evolution of light; while the temperature of the flame being reduced by the superfluous air which brushes along its surface, the intensity of ignition, and with it the splendor of the light, is proportionally diminished. This explanation is well illustrated by partially shutting the central part of the burner, and thus interrupting the supply of air to the internal surface of the flame; the moment this is done, the length of the flame is increased, and a visible improvement of the light takes place, thus indicating that more air was previously brought in contact with the gas than was requisite for its perfect combustion.

The second point to which Drs. Christison and Turner directed their attention was the construction of the burner itself, particularly the magnitude and position of the orifices at which the gas is emitted during the combustion. The same principles which explained the relation between the light and the expenditure in the case of flames of different lengths, suggested the rule for regulating the dimensions of the orifices; and accordingly they justly inferred that, in a single jet, the diameter of the aperture ought to be such as to ensure the complete combustion of the gas, without rendering it more vivid than is necessary for that effect. If the orifice be too small, the greater portion of the gas is liable to be consumed without suffering a previous decomposition, and thus the light is extremely feeble; and, on the other hand, if the orifice be too large, the surface of flame exposed to the action of the air being too small in comparison of the discharge of gas, the combustion is imperfect, and the carbon, after being separated from the hydrogen, either burns at a low temperature with a dusky flame, or, what is still worse, a large portion of it passes off in the state of smoke. In conformity with these views, they recommend, as we have already stated, a twenty-eighth of an inch for coal-gas, and a forty-fifth for oil-gas, as the most suitable dimensions for single jets. They acknowledge, however, that their experiments with coal-gas were too limited to justify them in using very confident language on the subject; and we have therefore the less hesitation in stating that we consider an orifice varying in diameter from a thirty-second to a thirty-sixth of an inch as better adapted to coal-gas of a specific gravity between ·62 and ·70.

In Argand burners the diameter of the orifices ought to be a little smaller. Drs. Christison and Turner state that the diameter which appeared to answer best for coal-gas of the specific gravity ·6, when the holes are ten in a circle of three-tenths of an inch radius, was a thirty-second of an inch. We consider this, however, to be too great for coal-gas of a better quality, and would recommend, in preference, apertures varying in diameter from a thirty-sixth to a fortieth of an inch.

The distance between the jet-holes of Argand burners is a matter of no less importance than the diameter of the orifices, and must be regulated by the same principles. When they are so far asunder that the flames of the separate jets do not coalesce, no advantage is derived from the Argand form; but when they unite, and compose a uniform and unbroken surface of flame, the light is considerably greater, compared with the expenditure of gas, than is obtained from detached jets. In order to determine the most suitable distance at which the orifices of Argand burners should be placed, Drs. Christison and Turner employed burners six-tenths of an inch in diameter, which they caused to be drilled with eight, ten, fifteen, twenty, and twenty-five holes, a fiftieth of an inch in diameter; and having determined with each of these burners the light and expenditure in the case of oil-gas, they obtained the following results:

Burners.	VIII.	X.	XV.	XX.	XXV.
Light	360	360	391	409	382
Expenditure	367	318	296	289	275
Ratio of light to expenditure	98	113	132	141	139

As the standard of comparison was a single jet, burning

with a four-inch flame, the ratio of the light yielded by which to the expenditure was expressed by 100, it was inferred that no advantage is gained by giving the jets the Argand arrangement with a burner of the dimensions above-mentioned if the holes are only eight in number; and that the gain does not increase after the number reaches to twenty. In the former case the distance of the holes must have been ·2356 inch, or nearly one-fourth of an inch, and, in the latter, ·0945; so that the most advantageous distance for jet-holes of a fiftieth of an inch in diameter would seem to be about $\tfrac{2}{100}$ths of an inch. For coal-gas burners, however, the distance between the jet-holes ought to be increased in a ratio varying inversely with the quality of the gas, or directly as the diameters of the orifices themselves. Hence, if the coal-gas were of an ordinary quality, the jet-holes should not be less than one-eighth nor more than one-sixth of an inch from each other.

The difference between the orifices being once assumed, serves to determine the diameter of the circle of holes. Thus, in a burner of eighteen holes, each a seventh of an inch asunder, the circumference ought to be $18 \times \tfrac{1}{7} = 2\cdot 57$ inches, and consequently the diameter of the circle of holes should be $\tfrac{2\cdot 57}{3\cdot 1416} = \cdot 818$ inch. If the breadth of the rim be supposed to be a tenth of an inch, and perhaps it ought not to exceed that quantity, it may be proper, in the case of the larger burners, to contract the lower part of the central air-hole, on account of the supply of air to the inside surface of the flame increasing in a faster ratio than the number of jets.

The only remaining point to be considered with respect to the burner is the glass chimney, which serves at once to protect the flame from irregular currents of air, and to convey to the gas a due supply of it during combustion. When the interval between the chimney and the external part of the flame is too great, the tendency of the air to flow through the air-hole is diminished, and the flame contracts toward the top, where it yields a dusky light, and indicates a disposition to smoke. The diameter of the chimney should therefore be reduced until it is perceived that the upper part of the flame is enlarged and acquires the same diameter as the lower part. When this is the case, the color of the flame is improved in brightness, and none of the carbon is uselessly wasted in the formation of smoke. On the other hand, if the supply of air to the external surface of the flame be diminished beyond a certain extent, either by reducing the diameter of the glass chimney, or by any other means, the flux of air through the central air-hole is unduly increased, the flame diverges in the form of a tulip till it touches the chimney, and the supply of air to the outside of the flame being thus interrupted, smoke is again produced. Hence the greatest degree of light, in relation to the expenditure of gas, may be expected to be obtained when the supply of air to the external and internal surface of the flame is so adjusted by the diameter of the chimney that the flame is perfectly cylindrical, neither burning with too much vivacity, nor showing any tendency to smoke. The length of the glass chimney is of much less importance than its diameter, and may vary from five to six inches.

A cylindrical chimney, however, is the least advantageous form that can be adopted. If the chimney be tall and narrow, and contracted toward the top, as in a, fig. 11, or suddenly contracted near the bottom, as in b, the draught is increased and the light improved. It is also useful to contract the diameter of the glass chimney about a couple of inches above the burner, as at c, so as to form a shoulder a few lines in width, the effect of which is to change the direction of the draught and project it on the flame at a certain angle.

In the Bude light proposed by Gurney, oxygen gas instead of air was passed through the flame, the effect of which was greatly to increase its brilliancy. In the Bude light as now constructed, there are two, three, or more concentric burners with chimneys supplied with common air, and a dioptric apparatus.

Attempts have been made of late years to ventilate gas-burners so as to get rid of the injurious products of combustion. One part by weight of good coal-gas produces nearly three parts by weight of carbonic acid, which produces many distressing symptoms when breathed with the air of the room. Sulphurous acid, and other compounds which are not entirely removed in the purification of the gas, form deleterious products during the combustion of the gas. The sulphurous acid forms sulphuric acid, which exerts a corrosive action on the walls and furniture, books, pictures, etc., while the hydrogen of the gas produces vapor of water which serves as a vehicle for some of the other products. To get rid of these noxious fumes a bell-shaped vessel is sometimes suspended over the chimney, and is connected with a tube leading into the open air. Unless this tube be judiciously arranged the condensed water may accumulate in it, and cause inconvenience. By a contrivance of Dr. Faraday, a copper tube of about the same diameter as the flame is conducted from its summit out of the apartment; the heat of this tube establishes a rapid current, which serves to convey away the products. The same distinguished chemist invented another contrivance, by which the ventilating current is made to descend between two concentric glass chimneys of different heights, the outer one being the taller, and this is covered with a disc of talc. When the current reaches the bottom of the space between the two glasses, it is conveyed away by a ventilating tube which bends upward. The descending current is first established by applying heat to the bend of the ventilating tube where it begins to ascend; when this current is established the gas is lighted, and the plate of talc is put on: the products of combustion are conveyed into a box, from which proceeds a pipe for conveying the vapors outside. A globe of ground glass open only at the bottom is placed over the lamp. The accumulation of condensed water in different parts of this apparatus is said to have greatly interfered with its successful action.

Mr. R. Brown of Manchester has a contrivance for ventilating by means of gas. Through an opening in the ceiling a wide tube is passed, one end of which conveys the foul air outside, and the other projects a little below the level of the ceiling. The gas-pipe enters on one side, and is bent so as to hang perpendicularly in the center of the tube, and has an annular burner at the lower extremity, surrounded by a glass chimney, which is supported on the top on a metal cone piece, secured to the lower extremity of the tube by screws. This arrangement is surrounded by a hemispherical glass shade with its mouth uppermost, and a few inches below the level of the ceiling. The air of the apartment passes off in the strong draught occasioned by the burner, and a fresh supply of air is admitted at the lower part of the room.

Oil-Gas, Resin-Gas, and Water-Gas.

When tallow or oleaginous matter of any kind is raised to a certain temperature, it is resolved into various gases, of which the compounds of carbon and hydrogen, viz., olefiant gas or bicarburetted hydrogen, and lighted carburetted hydrogen, are the principal, both in point of quantity and quality, for the purposes of illumination. As oil contains in its composition a portion of oxygen, existing most probably in union with hydrogen in the state of water, that substance also yields, during its destructive distillation, a considerable quantity of carbonic oxide, as well as traces of carbonic acid, hydrogen, and even nitrogen. With these products, all of which are of a determinate character, is found in greater or less abundance a quantity of a very inflammable vapor,* which seems to be a compound of carbon and hydrogen.

Oil-gas owes its illuminating power chiefly to the proportion of olefiant gas which it contains, and the oleaginous vapor which is diffused through it; and as both of these ingredients vary in quantity with the temperature at which the decomposition is effected, the quality of the oil-gas is extremely fluctuating. When the temperature is too high, a portion of the olefiant gas and oleaginous vapor is resolved, by the deposition of carbon, into light carburetted hydrogen; and though the quantity of gas from a given portion of oil is thus increased, the quality of it is diminished in a still higher ratio. On the other hand, if the temperature be rather too low, a larger quantity of olefiant gas, mixed with a greater proportion of oleaginous vapor, is obtained; but as the latter is gradually and rapidly condensed when the gas is allowed to stand over water, the higher

---

* The oleaginous vapor alluded to consists, according to the experiments of Mr. Faraday, of two distinct compounds of carbon and hydrogen. One of these he terms *bicarburet of hydrogen*, which, by his analysis, is composed of six proportions of carbon and three of hydrogen. The other compound, to which Dr. Thomson has given the name of *quadro-carburetted hydrogen*, consists of four proportions of carbon and four proportions of hydrogen, existing in a different state of aggregation from that in which they exist in olefiant gas, the elementary constituents of which are in the same proportion.

illuminating power of this richer gas is more than counterbalanced by the deficiency in its quantity, and the deterioration to which it is liable by keeping.

We are indebted to Dr. Henry of Manchester for the first analysis of the aëriform compounds obtained by the decomposition of oil by heat; and though his elaborate researches can scarcely be said to have led to the determination of the precise products of that decomposition, they furnish data from which their true nature may be inferred, with a probability nearly as great as that which belongs to the results of direct experiment. The principal difficulty of the analysis consists in determining the condition in which the elementary principles of carbon and hydrogen exist in union with each other, and reconciling the various suppositions that may be made respecting the compounds thus formed, with the specific gravity which belongs to the original gas, supposed to be produced by their mixture.

The results of Dr. Henry's first experiments were published in 1805; but it was not till about ten years after that period that an apparatus for decomposing oil, on a large scale for economical purposes, was constructed.

Oil being decomposed at a loss of nearly fifty per cent., the conversion of it into gas, after a protracted but ineffectual competition with coal, was gradually abandoned on the large scale, even in those places where, from the interest they had in the whale-fisheries, there was the strongest inducement to foster the prejudices which prevailed for some time against the use of coal-gas. The exaggerated advantages which it was pretended would be derived from compressing oil-gas and thus rendering it portable, served to prolong the delusions on the subject; nor were these delusions fully removed until a demonstration was given of the failure of the scheme, in the decay of the costly edifices and expensive apparatus which had been constructed for carrying it into effect. The late Professor Daniell of King's College, London, also contrived an ingenious form of apparatus for making gas from resin; but the plan did not succeed on account of the impossibilty of competing with the coal-gas works.

Of late years a new process of gas-making has been much discussed, and has formed the subject of a variety of patents. It is known as the *hydrocarbon process of gas-making*, or more briefly *water-gas*. The principle of the manufacture is to pass steam over red-hot coke, by which it is resolved into hydrogen and carbonic oxide, and then to supply these inflammable gases with the carbon required for their illuminating power, by passing them through a retort in which oil, resin, tar, naptha, cannel coal, or some other carbonaceous substance, is undergoing decomposition by heat. The process does not appear to have been successful with resin, but better results seem to have been attained with cannel coal.

Methods for determining the Illuminating Power of the Gases.

Having described the various manipulations by which gas is prepared, both from coal and oil, we now proceed to explain the methods which have been adopted for determining their respective illuminating powers; it being by these methods that we acquire a knowledge of one of the most important tests by which the comparative value of the gases can be ascertained.

The first and most obvious of these tests is to determine the intensity of the light which the gases are capable of diffusing during their combustion, upon a white and smooth surface directly exposed to its emanations. The determination of that intensity is obtained with a considerable degree of accuracy, not by a direct comparison of the degree of illumination shed on two separate surfaces, but by means of a contrivance, first proposed by Count Rumford, which allows the illuminated surfaces to be contrasted with each other on the same ground, and so closely adjoining that the eye can readily detect a slight difference between them. This contrivance is as follows: Let A and B, fig. 12, be two luminous objects; EF a smooth and white surface, having the same inclination to the rays of light emitted by A and B; and CD an opaque cylindrical rod parallel to the surface EF; then it is evident that *aa* and *bb* will be the shadows of CD, in reference to the lights A and B. But the shadow *aa* being illuminated by the light B, and the shadow *bb* by the light A, it follows that if these shadows be perfectly the same in point of intensity of shade, the light yielded by A and B must be the same in degree. If the shadows, however, be different, one of the lights must be removed either further from EF or

[GAS.] [PLATE 4.

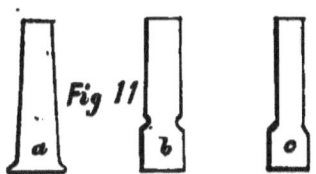
Fig 11

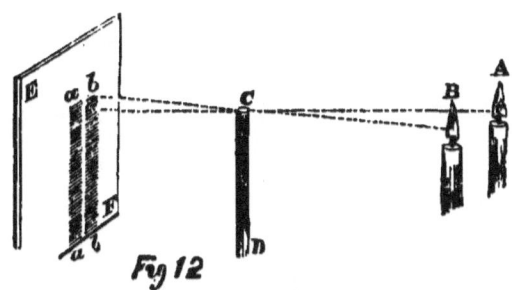
Fig 12

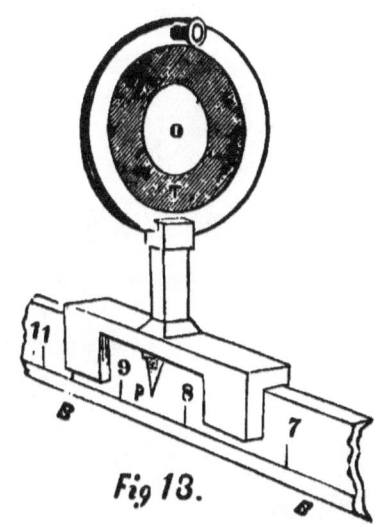
Fig 13.

brought nearer to it, until the shades seem to be exactly alike, when the light shed upon EF by A and B must, in point of intensity, be, as before, the same. But the intensity of light, like that of other emanations proceeding in straight lines from a central point, being inversely as the square of the distance, the relative degrees of light emitted by A and B must, in conformity with that principle, be proportional to the squares of their respective distances from the surface on which the shadows are projected. Thus, if the light A were at the distance of fifteen feet, and the light B at the distance of twenty-five feet, their relative illuminating powers would be as the square of fifteen to the square of twenty-five; that is, as 225 to 625, or as 9 to 25. As the quantity of gas consumed in the same time to yield the supposed lights might be different, it is evident that a correct estimate of the absolute value of the gases for the purpose of illumination would not be duly determined unless that circumstance were also taken into account. But the economical value of the gases, yielding equal degrees of light, being inversely as the quantities consumed, it follows that that value will be directly as the squares of the distances at which the shadows are the same, and inversely as the rate of consumption. Thus, if we now suppose that the gas yielding the light A consumed three cubic feet in the same time that the gas yielding the light B consumed five cubic feet, the value of the former would be to that of the latter as $\frac{9}{3}$ is to $\frac{25}{5}$, or as three to five. In obtaining the necessary data for determining the ratio of the lights, it may be proper to add that the screen on which the shadows are projected should be guarded with the utmost care from all extraneous light. If it be desired to contrast the illuminating power of a gas-light with that of a candle, the comparison is easily made. If, for example, the gas-light give a shadow equal to that of a candle placed at one-third the distance, the light of the gas is equal to the light of nine candles. If the candle be placed at one-fourth the distance of the gas-light, the latter is equal to sixteen candles, and so on.

Professor Bunson of Marburg has contrived a photometer which is now in common use in gas-works. The principle of this instrument is not the comparison by shadows, which forms a delicate experiment, but a comparison of light transmitted through a translucent surface with light reflected from an opaque surface. For this purpose a disc of paper, TO, fig. 13, is placed between the two lights to be compared; an annular portion of this paper T is made translucent by means of melted spermaceti, or that substance dissolved in oil of naptha, while a central disc of the paper O being left untouched by the composition, remains opaque. Fine cream-colored letter-paper answers the purpose very well, and the central opaque disc may be about the size of half-a-crown. Now it is evident that the translucent ring will be illuminated by a light behind the disc, while the opaque portion is illuminated by a light in front. The frame on which the disc is mounted is moved backward and forward on a graduated bar BB between the two lights until the transmitted and reflected lights appear of the same intensity. The pointer P then shows the division over which the disc stands. Under such circumstances, the lights are to each other in the ratio of the squares of their distance from the disc

The determination of the intensity of light by the above simple means is capable, under careful management, of all the precision which the nature of the problem requires; it is even preferred by engineers to the more elaborate method of chemical analysis. The latter method has for its object to ascertain the relative value of the gases used for illumination, by finding the quantity of olefiant gas which they contain under equal volumes; it being assumed that the illuminating power of the compound combustible gases derived from the decomposition of oil and pit-coal is directly proportional to the quantity of that gas existing in their constitution. Though that supposition is by no means a matter of certainty, or even of probability, we shall nevertheless briefly explain the mode of analysis which has been recommended. According to the experiments of Dr. Henry, chlorine has no action upon any of the gases obtained from oil or coal when the influence of light is carefully excluded, with the exception of olefiant gas; and as chlorine and olefiant gas unite together in equal volumes, this property affords an easy mode of determining the quantity of the latter which may exist in any compound gas of which it forms a constituent part. All that is required for the purpose is to add somewhat more chlorine than is absolutely necessary for uniting with the olefiant gas, and to allow the mixture to remain about fifteen minutes completely excluded from light. The extent of absorption being thus observed, half the quantity of the gas which has disappeared of the whole mixture will be olefiant gas. Thus, if twenty parts of chlorine by measure were added to twenty-five of coal-gas, and if the mixture, after being allowed to remain a sufficient length of time in the dark, were found to occupy thirty-six measures, the absorption would be nine measures, and consequently the coal-gas must have contained four and a half measures of olefiant gas, or eighteen per cent. The quantity per cent. of olefiant gas is determined without calculation, by adding to fifty measures of the gas to be analyzed an equal volume of chlorine; when the diminution of volume in the graduated jar, is the quantity which the gas contains per cent. of olefiant gas. Dr. Fyfe states that the illuminating power of the different specimens of oil and coal-gas which he subjected to this test bore a pretty exact ratio to the quantity of olefiant gas which they contained. One great advantage to be derived from this method of testing the quality of any species of carburetted hydrogen containing olefiant gas in its composition, is, that it admits of a comparison being made between gases in different places and at different times, without the necessity of transporting them to a distance, and making a simultaneous examination of their illuminating properties.

Of late years, bromine has been substituted for chlorine in the above analysis. The gas is passed up into a eudiometer tube, and the carbonic acid is removed by means of caustic potash: a small portion of bromine is dropped in and shaken in contact with the gas. Potash is again added to remove the bromine vapors, and the absorption is then noted. It is stated that some of the highly illuminating cannel-coal gases are condensed by this process as much as 12 or 14 per cent.; while some of the poorer gases not more than 4 or 5 per cent.

The specific gravity of oil and coal gas, and the quantity of oxygen which they require for their perfect combustion, have also been proposed as means of ascertaining their illuminating powers. The latter, however, even if it were a correct test, is determined with considerable difficulty; and that little reliance can be placed on the former may be inferred from

the fact that some of the gases which are component parts of oil and coal gas have a great specific gravity without possessing any illuminating power. This will readily be perceived from the subjoined table:

Gases.	Specific Gravity.	Quantity of Oxygen for 100 volumes.
Olefiant gas	·970	300
Carburetted hydrogen	·556	300
Hydrogen	·069	50
Carbonic oxide	·972	50
Carbonic acid	1·538	None.

Of these gases, carbonic oxide and carbonic acid possess the greatest specific gravity; while the latter is not only destitute of illuminating property, but calculated, as we shall afterward show, to deteriorate to a great extent the quality of the luminiferous gases with which it may happen to be mixed.

There are cases, however, in which it is necessary to determine accurately the composition of a sample of coal-gas, and the following is the now generally adopted method of conducting the analysis.* The ingredients or impurities which may be present in the gas are—1, Common hydrogen; 2, olefiant gas and other hydrocarbons; 3, light carburetted hydrogen; 4, carbonic oxide; 5, carbonic acid; 6, sulphuretted hydrogen; 7, ammonia; 8, oxygen and nitrogen derived from the atmosphere. A qualitative examination is made thus—the proportion of ammonia and of sulphuretted hydrogen is usually very minute, and in most cases these gases must be sought for by placing the tests for their presence for some time in a current of the gas. In searching for ammonia a piece of moistened litmus paper feebly reddened is placed for a minute in a jet of the issuing gas. If the blue color be restored, ammonia is present. Paper soaked in a solution of acetate of lead may be subjected to a similar trial. If it turn brown, sulphuretted hydrogen is present. The presence of oxygen is detected by admitting a bubble of the deutoxide of nitrogen into a tube filled with the gas under trial, and looking through the tube obliquely upon a sheet of white paper; very small traces of oxygen may thus be de-

---

* Abridged from *Elements of Chemistry*, by Professor Miller, of King's College, London.

tected by the red tinge produced, owing to the formation of peroxide of nitrogen. The presence of carbonic acid may be readily detected by throwing up a little lime water, or solution of sub-acetate of lead, into the gas whilst standing in a tube over mercury. The existence of the other gases may be assumed, as they are certain to be present in greater or less quantity. The sulphuretted hydrogen and ammonia being neglected, and supposing that oxygen and carbonic acid are found to be present, seven different gases are therefore supposed to exist in the mixture. The following method may be adopted for their quantitative determination:—
1. *Oxygen.*—A volume of the gas is confined over mercury, and its bulk is measured with due attention to temperature and pressure. A piece of moist phosphorus, which has been melted upon the end of a long platinum wire to serve as a handle, is introduced from below through the mercury into the tube. After twenty-four hours the phosphorus is withdrawn, when the amount of absorption indicates the proportion of oxygen which was present. 2. *Carbonic Acid.*—This gas is determined in a similar manner, substituting a ball of caustic potash for the phosphorus; the second diminution in bulk shows the proportion of carbonic acid. 3. *Olefiant Gas and Heavy Hydrocarbons.*—These gases are absorbed by introducing a third ball, consisting of porous coke, moistened with fuming sulphuric acid. It is necessary, however, before reading off the volume of the gas, to introduce a ball of potash a second time, to withdraw the vapor of anhydrous sulphuric acid, which possesses sufficient volatility to introduce a serious error by dilating the bulk of the gas, unless it be completely removed. The total amount of absorption will indicate the proportion of olefiant gas, together with the vapors of condensible hydrocarbons. 4. *Carbonic Oxide.*—The separation of carbonic oxide from the other gases is not easily done with accuracy. The gas may be divided into two portions, one of which is to be carefully measured as it stands over mercury, and a small quantity of a solution of subchloride of copper in hydrochloric acid is to be added, and the mixture briskly agitated; the gas is then transferred to a second graduated tube, also standing over mercury, and a ball of potash is introduced for the purpose of absorbing the vapors of hydrochloric acid with which the

gas is saturated; the bulk of the gas may then be read off, and the volume of carbonic acid may be known by the loss in bulk. 5. *Nitrogen, Carburetted Hydrogen, and Hydrogen.*—In determining the proportion of these gases, that of carbonic oxide may also be ascertained, for which purpose a portion of coal-gas, in which the carbonic oxide is still present, is transferred to a siphon-eudiometer, and its bulk is measured: it is then mixed with twice its volume of oxygen, and the bulk of the mixed gases is again measured: the mixture is then exploded by means of the electric spark, and the bulk is a third time measured: call this diminution in bulk a, next inject a small quantity of a strong solution of potash, and the resulting condensation due to the absorption of carbonic acid may be called b; the remaining gases, c, consist of oxygen in excess and nitrogen; the quantity of oxygen in excess is ascertained by mixing the residual gas with twice its bulk of pure hydrogen, and a second time causing the electric spark to pass; one-third of the condensation observed will be due to the excess of oxygen; on deducting this excess from the residue c, the difference gives the quantity of nitrogen. The difference between the amount of the oxygen thus found to be in excess, and that originally introduced, will of course represent the quantity of oxygen consumed; call this d. We have now all the data for calculating the proportion of carburetted hydrogen, of hydrogen, and of carbonic oxide, which are present in the mixture. Let x represent the quantity of light carburetted hydrogen; this gas requires twice its own volume of oxygen for complete combustion, and furnishes its own volume of carbonic acid, which requires an equal volume of oxygen for its formation, or half the amount consumed; the other half of the oxygen being required by the hydrogen, which condenses in the form of water, $2x$ will be the diminution in bulk of oxygen which occurs on dotonation. Again, when hydrogen is converted into water, it requires half its bulk of oxygen, and both are condensed entirely. If y represent the bulk of the hydrogen, $\frac{3y}{2}$ will be the diminution in bulk of the mixed gases on detonation, which is occasioned be the hydrogen in the mixture. Let z represent the volume of carbonic oxide present; carbonic oxide, for conversion into carbonic acid, requires half its bulk of oxygen, the carbonic acid produced occupying the same bulk as the carbonic oxide. $\frac{z}{2}$ will therefore indicate the condensation which occurs on firing the mixture. The total condensation in bulk (a) which occurs on firing a mixture of light carburetted hydrogen, hydrogen, and carbonic oxide, will consequently admit of thus being represented—

(1.) $a = 2x + \frac{3y}{2} + \frac{z}{2}.$

Further, the quantity of the carbonic acid formed by detonation, b, is composed of a volume of carbonic acid equal in bulk to the light carburetted hydrogen, and a volume equal to that of the carbonic oxide, so that the quantity of carbonic acid may be thus indicated—

(2.) $b = x + z.$

And lastly, the oxygen consumed, d, will be composed of the following quantities: Light carburetted hydrogen, twice its bulk, $2x$; hydrogen half its bulk, $\frac{y}{2}$; carbonic oxide, half its bulk, $\frac{z}{2}$; or the total quantity of oxygen consumed will be the following:

(3.) $c = 2x + \frac{y}{2} + \frac{z}{2}.$

From these three equations the values of x, y, z, are determined:

$$x = c - \frac{a+b}{3}$$

$$y = a - c$$

$$z = \frac{a + 4b}{3} - c.$$

Hints respecting the Improvement of Coal-Gas.

Of all the combustible bodies having an elementary character, carbon and hydrogen are not only the most widely and copiously diffused throughout the three kingdoms of nature, but best adapted for the evolution of light during their combustion. It is only, however, when they are united together in due proportion that they answer the purpose most effectually; and, indeed, in a separate state their illuminating powers are so feeble, that even when their combustion is accelerated and rendered more perfect by the presence of oxygen, the light which they yield is yet unfit for many of the useful ends to which light is subservient. The substances in which carbon and hydrogen are united in the best proportion for the production of light are pit-coal in the mineral kingdom, and oils and fatty matter in the animal and vegetable.

The great abundance of coal, and the comparative cheapness at which it can be obtained, give it a decided advantage in point of economy over oleaginous matter, whether of animal or vegetable origin; while the processes of decomposing it, with the view of converting it into a volatile and elastic product, have been so much improved as to render the gas which it yields equally fit for the purposes of illumination with the more costly gases obtained from the oils.

The gas produced by the decomposition of coal and oleaginous matter at a high temperature is a compound of carbon and hydrogen, and consists chiefly of two gases, in which these elementary substances exist in definite proportions. One of these gases is termed carburetted hydrogen, and the other olefiant gas or bicarburetted hydrogen. The former contains one atom of carbon united with two atoms of hydrogen, and the latter an atom of each of these elements.

Of these two compounds of hydrogen and carbon, that which contains the largest proportion of the latter element is found to yield during its combustion the most brilliant light, and that too for a longer period of time. And, indeed, so great is the difference in these respects, that the hydrogen may not improperly be regarded as the mere solvent or vehicle of the carbon, acting the part of wick, and thus presenting that substance in a state sufficiently comminuted for its more perfect combustion. Accordingly, the more abundantly the hydrogen is impregnated with carbon the greater may we expect to be its illuminating power, and the fitter in every respect for yielding artificial light. These views are fully supported by experiment; for not only is the brilliancy of the light modified by the quantity of carbon held in solution by the hydrogen, but the time which a given portion of the gas takes to consume away by combustion is affected by it in a still greater degree.

To determine in what ratio the illuminating power of the gases obtained both from oil and coal was reduced by diluting them in various proportions with hydrogen, we instituted a series of experiments, the results of which are of importance inasmuch as they indicate not only that the mixture is deteriorated, but that the same quantity of carbonaceous matter yields less light the more largely it is diluted with hydrogen.

In the first experiment we took a portion of coal-gas of the specific gravity ·67, which we found to consume at the rate of 4400 cubic inches per hour, and yielded the light of eleven candles, being 400 cubic inches per hour for the light of one candle. This gas being diluted with a fourth part of its bulk of pure hydrogen, acquired the specific gravity ·55, and wasted away at the rate of 6545 cubic inches per hour, yielding the light of ten candles. As a fifth part of the compound gas was hydrogen, the remaining four-fifths, amounting to 5236 cubic inches, was the quantity of the coal-gas which in its diluted state gave the light of ten candles for an hour; so that 524 cubic inches of the original coal-gas were requisite to give the light of one candle for the same time. But in its unmixed state, 400 cubic inches were sufficient to give the light of one candle for an hour; and, consequently, the deterioration occasioned by the dilution was in the ratio of 524 to 400, or of 100 to 76, being 24 per cent. It must be distinctly kept in view that the deterioration has been reckoned, not with respect to the whole volume of the mixture (in which case it would have been 39 per cent.), but simply in reference to the coal-gas itself; and therefore the experiment, so far as it goes, justifies us in adopting the conclusion, that had the hydrogen existed originally in union with the coal-gas, the latter would have improved in quality 24 per cent. by its abstraction; because the residuary portion would not only have lasted longer, but yielded during its combustion a superior light.

In a second experiment, conducted in a similar manner, in which the proportion of hydrogen was one-third of the quantity of the coal-gas, the deterioration was 27 per cent.; in a third experiment, the proportion of hydrogen being a half of the volume of the coal-gas, the deterioration amounted to 31 per cent.; and in a fourth experiment, the quantity of hydrogen being exactly equal to that of the coal-gas, the deterioration extended to 36 per cent.

These results indicate a progressive deterioration in the quality of coal-gas by the admixture of hydrogen; and the important conclusion to which they lead is, that the abstraction or removal of the latter, though diminishing the entire volume, would improve the nature of the residuary portion not only in a higher ratio than the loss which the whole sustained in its bulk, but render that portion capable of yielding, for a longer period of time, a greater light than it could have done in its original state. Hence it may be inferred that the illuminating power of coal-gas, whether considered with respect to the cost of its production or the intensity of its light, admits of being improved; first, by impregnating the hydrogeneous element more largely with carbon; secondly, by preventing the disengagement of hydrogen in a free state during the carbonization of the coal; and, lastly, by detaching a portion of that gas from coal-gas when it already exists in admixture with it.

The first of these modes of improvement seems to be practicable, at least to a certain extent, by thoroughly drying the coal before it is introduced into the retorts, and modifying the pressure under which the gas is generated; the second, by preventing the gas after its formation from being exposed to a high temperature by allowing it to pass over very hot surfaces, the effect of which is to deprive it of carbon. The second object may also be assisted by arresting the process of distillation at an earlier period than is usually practiced, hydrogen and carbonic oxide being the products which predominate during the last periods of decomposition. On this point, however, the interests of the public and of the manufacturer are at variance. The consumer pays by measure, and hence it is the interest of the manufacturer to carry on the process of distillation as long as possible, for, by so doing, not only does he increase the quantity of gas but he improves the quality of the coke. With respect to the third mode of improvement, we are unfortunately, in the present state of our knowledge, acquainted with no method of detaching hydrogen from the gases with which it is mixed in oil or coal gas that would not impair the illuminating power of these gases to a greater extent perhaps than the benefit that would be derived from the removal of the hydrogen. A plan has been proposed by Mr. Lowe to increase the quantity of carbon in the gas by impregnating it with the vapor of coal naptha; for which purpose it was proposed to fill the wet gas meter at the house of the consumer with purified naptha, and to maintain it at the same height by means of a reservoir connected with the meter, by which means the gas would be measured and saturated with naptha at the same time. A more practical plan was to pass the gas through an ornamental vase containing a sponge saturated with naptha, and placed at some point between the meter and the burner.

To determine the diminution of the illuminating power produced by separating the particles of the inflammable gas during its combustion, and thus diminishing the temperature of the flame, it occurred to the writer of the present article that nitrogen, having neither the property of supporting combustion nor of adding to the quantity of combustible matter submitted to that process, was well fitted to answer for the intended purpose; and, accordingly, on mixing coal-gas of ordinary quality (which, when burnt alone, yielded the light of twelve candles when it consumed 5400 cubic inches per hour) with varying portions of nitrogen, results were obtained which implied that the diminution of the intensity of the light proceeded in a ratio much more rapid than was observed when the gas was diluted with hydrogen. Thus, when six volumes of the coal-gas were mixed with one volume of nitrogen, the expenditure per hour was 6000 cubic inches, and the light equivalent to that of nine candles, being 667 cubic inches per hour for the light of one candle. But one-sixth of the whole being nitrogen, the remaining five-sixths, amounting to 556 cubic inches, was the quantity of the coal-gas which, in its diluted state, afforded the light of a candle for an hour. On the other hand, the quantity of the coal-gas requisite, in its unadulterated state, to give an equal degree of illumination being $5\frac{40}{12}°$, or 450 cubic inches, it follows that the deterioration was in the ratio of 556 to 450, or 100 to 81 nearly.

By diluting the same coal-gas with other proportions of nitrogen as subjoined, and afterward applying to each of the results the same kind of reduction as that which we have already made, we have deduced the following table, which exhibits the gradual deterioration of the illuminating power of the same quantity of coal-gas, produced by the mere separation of the atoms of the gas during its combustion.

Volumes of Coal-gas.	Volumes of Nitrogen.	Illuminating Power.
60	0	100
60	10	81
60	12	69
60	15	55
60	20	37
60	30	29
60	60	4

When carbonic acid was used instead of nitrogen, similar results were obtained; only the deterioration was considerably greater. Thus, when five volumes of the coal-gas were mixed with one volume of carbonic acid, the illuminating power was reduced from 100 to 30, whereas in the case of the nitrogen it was from 100 to 69. It is therefore a fortunate circumstance that carbonic acid, which is so apt to be generated during the production of coal-gas, and has so debasing an influence upon its illuminating power, is readily absorbed by a variety of substances; while nitrogen, the less injurious as well as the less abundant accompaniment, cannot be separated from the other gases with which it may exist in mixture by any process yet known.

Deterioration of Gas by keeping it after it is prepared.

Both oil and coal gas suffer, by keeping, a gradual loss in their power of illumination, which seems to increase in a more rapid ratio than the time they are kept. The deterioration, though greatest when the gases are allowed to stand over water, takes place in a considerable degree even when they are kept over oil, or in air-tight vessels. Hence it may be presumed that the carbon held in solution by the hydrogen is separated from that element, partly by its own gravity, and partly perhaps by solution in the water, or by condensation in the liquid form.

To whatever cause the deterioration is owing, the fact itself is undoubted. Thus, an oil-gas which, when newly prepared, had the specific gravity 1·054, gave the light of a candle for an hour when it consumed 200 cubic inches; kept two days, it gave the same light with a consumpt of 215 cubic inches per hour; and kept four days, it required for the same light 240 cubic inches per hour. In the case of a portion of coal-gas, which, when newly prepared, required 404 cubic inches to yield the light of a candle for an hour, the same gas kept two days required 430 cubic inches; and kept four days, 460 cubic inches to yield the same light. These results indicate a progressive deterioration in the quality of the gases, increasing with the length of time they are kept; and it is deserving of remark, that in both gases the diminution of the illuminating power decreases in a faster ratio than the time increases. After being kept three weeks, the oil-gas was so much debased in quality that it required 606 cubic inches of it to yield the light of a candle for an hour; and hence its illuminating power was reduced to one-third of what it was when the gas was newly made. From these experiments it may obviously be inferred that both oil and coal gas should be used as soon as possible after they are prepared.

Economy of Coal-Gas.

Among the advantages which have resulted from the introduction of coal-gas, we may reckon, first, its comparative cheapness; and, secondly, its superiority to all the other modes of artificial illumination.

In forming a comparative estimate of the cost of coal-gas and that of the other means employed for procuring artificial light, we may contrast it with the expense of wax, tallow, and oil, the ordinary substances used for the purpose. It deserves to be remarked, however, that while the price of coal, in consequence of the regular and abundant supply of that article, is liable to little fluctuation, the cost of wax, tallow, and oil, on account of the more precarious nature of the sources from which they are obtained, varies exceedingly in different seasons. The very extensive use, too, into which coal-gas has been brought has produced a considerable effect upon the price of oil and tallow, as well as of wax; so that a comparative estimate of the expense of procuring the same extent of illumination from coal-gas and from these substances must appear less favorable to the former than would have been the case had the comparison been made when gas was first introduced. But by way of illustration, the approximative economy of the substances commonly employed for illumination may be contrasted as follows :—Supposing that 5 cubic feet of gas per hour give a light equal to that of 12 candles, then 1000 cubic feet, if burnt at the rate of 5 feet per hour, would give a light equal to that of 12 candles for 200 hours, at the cost of 4s. 6d., which is about the average price of gas in London per 1000 feet at the present time (1855). Suppose the candles to cost 9d. per lb., then 2 lbs. of candles, 6 to the lb., would burn for 6⅔ hours at the cost of 1s. 6d., or 60 lbs. would burn 200 hours, at the cost of 2l. 5s. Assuming wax to be three times the price of the candles, the cost of wax candles for 200 hours would be 6l. 15s.; and taking sperm oil at 8s. per gallon, 4 gallons would give a light equal to that of 12 candles for 200 hours, at a cost of 1l. 12s. So that, by comparing the cost of these various sources of light for equal periods of time, we have—

	L.	s.	d.
For wax candles, the cost of	6	15	0
For tallow candles, "	2	5	0
For sperm oil, "	1	12	0
For gas, "	0	4	6

The expense of gas, as compared with that of the other sources of light, will be—

Gas	1·0	Candles	10·0
Oil	7.1	Wax	30·0

In the above comparison we have taken London gas as the standard, which is scarcely fair, seeing that this gas is inferior in illuminating power to that of most other towns.

But the light obtained from coal-gas is not only procured at a smaller expense; it is also more convenient for most purposes than the light yielded by other substances. In the ordinary mode of lighting by tallow and oil, the light derived from their combustion cannot be diminished in intensity without considerable disadvantage and trouble; whereas in the case of gas, it may be reduced in an instant from the most perfect splendor to the feeblest degree of illumination by the simple adjustment of the stop-cock. The advantages arising from this easy method of regulating the light of gas, when it is used in the chambers of the sick, and indeed in all apartments where a variable but uninterrupted supply of light must be kept up, can only be duly estimated by those who have experienced them. To every branch of manufacturing industry which requires a steady and powerful light, the benefits which have resulted from the introduction of coal-gas are not less important. In many operations the light may be conveyed by means of flexible pipes, connected together with ball-and-socket joints, so as to be almost in contact with the fabric it is intended to illuminate, without the slightest risk of injury; and it may be kept in the same state for many hours in succession, or altered, as circumstances may render necessary.

For lighting churches, theaters, and other public buildings, where a strong and uniform light is required, gas answers the purpose more effectually than any other mode of illumination; partly from the facility of its application, and partly from the diversified and tasteful manner in which the jets of flame may be exhibited in various kinds of burners.

As a street light, its superiority is universally admitted; and from that application of gas it cannot be doubted that the metropolis, and other large towns, have derived great additional security against the perpetration of nocturnal crimes, as well as the means of carrying on the ordinary business of life, during the evening with nearly the same convenience as during the full light of day.

Secondary Products.

The chemistry of the gas manufacture has been for some years in a state of mutation, the effect of which has been to bring about important changes in the nature and amount of the secondary products. We may, however, refer to the methods of disposing of the usual secondary products, namely, the coke, the tar, and the ammoniacal liquor. A ton of Newcastle coals of the average weight of 2240 lbs. yields—

1 Chaldron of coke	=	1494 lbs.
12 Gallons of tar	=	136 "
10 Gallons of ammoniacal liquor	=	100 "
9000 to 10,000 Cubic feet of Gas	=	291 "
Loss	=	220 "
		2249 lbs.

It is found, on an average, that 1 cwt. of coals yields about 2 bushels of coke. About one-fourth of the quantity of coke produced is used as fuel for heating the retorts, and the remainder is sold. The tar and ammoniacal liquor or gas-water separate in the tar cistern, the tar forming the lower stratum. This is used in the manufacture of patent fuel and of creasote, and as a rough paint for out-door work, 100 lbs. of tar yield by distillation about 26 lbs. of an oily liquid known as *coal-oil*. A light product first distils over, which is called coal-naptha; the remaining pitch is used for paying the bottoms of ships, wooden piles, etc. The coal-naptha is used for dissolving caoutchouc, and for burning in the naptha-lamp. The ammoniacal liquor is used in the manufacture of sal-ammoniac, carbonate of ammonia, and prussian-blue. The presence of cyanogen in the ammoniacal liquor has led to its employment in the manufacture of ferrocyanide of iron or prussian-blue. It is stated that a gallon of ammoniacal liquor, when saturated with sulphuric acid, contains enough of cyanogen and cyanates to form, with a salt of iron, 24 grains of prussian-blue.

The secondary products of the Edinburgh gas-works are turned to account at the chemical works, situate at a distance of about two miles from them, the gas-works being on a lower level. They are, however, connected by a line of pipes, and the gas-liquor is lifted over the shoulder of the Calton Hill by means of a force-pump. The difference of level is then sufficient to carry it to the chemical works. The liquor is left for the tar to subside, but the ammoniacal liquor, consisting of an impure solution of carbonate and hydrosulphuret of ammonia, still contains a portion of tar, which is got rid of by distillation. The larger portion of the distilled liquid is converted into sal-ammoniac, and a portion into sulphate of ammonia. In order to obtain the sal-ammoniac, the liquor is neutralized with hydrochloric acid, and is then pumped into large cauldrons and evaporated to the crystalizing point, when it is drawn off into large vats, and on cooling deposits small feathery crystals; these are transferred to a stone chest, and are dried by the heat of a furnace below. The salt then resembles brown sugar; it is mixed with charcoal powder for the purpose of reducing any oxide of iron which may be present, and thus to get rid of the brown tint in the process of sublimation. The subliming vessels resemble a man's hat, and are arranged in the furnace with the crown downward; they are about three feet in depth, and two and a half in diameter, and they contain sufficient for a week's charge. Each pot is covered with a leaden cupola, luted on with clay, and the salt is at first allowed to sublime away through a hole in the center. This occasions some loss, but it appears to be a necessary precaution to prevent porosity in the sublimate. The central hole is then plugged with clay, and the sublimation is continued for a week. In this way hemispherical cakes of sal-ammoniac are produced; they are rasped on the surface to remove crust or coloring matter, and are broken into wedges, which are packed in barrels for exportation.

In preparing sulphate of ammonia the distilled ammoniacal liquor is saturated with sulphuric acid, and concentrated until small crystals are formed, which are removed by perforated ladles, dried, and packed in barrels lined with paper.

The tar, which contains a considerable portion of water, is transferred to a still, where crude naptha and vapor of water distil over. They separate in consequence of their different densities, and the naptha is digested with sulphuric acid in a leaden trough. This separates ammonia and other substances; the acid is removed by means of quick-lime, the naptha is washed with water, distilled, and is ready for the market. The remaining tar is raised to a higher temperature, and a liquid less volatile than naptha is produced; it is termed *pitch-oil*, and is used for impregnating wood, etc. The pitch in the still is then run out, when it settles into a soft solid, for which at Edinburgh no market has yet been found, but it may probably be turned to account as a cheap fuel.

Scarcely any market is found for the tar, which was formerly largely consumed at Continental seaports. The increase of gas-works on the Continent, and the absence of duty on foreign tar as distinguished from British tar, has greatly retarded the sale of the latter abroad.

Since the introduction of the Boghead cannel coal, a new secondary product has been obtained in the form of paraffine. It is separated at the Westminster gas-works as paraffine-oil, and is used for lubricating the machinery.

IRON.

HISTORY OF ITS MANUFACTURE,

WITH

AN ACCOUNT OF ITS PROPERTIES AND USES.

IRON.

Iron, on account of its abundance, working qualities, and tenacity, is probably the most useful and valuable of metals. According to Dr. Ure, "it is capable of being cast into moulds of any form, of being drawn into wire of any desired length or fineness, of being extended into plates or sheets, of being bent in every direction, of being sharpened, or hardened, or softened at pleasure. Iron accommodates itself to all our wants and desires, and even to our caprices; it is equally serviceable to the arts, the sciences, to agriculture, and war; the same ore furnishes the sword, the plowshare, the scythe, the pruning-hook, the needle, the graver, the spring of a watch or of a carriage, the chisel, the chain, the anchor, the compass, the cannon, and the bomb. It is a medicine of much virtue, and the only metal friendly to the human frame." In its primitive position it is commingled with the earth's strata in bountiful profusion; it is found in various combinations and conditions in every formation, and it is a constituent element of both animals and vegetables.

HISTORY OF THE IRON MANUFACTURE.

Malleable iron appears to have been known from a remote antiquity. Its obvious utility and great superiority over the softer metals, then commonly used, combined with the expense of its reduction, caused it to be highly prized, though the extreme difficulty of working it by the rude methods then employed greatly restricted its application.* There are notices in Homer and Hesiod of the arts of reducing and forging iron, but cast-iron was then unknown, an imperfectly malleable iron being produced at once from the ores in the furnace.

* This is shown by the epithet *much-wrought*, applied to it by Homer—*Iliad*, vi. 48.

It is probable that the Greeks obtained most of their iron through the Phœnicians from the shores of the Black Sea, and from Laconia.

It would be interesting to trace the gradual advances which have been made in the reduction of iron from its discovery to the present time ; to inquire into the circumstances which led to the successive changes in the processes, and into the principle on which those changes were founded ; to examine into differences in the products which from time to time ensued, and to notice the influence of these conditions on the extent and progress of the manufacture. Our knowledge of these changes, however, is scanty and imperfect, and we can only conjecture what was probably its early progress.

The furnaces which were first employed for smelting iron were probably similar to those now called *air-bloomeries*. They were probably simple conical structures, with small openings below for the admission of air, and a large one above for the escape of the products of combustion, and would be erected on high grounds in order that the wind might assist combustion. The fire being kindled, successive layers of ore and charcoal would be placed in it, and the heat regulated by opening or closing the apertures below.

The process of reduction would consist of the de-oxidation of the ore and the cementation of the metal by long-continued heat. The temperature would never rise sufficiently high to fuse the ore, and the product would therefore be an imperfectly malleable iron, mixed with scoriæ and unreduced oxide. It would then be brought under the hammer, and fashioned into a rude bloom, during which process it would be freed from the greater portion of the earthy impurities.

By such a process as this the Romans probably worked the iron ores of our own island; scoriæ, the refuse of ancient bloomeries, occur in various localities, in some cases identified with that people by the coincident remains of altars dedicated to the god who presided over iron. Mungo Park saw a rude furnace of this kind used by the Africans, and, indeed, with some modifications, it is still retained in Spain, and along the coast of the Mediterranean, where rich specular ores are worked.

The advantages of an artificial blast would soon become manifest, and a pair of bellows or a cylinder and piston would

soon be applied to the simple construction mentioned above.
Homer represents Hephæstus as throwing the materials from
which the shield of Achilles was to be forged into a furnace
urged by 20 pairs of bellows ($\varphi \tilde{\upsilon} \sigma \alpha \iota$). The inhabitants of
Madagascar smelt iron in much the same way, their blowing
apparatus, however, consisting of hollow trunks of trees,
with loosely fitting pistons.

The furnace corresponds to the *blast*-bloomery, and has by
successive improvements developed into the blast furnace,
now almost universally used, and into the *Catalan forge*, still
employed in some districts. The application of the blast
would offer considerable advantages; it would obviate the
necessity of an elevated site, place the temperature more
immediately under the direction of the smelter, and render
the whole process more regular and certain. The method of
reduction remained the same as before, but the product would
differ considerably, for whenever the blast was sufficiently
powerful, the iron would be *fused*, a partial carburation
would take place, and the resulting metal would be a species
of steel, utterly useless to the workmen of those days; hence,
it seems necessary to infer, that a rude process of refining
was invented, the metal being again heated with charcoal, and
the blast directed over its surface, the carbon would be burnt
out, and the iron become tough and malleable. The processes
might perhaps form two successive stages of one operation, as
at present practiced with the Catalan forge.

The increasing demand for iron, and the progress of internal communication, would lead the smelter to increase the
size and height of his bloomery, and this probably would
lead to a very unexpected result. The greater length through
which the ore had to descend would prolong its contact with
the charcoal, and a higher state of carburation would ensue,
the product being cast-iron—a compound till then perhaps
unknown.

From the time that cast-iron became the product of the
smelting furnace, the refining would be made a separate process, requiring a separate furnace and machinery. It would
soon be found also that, as the furnace increased in height, the
pressure of the superincumbent mass would render the materials so dense as to retard the ascent of the blast, and thus
cause it to become soft and inefficient; hence the internal

buttresses called *boshes* were first introduced to support the weight of the charge, relieving the central parts from the pressure, and permitting the free ascent of the blast. Whilst the good quality of the iron and the regularity of the process were thus insured, increase of quantity was the result of improvements in the blowing apparatus, which was now enlarged and worked by water-power. With these modifications, the furnace was the same essentially as the blast-furnace now employed, though not so large; indeed, until the introduction of coke at a much later period, the blast-furnace seldom exceeded 15 feet in height by 6 at the widest diameter. The more perfect operation of the blast-furnace allowed the reduction of the heaps of scoriæ which had been gradually accumulating during the period that the blast bloomeries had been in operation, and which contained 30 to 40 per cent. of iron. A new species of property was thus created, extensive proprietorships of Danish and Roman cinders were formed; large deposits of scoriæ which for ages had lain concealed beneath forests of decayed oak, were dug up, and in Dean Forest it is computed that 20 furnaces, for a period of upward of 300 years, were supplied chiefly with the bloomery cinders as a substitute for iron ore.

At what period the complete transformation of the blast-bloomery into the blast furnace was effected, it is impossible to say. It was probably in the early part of the 16th century, as we find that in the 17th the art of casting had arrived at a considerable degree of perfection, and in the reign of Elizabeth there was a considerable export trade of cast-iron ordnance to the Continent. In the forest of Dean are the remains of two blast furnaces, which formerly belonged to the kings of England, but they have been out of blast since the commencement of the struggle between Charles I. and his Parliament. Calculating from the quantity of scoriæ accumulated in their immediate neighborhood, which appear to have lain undisturbed for the last two centuries, Mr. Mushet has attempted to deduce the period of their erection, which he conceives to have been about the year 1550, in the time of Edward VI.

Up to this period wood charcoal was the only material employed in smelting operations, but the wants of a constantly increasing population, not less than the great consumption of the blast furnaces themselves, created a scarcity of this essential material, and gave a check to the manufacture. To such an extent had the wood been destroyed, that the cutting down of timber for the use of the iron-works was prohibited by special enactments; and the forests of Sussex alone appear to have been exempt from the general decree of conservation. The number of furnaces in blast decreased three-fourths, and the annual production, which but a short time before is said to have been 180,000 tons, was in 1740 reduced to only 17,350 tons.

James I. granted patents to ironmasters in various parts of the kingdom for using pit-coal in the manufacture of iron. The obstacles to its introduction, however, were numerous, and not easily overcome. The comparatively incombustible nature of coke, and its feebler chemical affinities, rendered a more powerful blast and a longer subjection to the heat indispensable to its successful adoption. Ignorance of the causes of failure operated long and seriously, but all difficulties were at length surmounted. An enlargement of the height of the furnace prolonged the contact of the ore and coke, and at last the employment of the steam-engine and improved blowing apparatus rendered the blast much more powerful and regular, and gave that impetus to the manufacture which has caused Great Britain to take the first rank in this branch of industry.

The first great improvement on the blowing apparatus was the substitution of large cylinders, with closely fitting pistons, for the bellows. The earliest of any magnitude were probably those erected by Smeaton at the Carron Iron-Works, in 1760.

In 1783–4, Mr. Cort of Gosport introduced the processes of puddling and rolling, two of the most important inventions connected with the production of iron since the employment of the blast furnace.

About this time the steam-engine of James Watt came into use, and along with it commenced a new era in the history of the iron trade and every other branch of industry. Its immense power, economy, and convenience of application, brought it at once into general employment. It was soon applied to pumping, blowing, and rolling; it enabled the mines to be sunk to a greater depth; refractory ores to be reduced with facility, and the processes of rolling, forging, etc., to be effected with a rapidity previously unknown.

Of late years, Scotland has made considerable progress in the iron manufacture. The introduction of railway communication, and the invention of the hot-blast, have given a stimulus to the trade which has raised Glasgow into importance as an iron district, and few towns possess greater facilities for the sale of their produce, than this central depot of the mineral treasures of the country by which it is surrounded.

The hot-blast process, for which a patent was taken out by Mr. Neilson in 1824, has effected an entire revolution in the iron industry of Great Britain, and forms the last era in the history of this material. This simple but effective invention has given such facilities for the reduction of refractory ores, that between three and four times the quantity of iron can be produced weekly, with an expenditure of little more than one-third the fuel ; and, moreover, the coal does not require to be coked, or the ores to be calcined.

In conclusion, we may add that there appear to have been five distinct epochs in the history of the iron trade.

The *first* dating from the employment of an artificial blast to accelerate combustion.

The *second* marked by the employment of coke for reduction, about the year 1750.

The *third* dating from the introduction of the steam-engine, and on account of the facilities with which that invention has given for raising the ores, pumping the mines, supplying the furnace with a copious and regular blast, and moving the powerful forge and rolling machinery, we may safely attribute this era to the genius of James Watt.

The *fourth* epoch is indicated by the introduction of the system of puddling and rolling, very soon after the employment of the steam-engine.

The *fifth*, and last—though not the least important epoch in the history of this manufacture—is marked by the application of the hot-blast—an invention which has increased the production of iron fourfold, and has enabled the ironmaster to smelt otherwise useless and unreducible ores; it has abolished the processes of coking and roasting, and has given facilities for a large and rapid production, far beyond the most sanguine anticipations of its inventor. Manufacturers taking advantage of so powerful an agent, have not hesitated to reduce improper materials, such as cinder-heaps and impure ores, and by unduly hastening the process, and attending to quantity more than to quality, have produced an inferior description of iron, that has brought the invention into unmerited obloquy.

THE ORES.

The ores of iron are found in profuse abundance in every latitude, imbedded in or stratified with every formation. They occur both crystallized, massive, and arenaceous; lying deep on strata of vast extent, filling veins and faults in other rocks, and scattered over the surface of the ground. Sometimes, but rarely, found native; usually as oxides, sulphurets, or carbonates, more or less mingled with other substances Of these ores there are perhaps twenty varieties, many of which are, however, rare; others are combined with substances which unfit them for the manufacture of iron, so that the remainder may be classed under the following general heads; their composition, however, varies greatly:

1. The magnetic oxides, in which the iron occurs, as $Fe_3 O_4$ or $Fe_2 O_3 + Fe O$. This is the purest ore which is worked; the best Swedish metal is manufactured from it. It is found in primitive rocks, and is widely diffused over the globe.

2. Specular iron ore, peroxide of iron, $Fe_2 O_3$. This is rich and valuable ore, and has been worked from a remote antiquity in Elba and Spain. It is found chiefly in primary and transition rocks.

3. Red and brown hæmatites, hydrated peroxide of iron. These ores occur in botyroidal radiating masses, in Cumberland, Ireland, America, and other places.

4. Carbonate of iron. This ore occurs mixed with large quantities of argillaceous, carbonaceous, and silicious substances, forming the large deposits of clay-ironstone and blackbands, from which most of the iron of this country is obtained. These strata are generally found in close proximity to the coal measures.

All the above ores are more or less mixed with silica, alumina, oxide of manganese, etc., and it may not be uninteresting to glance at their geographical distribution in Europe and America.

This country possesses peculiar and remarkable advantages for the manufacture of iron. The ores are found in exhaustless abundance, usually interstratified with the coal for their reduction, and in close proximity to the mountain limestone, which is used as a flux. In few countries do these three essential materials occur in such abundance, or so near together as to give the necessary facilities for a large and profitable production.

The ores principally employed are the clay-ironstones and carbonates of blackbands, which are found interstratified with the coal fields of Ayrshire, Lanarkshire, Shropshire, South Wales, and other parts, and these vary in richness in different localities, according to position and the amount of silica clay and other foreign matter with which they are associated. The chemical composition of three varieties of the ore used in Lanarkshire is given by Dr. Colquhoun, as follows:

	No. 1.	No. 2.	No. 3.
Protoxide of iron	53·03	47·33	35·22
Carbonic acid	35·17	33·10	32·53
Silica	1·40	·6·63	9·56
Alumina	0·63	4·30	5·34
Lime	3·33	2·00	8·62
Magnesia	1·77	2·20	5·19
Peroxide of iron	0·23	0·33	1·16
Bituminous matter	3·03	1·70	2·13
Sulphur	0·00	0·22	0·62
Oxide of Manganese	0·00	0·13	0·00
Moisture and loss	1·41	2·26	0·00
	100·00	100·00	100·37

The carbonic acid in the above ores may be partly combined with the lime as carbonate of lime, as well as with the protoxide of iron.

M. Berthier gives, according to Dr. Ure, the following analyses of the English and Welsh ironstones of the coal measures:

	Rich Welsh Ore.	Poor Welsh Ore.	Dudley Rich Ore or Gubbin.
Loss by ignition	30·00	27·00	31·00
Insoluble residuum	8·40	22·03	7·66
Peroxide of iron	60·00	42·66	58·33
Lime	0·00	6·00	2·66
	98·40	97·69	99·65

Calculating the amount of carbonate of iron and metallic iron indicated by the above analyses, we have:

Carbonate of iron	88·77	65·09	85·20
Metallic iron	42·15	31·38	40·45

The richness of the above ironstones would be about 33 per cent. of iron. In the process of roasting, 28 per cent. of the ore is dissipated.

Mr. Mitchell gives also the following assays of clay-iron-stone and blackband ore, as under:

	Clay iron-stone, Leitrim, Ireland.	Blackband Carbonate Ore.
Protoxide of iron	51·653	20·924
Peroxide of iron	3·742	·741
Oxide of Manganese	·976	1·742
Alumina	1·849	14·974
Magnesia	·284	·987
Lime	·410	·881
Potash	·274	trace.
Soda	·372	trace.
Sulphur	·214	·098
Phosphoric acid	·284	·114
Carbonic acid	31·142	14·000
Silica	6·640	26·179
Carbonaceous matter	2·160	16·940
Loss		2·420
	100·000	100·000

In North Lancashire and Cumberland, the red hæmatite ores

are now extensively worked, and great quantities are yearly shipped from Whitehaven, Ulverstone, etc., to Staffordshire, South Wales, and Scotland, for mixing with the poorer argillaceous and blackband ores. In Cumberland and North Lancashire, no less than 546,998 tons were raised in 1854 for this purpose, and the greater portion was exported from those districts.

In addition to these exports, about 25 to 30,000 tons are smelted by the hot blast at Cleator, in the neighborhood of Whitehaven. It produces a strong and ductile iron, considered highly valuable for mixing with the weaker irons. These ores have been carefully analyzed, and contain:

Peroxide of iron	90·3
Silica	5·0
Alumina	3·0
Lime	trace.
Magnesia	trace.
Water	6·0
	104·3

Or about 62 per cent. of metallic iron.

In Ireland there are vast deposits of iron ore of great richness, though as yet but little worked. Some of these, such as the ores worked at the Arigua mines, and the Kidney ores of Balcarry Bay, yield as much as 70 per cent. of iron. If these mines were worked more extensively, and if peat fuel were used in the smelting operations, the iron would probably be of the very best quality, and might rival the famed Swedish charcoal metal. Of this there is now every reason to hope, as the establishment of railway communication, with almost every part of Ireland, will open out the immense peat bogs of that country, and facilitate the introduction of vegetable fuel for the reduction of the ores, and create a large and important addition to other branches of Irish industry. In a communication to the writer from Mr. M'All, dated Scrabby, he states—"I have sent you samples of two kinds of iron ore, one is the red, the other the purple hæmatite. There are strata which are inexhaustible, and the ore can be raised and delivered at the furnace for less than a shilling a ton; the peat or vegetable carbon is equally cheap and abundant. Limestone of the purest quality is also close at hand, and can be delivered at the furnace at ninepence per ton. On account of the purity of these materials, iron of the greatest strength and ductility can be made, which, from its non-liability to corrode, would be admirably adapted for naval and marine purposes." Ireland is, therefore, according to Mr. M'All and others, in a condition to supply large quantities of excellent iron.

France possesses an abundant supply of iron ore, but on account of the scarcity of coal, the manufacture has been greatly restricted in extent. The introduction of railway communication is, however, rapidly removing the difficulty, and the operations of smelting are greatly on the increase. The railroad has enabled the French ironmaster to substitute coal for charcoal in the reduction of the iron ores, and in consequence an immense increase has taken place in the production of pig and manufactured iron. The ores are found in beds or strata in the Jura range; accumulated in kidney-shaped concretions in the fissures of the limestone; or dispersed over the surface of the ground, and but slightly covered with sand or clay.

They are found in the departments of the Yonne, the Meuse, and the Moselle, and indeed may be traced from the Pas de Calais on the north to the Jura on the south, indicating throughout an abundant and ample supply.

The present increased production of iron in France is chiefly due to the introduction of coal in smelting, but it may also be traced in some measure to the encouragement given by the Government to that branch of industry, and to the enterprise of such men as M. de Gallois and M. Dufrènoy, who have exerted themselves to extend its manufacture in that country. M. de Gallois resided in England for several years, immediately subsequent to the peace of 1815, and having obtained admission into the different iron-works here, he returned to France and established the works at St. Etienne, now probably the largest and most extensive in that country.*

---

* The universal exhibition of last year (1855) fully justifies the remarks in reference to the great increase of the iron trade of France. Any person in the least conversant with the imperfect machinery and processes of the iron manufacture as it existed in France some years since, could not have been otherwise than struck with the improved character of those exemplified in the Paris Exhibition. In no country (probably not excepting even this) has so great progress been made in so short a time, in advancing from a state of comparative rudeness to one of considerable perfection, as in France.

The production of crude pig-iron in France is now little short of 1,000,000 tons annually, but the demand for railways, rolling-stock, bridges, iron ships, girders, and other constructions is so great that large quantities of iron are still annually imported into this country.

Valuable deposits of the blackband and clay carbonate ores are found interstratified with the great coal-field of Ruhr; and the bog-iron and hæmatite ores are found in considerable profusion in Rhenish Prussia and other parts. In Upper Silesia, on the Vistula and the Oder, large deposits of coal and iron are found in juxtaposition, and are worked to a considerable extent.

The consumption of iron is not so great as in France, though it is increasing rapidly, as may be seen from returns recently given by the British Chargè d'Affaires at Berlin. These returns show that the amount of iron ore raised in Prussia has increased from 1,495,516 tons in 1853, to 2,144,509 tons in 1854; this has taken place in nearly all the producing districts, but chiefly on the Rhine, where the demand has increased from 719,684 to 1,068,656 tons; in Westphalia, from 146,320 to 380,014 tons; in Silesia, from 563,739 to 650,369 tons; in Lower Saxony and Thuringia, from 51,963 to 70,676 tons; in Prussian Brandenburgh, from 8084 to 12,731 tons; and in the Upper Zollverein, from 6736 to 13,063 tons.

In AUSTRIA, all the iron is smelted with charcoal or carbonized peat, and is in consequence of the finest quality; it may be applied to every description of manufacture, from the most ductile wire to the hardest steel. The production is, however, small. The ores are found in Hungary, Styria, Moravia, and Upper Silesia.

In BELGIUM, both coal and iron are found in equal abundance, and are worked at Charleroi, Liege, and at other places. The ores which are chiefly hæmatite, are derived from the limestone at the base of the coal measures.

The superiority of the SWEDISH iron has long been acknowledged, and till recently it has been unrivaled. This arises not only from the purity of the ore — the magnetic oxide of iron — but in consequence of its being smelted with charcoal only. The quantity is however restricted, as the ironmasters are allowed by law only a certain number of trees per annum, in order that the forests may not be totally destroyed. Coal does not exist in either Sweden or Norway.

In 1844 some experimental researches were undertaken by Mr. Fairbairn of Manchester, at the request of the Sublime Porte, in regard to the properties of iron made from the ores of Samakoff in Turkey. The ores were strongly magnetic, and contained, according to Dumas and others, 62 to 64 per cent. of iron. They consist of:

$$\begin{array}{ll} \text{One atom iron } 28 + \text{one atom oxygen } 8 = 36 \\ \text{Two atoms iron } 56 + \text{three atoms oxygen } 24 = 80 \\ \hline \text{Iron } \ldots 84 \quad\quad \text{Oxygen} \ldots 32 \quad 116 \end{array}$$

Some of these ores have been smelted with charcoal, and some very fine specimens of iron and steel produced. The manufacture is, however, in a languid state in Turkey, and although smelting furnaces, blowing apparatus, forges, rolling mills, etc., were prepared and sent out from this country, they are to a great extent useless among a people who have deeply rooted prejudices and habitual inactivity to overcome, and every thing to learn in all those habits of industry which indicate the rising prosperity of an energetic and an active people.

Both the magnetic, hæmatite, and clay-ironstones abound in the United States. The magnetic ores worked in New England, New York, and New Jersey; the hæmatite in Pennsylvania, New York, New Jersey, and other localities; but the greater part of the manufacture must eventually establish itself in the valley of the Mississippi west of the Alleghany range, where vast deposits of coal and iron exist, though at present but imperectly known or developed.* The ores in most of these districts are smelted with a mixture of charcoal and anthracite, and the usual limestone flux, and produce a very excellent quality of iron.

In Nova Scotia some of the richest ores yet discovered occur in exhaustless abundance. The iron manufactured from them is of the very best quality, and is equal to the finest Swedish metal. The specular ore of the Acadian mines, Nova Scotia, is said by Dr. Ure to be a nearly pure peroxide of iron, containing 99 per cent. of the peroxide, and about 70 per cent. of iron. When smelted, 100 parts yield 75 of iron, the increase in weight being due to combined carbon. The red

---

* Especially in Ohio and in Missouri.

ore Dr. Ure states to be analogous to the kidney ore of Cumberland, and to contain:

	(1)	(2)
Peroxide of iron	85·8	84.4
Silica	8.2	8·0
Water	6·0	7·6
	100·0	100.0

The Acadian ores are situated in the neighborhood of large tracts of forests, capable of supplying almost any quantity of charcoal for the manufacture of the superior qualities of iron and steel. Several specimens of iron from these mines have been submitted to direct experiment, and the results prove its high powers to resist strain, ductility, and adaptation to all those processes by which the finest description of wire and steel are manufactured.

The difficulties which the Government have had to encounter, during the last two years, in obtaining a sufficiently strong metal for artillery, are likely to be removed by the use of the Acadian pig-iron. Large quantities have been purchased by the War Office, and experiments are now in progress, under the direction of Lieutenant-Colonel Wilmot, Inspector of Artillery, and of Mr. Fairbairn, which seem calculated to establish the superiority of this metal for casting every description of heavy ordnance.

There are also some very rich ores at the Nictau mines, as the following analyses by Dr. Jackson show. They contain impressions of Silurian tentaculities, spirifers, etc.:

	Brown Ore somewhat magnetic.	Red Iron Ore.
Peroxide of iron	70·20	64·40
Silica	14·40	19.20
Carbonate of Lime	5·60	5·40
Carbonate of Magnesia	2·80	3·20
Alumina	6·80	1·20
Oxide of Manganese	·40	4·40
Water	·00	2·40
	100·20	100·20
	·20*	·20†
	100·00	100·00

* Gain from oxygen.
† Over-run, probably carbonic acid from carbonate of lime.

As our limits are circumscribed, it will not be necessary to extend this section further; suffice it therefore to observe, that in all countries nature has, with a beneficent purpose, interlaid and interstratified the whole surface of the globe with this useful and indispensable material, and it would ill bespeak that high intelligence with which man is endowed if he did not avail himself of, and turn to good account, the immense stores of mineral treasures which are so profusely laid at his feet.

THE FUEL.

The inquiry into the properties and composition of the ores of iron, and the processes employed for their reduction and subsequent conversion into bars and plates, would be incomplete unless accompanied by a descriptive analyses of the fuel by which they are fused. Indeed the results of the operations of smelting, puddling, etc., are so intimately dependent on the quality of the fuel employed, as to render a knowledge of its constituents essential to the manufacture of good iron.

Charcoal was at first universally employed in the manufacture of iron, and on account of its purity compared with other kinds of fuel, and its strong chemical affinities and consequent high combustibility, it is of very superior value where it can be obtained in large quantities at a moderate cost. This, however, is rarely the case, and hence its use is restricted within very narrow limits in most countries. Charcoal is the result of several processes, in each of which the object is to increase the amount of fuel in a given bulk. The wood being cut into convenient lengths, and piled closely together, in a large heap, the interstices being filled with the smaller branches, and the whole being covered with wet charcoal powder, is then set on fire. Care is taken that only sufficient air is admitted to consume the gaseous products of the wood, so as to maintain the high temperature without needlessly consuming the carbon. After the whole of the gaseous products have been separated, and the carbon and salts only are left, the access of air is prevented, and the heap allowed to cool.

Another and better process is to throw the wood into a large close oven or furnace, heated either by the combustion within it, or by a separate fire conducted in flues around it. By this process, not only is the yield greater and of better quality, from the slower progress of the operation, but the products of the distillation may be preserved and employed for a great variety of purposes. The following results of some experiments by Karsten, show the difference in yield of very rapid and very slow processes:

Wood.	Charcoal produced by quick carbonization	Charcoal produced by slow carbonization.
Young Oak	16·54	25·60
Old "	15·91	25·71
Young Deal	14·25	25·25
Old "	14·05	25·00
Young Fir	16·22	27·72
Old "	15·35	24·75
Mean..........	15·38	25·67

These, on the average, give for the quick process 15·3, and for the slow 25·6, being in the ratio of 1 : 1·67, or 0.67 in favor of the quick process.

PEAT.—This material seems likely to come into use for

smelting iron in countries such as Ireland, where neither coal nor wood are found in abundance. It is purer and less objectionable than coal, and if properly dried, compressed, and carbonized, would prove a very valuable fuel for the reduction of such ores as we have already described in the section on the iron ores of Ireland. It is carbonized in the same way as the charring of wood.

COKE.—Before the introduction of the hot-blast, this material was used to a very great extent in the manufacture of iron; it is prepared from coal in the same way that charcoal is prepared from wood, the operation being called the coking or desulphurizing process. The heaps do not require so careful a regulation of the admission of air as those of charcoal, on account of the comparatively incombustible character of the coke. Sometimes the heaps are made large, with perforated brick chimneys, to increase the draught through the mounds; at other times they are formed into smaller heaps, and the conversion takes place without the intervention of flues. The more usual and economical plan is, however, the employment of close ovens, by which process a great saving is effected, the yield being from 30 to 50 per cent. in the one case, and from 50 to 75 in the other, according to the nature and quality of the coal.

COAL.—The hot-blast has enabled the ironmasters to use raw coal in the blast furnaces, the great heat of the ascending current of the products of combustion coking it as it falls in the furnace. The sulphur, however, and other deleterious ingredients, do not appear to be so completely got rid of as when the coal is used in the shape of coke; and it appears probable that even with the hot blast, the separate process of coking might be advantageously used, on account of the greater purity of the iron produced.

The following tables, selected from various sources, give the composition of the different kinds of fuel, all of which are applicable to the reduction and fusion of the iron ores:

Fuel.	Locality.	Specific Gravity.	Carbon.	Hydrogen.	Oxygen and Nitrogen.	Ashes in 100 parts.	Authority.
Splint Coal		1 29	75 00	6.25	18 75	..	Thomson.
"		1.266	70 90	4 30	24.80	..	Ure.
"	Newcastle, Wylam	1.302	74.823	6.180	5.085	13.912	Richardson.
"	Glasgow.	1.307	82.924	6.491	10.457	1.128	
Cannel Coal		1 272	64.72	21.56	13.72	..	Thomson.
"		1.228	72.22	3.93	23.85	..	Ure.
"	Lancashire, Wigan	1.319	83.753	5.660	8.039	2 545	Richardson.
"	Edinburgh. *Parrot coal.*	1.318	67 597	5.405	12 432	14.566	
Cherry Coal		1 263	74.45	12.40	13.15	..	Thomson.
"	Newcastle, Jarrow.	1.266	84.846	5.048	8.430	1.676	
"	Glasgow.	1.286	81.208	5.452	11.923	1.421	Richardson.
Caking Coal	Newcastle, Garesfield.	1.280	87 952	5.239	5.416	1.393	
"	Durham, South Hetton.	1.274	83 274	5.171	3 036	1.519	
"		1.269	75 28	4.18	20.54	4.670	Thomson.
Anthracite	Swansea,	1.348	92.56	2 330	2.530	1.720	Regnault.
"	"	1.270	90 58	2.600	4. 00	..	Jacquelin.
"	South Wales,		94 05	3 38	2.570	..	Overman.
"	Pennsylvania,	1.462	90.15	2.430	2 45	4.770	Regnault.
"	"	..	94.89	2.550	2.560	..	Overman.
"	Massachusetts, Worcester.	..	28 35	0.920	2.150	68.65	
Peat	Vulcaire,	..	57.03	5 680	31.760	..	
"	Long,	..	58 09	0.980	31 370	..	Regnault.
"	Camp de Feu,	..	57 79	6.110	30.710	..	
"	Cappage,	..	51.05	6.85	39.555	2.55	
"	Kilbeggan,	..	61.04	6.67	30.46	1.83	Dr. Kane.
"	Kilbakan,	..	51.13	6.33	34.48	8.06	

According to Knapp, peat contains from 1 to 33 per cent. its weight of ash. In coal we have the following from Mr. Mushet's analyses:

	Specific gravity.	Carbon.	Ashes.	Volatile matter.
Welsh furnace coal	1·377	88·068	3.432	8·300
" " "	1·393	89·709	2.300	8·000
" slaty "	1·409	82·175	6 725	9·100
Derbyshire furnace coal	1·264	52·882	4·288	42·830
" cannel "	1·278	48·362	4·638	47·000

And again the analyses, from Overman, of the ash of coal, may be quoted, as showing the constituents contained in the ashes derived from combustion :

Sulphate of lime	80·3	3·6
Lime	3·8	2·5
Silex	14·2	85·7
Oxide of iron	1·7	0·0
Alumina	0·0	8·2
	100·0	100·0

The following table of the heating power of various kinds of fuel, from Knapp's Chemical Technology, is not without interest; in practice, however, only a portion of the absolute heating power is made available :

	Authority.	Lbs. of water heated from 0° to 100° centig. by 1 lb. of fuel.
Charcoal—		
Average	Berthier.	68·0
Peat from Allen in Ireland—		
Upper	} Griffith.	62·7
Lower		56·6
Pressed		28·0
Peat charcoal—		
Essone	..	50·7
Framont and Champ de Feu	Berthier.	58·9
Coke—		
St. Etienne		65·6
Besseges	} Berthier.	64·3
Rive de Gier		58·9
Brown coal—		
Mean of 7 varieties	Berthier.	50·3
Cannel coal, Wigan		64·1
Cherry, Derbyshire		61·6
Cannel, Glasgow	} Berthier.	56·4
" Lancashire		53·2
Durham		71·6
Gas coke, Paris—		50·3
Anthracite	} Berthier.	
Pennsylvania		69·1
Mean of 5 varieties		67·4

In concluding the observations on fuel, we may notice that the various kinds of coal are classed by mineralogists as the bituminous, and stone or anthracite coal. The first class is chiefly employed for the purpose of smelting, though, since the introduction of the hot-blast, anthracite is coming largely into use both in this country and America. Mr. Crane of South Wales was the first who attempted the reduction of iron ores by anthracite, and Mr. Budd, at his works at Ystalyfera, followed successfully in the same path. To these two gentlemen the public are indebted for having surmounted the obstacles to the employment of this fuel for smelting iron.

THE MANUFACTURE OF IRON.

The processes for the manufacture of iron, as we have already pointed out, are of two distinct kinds, those of cementation and those of smelting; the product of the former is imperfectly malleable iron, that of the latter, cast-iron, or iron combined with more or less carbon.

The first and older process is uncertain in its results, involves considerable expense, and as there are no efficient means of getting rid of the earthy impurities, it necessitates the employment of rich magnetic, specular, or hæmatite ores; on account of these defects, it is now seldom employed. The ores to be reduced by this process were heated with charcoal in open furnaces, the fire being urged by a blast. The oxygen, water, and volatile substances were driven off, and the iron—carburized and partly fused—sunk to the bottom of the hearth. The blast was then directed downward, so as to play over the surface of the iron, and oxidized the greater part of the combined carbon; during this operation the iron became tough and malleable, and fit for the hammer.

The process of smelting in the blast furnace is now almost universally adopted for the reduction of iron ores, and for the cheapness and working qualities of the metal produced, as well as for the rapidity of the manufacture, it is decidedly superior to all others.

Ores which contain much carbonic acid, water, or volatile matter, were at one time invariably subjected to a preparatory process of calcination, but since the introduction of the hot-blast, they are now frequently employed in the raw state.

[IRON.] [PLATE 1.

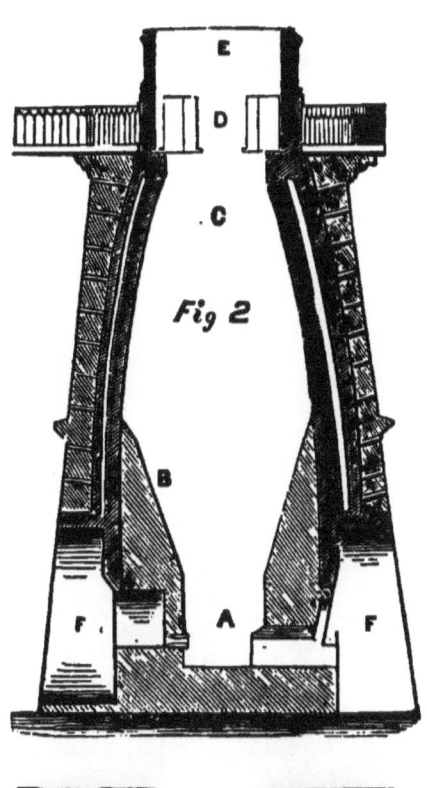

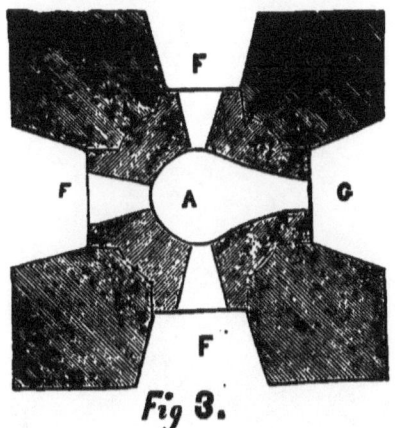

The calcination is sometimes effected in the open air, by stacking the ore with coal, setting fire to it, and allowing it to burn out; but this method is liable to serious objection. It is impossible to keep the temperature uniform throughout the heap, and in consequence, while some portions are scarcely affected, others are fused together into large masses, which cannot be smelted without difficulty, even when broken up. Apart from the irregularity and uncertainty of the open air process, it appears to be more expensive than the calcination in kilns, when the admission of air is entirely under command. These ovens or kilns are usually built of masonry, and are placed, if possible, on a level with the charging platform of the smelting furnace. The argillaceous ores lose, during this process, 20 to 30 per cent.; the carbonaceous, 30 to 40 per cent. of their weight.

The blast furnace consists of a large mass of masonry, usually square at the base, from which the sides are carried up in a slightly slanting direction, so as to form, externally, a truncated pyramid. In the sides there are large arched recesses, in which are the openings into the furnace for the admission of the blast, and for running out the metal and cinder; at the top of the furnace is a cylindrical erection of brickwork, called the tunnel-head, for protecting the workmen from the heated gases rising from the furnace, and having one or more doors through which the charges of ore, fuel, and flux are thrown into the furnace. In front, protected by a roof, is the casting-house, where the metal is run from the furnace into moulds.

Fig. 2 is a vertical section, and fig. 3 a plan of one of the furnaces at the Dowlais Iron Works. Mr. Truran, in a recently published and elaborate work on iron, has figured and described it. He states that it is one of the largest class, 38 feet square at the base, diminishing upward 3 inches for every vertical foot, till it attains a height of 25 feet, where the square form ends with a moulded cap; above this, the form is circular, diminishing in diameter at a similar rate, and finishing at top with a plain moulded cornice, as a support for the charging platform. In the section and plan A is the hearth, 8 feet high and 8 feet in diameter. BB the boshes, rising to the height of 15 feet, and 18 feet wide at their greatest diameter. From the top of the boshes the body of the furnace contracts, in a barrel-shaped curve, so that at the charging platform D, at a height of 50 feet, it is only ten feet in diameter; E is the tunnel-head, with doors of iron, to admit the charges of ore and fuel; FFF the tuyere-houses, arched over and spread outward, with the openings into the furnace for admitting the blast. G, the opening through which the iron is run from the furnace. The exterior is generally built of stone, and requires to be strongly bound with iron hoops, to prevent fracture from the expansion of the interior by the heat. The interior is lined with fire-brick set in fire-clay, a space of 2 or 3 inches being left between the two courses, to allow the expansion of the inner course. The hearth and boshes were usually constructed of refractory sandstone grit, or conglomerate, but fire-bricks are now chiefly used, and although they do not last so long, they are, in the end, more economical, and may be replaced whenever the furnace is blown out. The proper inclination of the boshes is a point of much importance, so that the materials, whilst smelting, may neither press too heavily downward, nor yet be so retarded as to adhere in a half-liquid state to the brickwork, and cool there, thus forming what are known by the name of *scaffolds*, the removal of which is a source of great inconvenience.

Another form of furnace is occasionally used for smelting, called the cupola, and built much more slightly than the blast furnace. Its form is circular, and from the boshes upward it is constructed of fire-brick, one, or sometimes two, courses in thickness. It is strongly bound together with wrought-iron hoops, and pillars of cast-iron, bolted at each end to imbedded rings of the same metal, rise through the foundation to the summit of the tuyere arches, giving considerable firmness and stability to the structure. Cheapness and facility of construction are much in its favor, and although objections have been made to the thinness of its sides, as permitting great loss of heat by radiation, it has met with very general adoption.

In addition to the cupola furnace, another of the same character has of late years been introduced. It consists of a truncated cone, composed entirely of boiler plates riveted together. On the four opposite sides recesses are cut to admit the tuyeres and the opening from the hearth into the casting-house. The interior of the furnace is lined with firebrick and fire-clay in the usual way, and this plate furnace is not only perfectly secure, as regards the expansion and contraction, but it is found to be economical and to answer every purpose in common with the large stone and iron-bound furnaces.

The blast is usually created by a steam-engine; a piston being attached to the extremity of the beam, working in a cylinder of large diameter, and forcing the air through proper valves into a large spherical reservoir, constructed of boiler-plate, whence its own elasticity causes it to flow in a regular unintermitting stream into the furnace. A cylindrical vessel, open at bottom, and immersed in a pit of water, has sometimes been used to regulate the pressure of the blast, but the water evaporated is detrimental to the working of the furnace. The nozzles by which the blast is directed into the furnace are made of cast or wrought-iron, and sometimes a current of water is conveyed round their extremities to keep them cool. The number of blow-pipe nozzles to each furnace varies at different works; the usual number is three, one for each of the tuyere houses, but sometimes six, eight, or twelve are employed; it, however, appears questionable whether this is not objectionable, as the density and penetrating power of the blast is considerably diminished by this system of diffusion. This, however, is a point which can only be decided by practice, and must be left to the judgment of the smelter. The usual pressure of the blast as it enters into the furnace is $3\frac{1}{2}$ lbs. per square inch, but in some cases it is as much as 5 lbs. per square inch.

The communication between the ground and the tunnel-head is effected in various ways. In South Wales the furnaces are usually built on a declivity, which affords ready means of access from behind; sometimes an incline is constructed, or other contrivances, such as the balance and pneumatic lifts, are resorted to for the elevation of the materials.

The dimensions and form of the blast furnace vary greatly, according to the fashion of the district, and the notions of the builder. Yet so much does the quantity and quality of the iron depend upon the size of the furnace and strength of the blast, that we may venture to assert that the production varies in the ratio of the cubical contents of the furnace, and the volume of air admitted. Mr. Truran gives the following particulars of the Dowlais Foundry iron furnace: "The capacity is 275 cubic yards. It is blown with a blast of 5390 cubic feet of [cold] air per minute. The materials charged at the top consist of calcined argillaceous ore, coal, and limestone. The yield or consumption averages 48 cwts. of calcined ore, 50 cwts. of coal, and 17 cwts. of broken limestone, to 20 cwts. of crude iron obtained. The weekly make of iron is occasionally over 190 tons. The weekly product of cinder amounts to 250 tons. For the production of white iron for the forge, in furnaces of the same capacity as the foregoing, a larger volume of the blast is employed, along with a different burden of materials. The blast averages 7370 feet per minute. The consumption of materials to one ton of iron averages 28 cwts. of calcined argillaceous ore, 10 cwts. of hæmatite, 10 cwts. of forge and finery cinders, 42 cwts. of coal, and 14 cwts. of limestone. With these materials the weekly produce amounts to 170 tons of crude iron, and 310 tons of cinder."

The action which takes place in the blast furnace is as follows: The contents being raised to an intense heat by the combustion of the fuel, are brought into a softened state; the limestone parts with its carbonic acid, and combining with the earthy ingredients of the ironstone, forms, with them, a liquid slag, whilst the separated metallic particles, descending slowly through the furnace, are deoxidized and fused; in their passage they imbibe a portion of carbon, and at last settle down in the hearth, from whence they are run off into pigs about every twelve hours; the slag, being lighter, floats upon the surface of the liquid metal, and is constantly flowing out over a notch in the dam-plate, level with the top of the hearth. This slag indicates, by its appearance, the manner in which the furnace is working; thus, if the cinder is liquid, nearly transparent, or of a light grayish color, and has a fracture like limestone, a favorable state of the furnace is indicated. Tints of blue, yellow, or green are caused by a portion of oxide of iron passing into the slag, and show that the furnace is working cold. The worst appearance of the cinder is, however, a deep brown or black color, the slag flowing in a broad hot rugged stream, and indicating that the supply of coke is not sufficient to deoxidize the whole of the iron.

During the process of smelting, the interior of the furnace requires to be very carefully watched. The stream of air constantly rushing in at the tuyeres, exerts a chilling agency on the melted matter directly opposed to it at its entrance. The consequence of this is the formation of rude perforated cones of indurated scoriæ, stretching from either side horizontally into the furnace, each one having its base directly over the embouchure of a blast-pipe. When these project only to a certain extent, they are favorable to the working of the furnace, as the blast is thrown into the center, and prevented from passing up the sides and burning the brickwork. Sometimes, however, when the furnace is driving cold and slow, these conduits of slag become so strong, and jut out so far as to meet in the middle, and thus cause a great obstruction to the entrance and ascent of the blast. When this happens, there is usually no remedy but to increase the burden, that is, to increase the quantity of *mine* or ore to the charge. This causes an intense heat, the furnace is said to work hot, and the conduits of slag drop off from the sides. This, however, is followed by bad as well as good consequences; the brickwork is frequently melted, and, for a time, the iron produced is small in quantity and of the worst quality. To bring the furnace again to its proper state, the burden must be reduced; the sides then become cool, new tubes of slag are formed, and the iron produced is good.

At the end of every twelve hours, more or less, the furnace is tapped, that is to say, the aperture in the dam-stone, which, at the commencement, had been stopped up with a mixture of loam and sand, is re-opened, and the metal contained in the hearth allowed to flow out into moulds, made in the sand of the cast-house floor, thus forming a cast or sough of pigs. When this operation ceases, the dam-stone is again secured, and the work proceeds as before. In this manner a furnace is kept continually going, night and day, and never ceases to work until repairs are necessary. Incessant action has even been thought necessary to the successful carrying on of an iron-work, but the example of perhaps the largest ironmaster in South Wales has shown, contrary to general practice in that district, that smelting may be discontinued for at least one day in the week without any very serious derangement of operations.

Thus far we have confined our observations to the production of iron by the cold-blast process; we have now to consider the changes introduced by the employment of a heated blast.

In the year 1828, Mr. J. Beaumont Neilson, a practical engineer at Glasgow, took out a patent for an "improved application of air to produce heat in fires, forges, and furnaces, where bellows or other blowing apparatus are required." Mr. Neilson proposed to pass the current of air through suitably shaped vessels, where it was to be heated *before it entered the furnace.* In this simple substitution of a hot-blast, heated in a separate apparatus, for a cold-blast heated in the furnace itself, consists the whole invention.

Like most other improvements, the progress of this was at first slow. Retarded by practical difficulties, which beset all new processes in their first use—stopped every now and then by the prejudices of custom and ignorance, which cling with inveterate tenacity to maxims of established practice, and repel indiscriminately innovations which improve and those which modify without improving—the invention was more than once on the point of being abandoned. A great part of the interest in its possible remuneration was transferred by the inventor to strangers, whose combined efforts and influence were necessary to insure its success. But though thus tardy in its first steps and feeble in its early efforts, the hot-blast process is now adopted at the greater number of the iron-works of Great Britain, and other parts of Europe and America.

It is perhaps not generally known that practical men, previous to Mr. Neilson's invention, universally believed that the colder the blast the better was the quality and quantity of the iron produced; and this opinion appeared to be confirmed by the fact that the furnaces worked better in winter than in summer. Acting on such views, the ironmaster actually resorted to artificial means of refrigeration, to reduce the temperature of the blast before it entered the furnace. The fact of the improved action of the furnace in winter may perhaps be explained as a consequence of the diminished amount of the aqueous vapor contained in the atmosphere in cold weather; and the opinion that the low temperature is the cause of the alleged increase of production has been shown to be wrong by the success of Mr. Neilson's invention.

This simple invention affects only the transit of the air from the blowing cylinder to the furnace, an oven or stove being interposed, through which, in appropriately shaped vessels, the air in conducted, and in which it is heated to 600° or 800° Fahr., or to any other temperature adapted for the purpose of smelting.

The earliest and simplest plan by which the blast was heated is shown in the sketch, fig. 4. In an oven of brickwork *000, with a fire fed by the door D, a large cylindrical tube or receiver $h\,h$, made of riveted boiler-plate, about 3 feet in diameter, and 8 or 10 feet long, was placed. The pipes, B and S, attached to the receiver $h\,h$ at the opposite ends, communicated with the blowing-cylinder and smelting-furnace respectively. Lunular partitions $p\,p\,p$, projecting from opposite sides on the interior of the receiver, caused the air passing through it to inpinge alternately first on one side and then on the other, in order that the temperature might be uniformly and effectively communicated from the metal to the blast. By this means a moderate current of air has been heated up to 300° or 400° Fahr.*

The figures of the transverse pipes vary considerably at different iron-works. Sometimes they rise up and form a large semicircular arch over the fire, 8 or 10 feet perpendicularly, and are then connected by an arch at the top; sometimes they cross the fire in the form of a pointed arch, variously acuminate, or a single large tube is used, traversing the furnace in a long spiral direction. Their cross-section is as various as the form in which they are bent; pipes of circular, flattened, elliptical, rectangular, heart-shaped, and other sectional forms have been employed, in order to increase the heating surface in proportion to the volume of the blast. All these forms of apparatus, although admirably adapted for heating the air, are liable to fracture, from the unequal expansion of the metal.

The more difficult the reduction of the ironstone the smaller must be the diameter of the hearth, so as to enable the blast to penetrate and circulate throughout the whole of its contents. In other conditions, where the ores are easily reduced,

---

* Various modifications of this plan are in use.

hearths of 9 feet diameter have been introduced with great advantage, and that without detriment to the quality of the iron produced. The diameter of the body of the furnace is likewise regulated by the quality of the materials used, and in cases where the coal is not bituminous, and the ore hard, a large diameter is found to work very irregularly; and the results have been, where furnaces have been erected 18 feet diameter, to have them reduced to only 6 feet.

The height of the furnace is also regulated by the nature of the materials and the strength of the blast by which they are reduced. Sometimes, when the coal is soft and crushed by the superincumbent pressure, it is bound or compressed to such an extent as to prevent the blast from penetrating the mass, and causes an irregular working of the furnace; and, moreover, under these conditions, it makes what is called white or silvery iron.

The pressure of the blast requires also to be regulated to suit the materials, and, according to the workings at Coltness Works, the pressure is about 4 lbs. on the square inch, and as much as 10,000 cubic feet of air is discharged into the furnace per minute. The temperature of the blast is 594° Fahr., and the area of the heating surface of the apparatus for raising that temperature is 3500 square feet.

The quantity of materials to make a ton of iron at these works varies in some relative proportion to their densities; but the following may be taken as a fair average of the consumption of fuel, ore, limestone, etc.:

Ton.	Cwt.		Ton.	Cwt.	
1	10	of raw coal.	0	4	of coal for heaters.
1	17	of calcined ironstone.	0	4	of " for blowing engine.
0	12	of broken limestone.			

With the above charges the furnaces will produce from 168 to 170 tons per week, or 8700 tons of good iron per annum.

With regard to the advantages and defects of the hot-blast process, much has been said on both sides, and the question does not appear by any means settled. It is asserted, on the one hand, that iron reduced by the hot-blast loses much of its strength, whilst, on the other, it is contended that the quality of the iron is richer, more fluid, and better adapted for general purposes than that produced by the cold-blast. The advocates of the hot-blast say that the process has in-

IRON.] [PLATE 2.

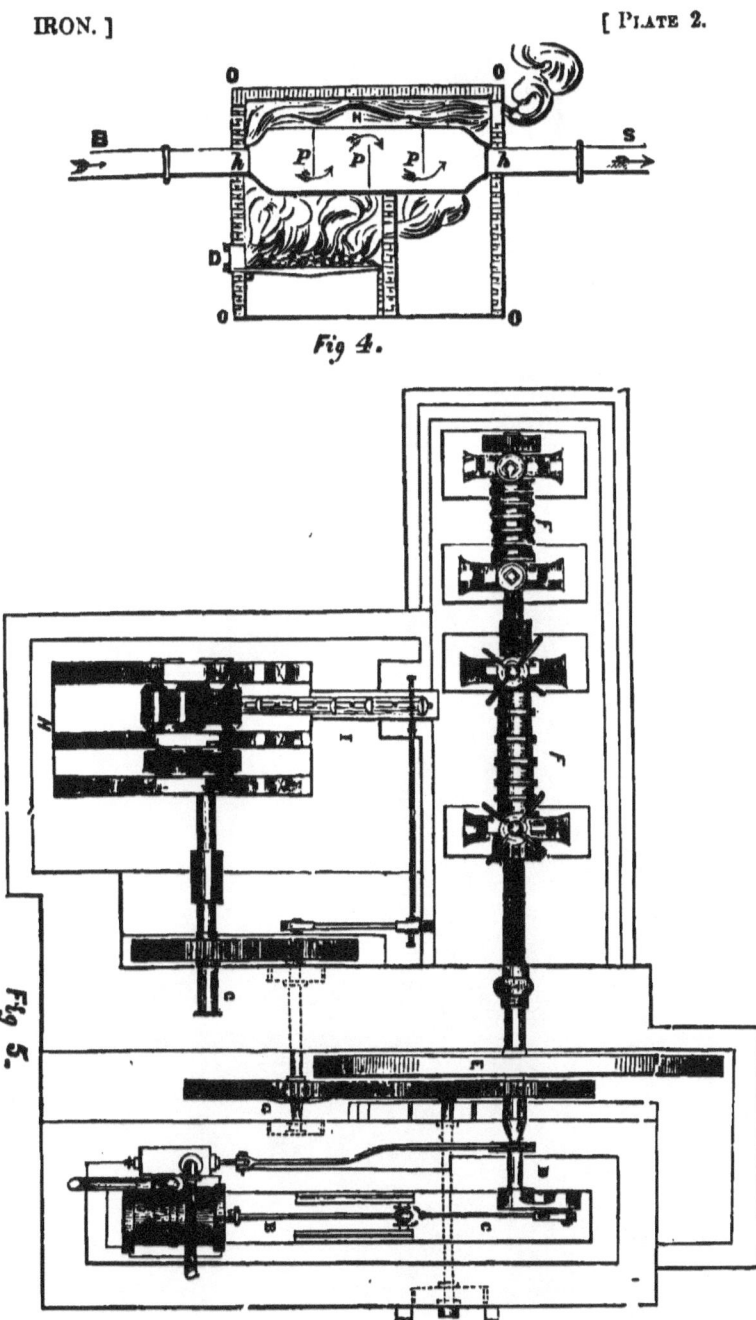

Fig 4.

Fig 5.

creased the production and diminished the consumption of coal three or four fold; and the upholders of the cold-blast maintain that the same effects may be produced, to almost the same extent, by a judicious proportion of the shape and size of the interior of the furnace, a denser blast, and greater attention on the part of the superintendent to the process.

On these points it appears to us that although the hot-blast has enabled the manufacturer to smelt inferior ores, cinder-heaps, and other improper materials, and to send into the market an inferior description of iron; this is no reason for its rejection, but rather an argument in its favor. It is true that when a strong rigid iron is required for such works as bridges or artillery, the somewhat uncertain character of hot-blast metal renders it objectionable, but this appears to be due rather to the carelessness or want of attention in the manufacture than to the use of heated air or defects in the process. On the other hand, the hot-blast, by maintaining a higher temperature in the furnace, insures more effectually the combination of the carbon with the iron, and produces a fluid metal of good working qualities, generally superior to cold-blast iron, in cases where great strength is not required; and, moreover, we have yet to learn why even the strongest and most rigid iron cannot be made by this process. The comparative strength of hot and cold-blast iron will, however, be given in another part of this article; for the present it is sufficient to observe that the results of the experiments are not unfavorable to the hot-blast iron, either as regards its resistance to a transverse strain, or its power to resist impact.

Dr. Clark, Professor of Chemistry in the University of Aberdeen, investigated the merits of the hot and cold-blast process in regard to the consumption of fuel, as early as 1834–5. He states, that after the hot-blast had been brought fully into operation at the Clyde Iron Works, "during the first six months of the year 1833, one ton of cast-iron was made by means of 2 tons $5\frac{1}{4}$ cwt. of coal, which had not previously to be converted into coke; adding to this 8 cwt. of coal for heating, we have 2 tons $13\frac{1}{4}$ cwt. of coal required to make one ton of iron. In 1829, when the cold-blast was in operation, 8 tons $1\frac{1}{4}$ cwt. of coal had to be used. This being almost exactly three times as much, we have from the change of the cold-blast to the hot, combined with the use of coal instead of coke, three times as much now made from the same quantity of coal." Dr. Clark adds the following statistics of the Clyde Iron Works:

"In 1829, the weekly produce of three furnaces, cold air and coke being used, was 110 tons 14 cwt.; and the average of coal to one ton of iron was 8 tons 1 cwt. 1 qr.

"In 1830, the weekly produce of three furnaces, coke, and air at 300° Fahr. being used, was 162 tons 2 cwt.; and the average of coal to one ton of iron was reduced to 5 tons 3 cwt. 1 qr.

"In 1833, the weekly produce of four furnaces, *raw coal*, and air heated to 600° being used, was 245 tons; and the average of coal to one ton of iron was reduced to 2 tons 5 cwt. 1 qr.

"On the whole then, the application of the hot-blast has caused the same fuel to reduce three times as much iron as before, and the same blast twice as much."

This decrease in the amount of fuel and blast required for the reduction of iron, Dr. Clark accounts for by showing that in an ordinary furnace, "2 cwt. of air a minute or 6 tons an hour are injected into the furnace." This he considers "a tremendous refrigeratory passing through the hottest part of the furnace," and to a great extent repressing the temperature which is necessary for the complete and rapid reduction of the iron.

Mr. Truran considers that "writers on the hot-blast have greatly exaggerated the effects of this invention on the iron manufacture of this country. If we are to believe the majority of them, the great reductions which have been effected within the last 25 years, in the quantities of fuel and flux to smelt a given weight of iron, and the large increase of make from the furnaces, is entirely owing to the use of this invention. That the hot-blast, under certain circumstances, has also effected a saving in the consumption of fuel, and also augmented the weekly make, we freely admit. But the saving of fuel, and increase of make due to its employment, is not generally one-fourth of the quantity which writers have asserted." Here Mr. Truran is at issue with Dr. Clark, and denies the cooling effect of a cold-blast. He attributes the effects of a heated-blast, "first to the caloric thrown into the furnace along with the blast, enabling a corresponding quantity of coal to be withdrawn from the burden of materials, with a proportionate reduction in the volume of blast, the effects of which are seen in an augmentation of the make, but do not result in the saving of fuel; secondly, to the reduced volume of blast and large proportion of caloric which it carries into the furnace, causing a diminished consumption of fuel in the upper parts of the furnace." Although we do not agree with all Mr. Truran's strictures on the hot-blast, the consumption of fuel in the throat is, nevertheless, a question well worthy of investigation. The combustion is of course largely increased by the narrow form of throat given to furnaces, which greatly increases the effect of the blast there, and accounts for the difficulty of using those kinds of coal, in the raw state, which splinter if rapidly heated. If Mr. Truran's conjectures be correct, and it be found, that by increasing the area of the throat, raw coal and anthracite can be advantageously used with the cold-blast, the superiority of the hot-blast will not be so decidedly marked. This must, however, be determined by practice; as at present, certainly, it is well known that the anthracite and splint coal can be used most effectively and economically with the hot-blast.

We quote from one more authority on this subject. M. Dufrènoy, in his report to the Director-General of Mines in France, states, that upon heating the air proceeding from the blowing cylinder up to 612° Fahr., a considerable saving in fuel was effected by the use of raw coal instead of coke, and that this caused no derangement of the working of the furnace or deterioration of the iron produced. On the contrary, "the quality of the metal was improved, and a furnace which, when charged with coke, produced only about half No. 1 and half No. 2 pig-iron, gave a much larger proportion of No. 1 after the substitution of raw coal. Besides this, the quantity of limestone was considerably diminished." This last circumstance, according to M. Dufrènoy, is due to the increased temperature of the furnace, which fuses more readily the earthy matter and other impurities in combination with the ores.

To show the saving effected, M. Dufrènoy gives the quantities used in each of the experiments at the Clyde Iron Works:

In 1829, the combustion being produced by cold air, the consumption for one ton of iron was—

	Tons.	Cwt.	Tons.	Cwt.
Coal—for fusion, 3 tons of coke, corresponding with	6	13		
" for blowing engine	1	0		
Total coal used			7	13
Limestone			0	10½

In 1831, the furnaces being blown with air heated to 450° Fahr.—

	Tons.	Cwt.	Tons.	Cwt.
Coal—for fusion, 1 ton 18 cwt. coke, corresponding with	4	6		
" for the hot air apparatus	0	5		
" for blowing engine	0	7		
Total coal used			4	18
Limestone			0	9

In July, 1833, the temperature of the blast being raised to 612° Fahr., and the fusion effected by *raw coal*, the consumption per ton of iron was—

	Tons.	Cwt.	Tons.	Cwt.
Coal—for fusion	2	0		
" for the hot air apparatus	0	8		
" for blowing engine	0	11		
Total coal used			2	19
Limestone			0	7

Since that time, the employment of a blast heated to 800° or 900° has still further increased the weekly production and saving of fuel.

The Waste Gases.—From the description that we have given of the smelting operations, it is evident that a large volume of gaseous products are constantly escaping at the top of the blast-furnace. These are found to contain a large proportion of unconsumed inflammable gas, capable of developing heat, and in countries where fuel is expensive, it is of great importance that these should be applied to useful purposes, and not be wasted in the atmosphere. Various contrivances have been adopted for this purpose, and in some places, particularly on the Continent, they have been utilized with great economy.

To enable the waste gases to be collected and applied to raising steam, heating hot-blast stoves, etc., without detriment to the working of the blast-furnace, it is necessary to withdraw them at an elevation where they have completed their work, yet at such a distance from the mouth of the furnace that they may be extracted in a dry state, and before they come into contact with the atmosphere, so as to cause combustion. This may be effected, either by increasing the height of the blast-furnace, withdrawing a portion of the gases through apertures in the side, or, if the furnace be not too large, by closing the top of the furnace with a movable door.

The Conversion of Crude into Malleable Iron.

The conversion of the carburized crude iron, obtained from the blast-furnace, into malleable or wrought iron is effected by several operations of an oxidizing character, in which it is sought to separate, in the gaseous state, the carbon contained in the iron, by combining it with oxygen, whilst the other metals alloyed with the iron and the phosphorus pass into the slag.

In reference to subsequent operations, the iron produced in the smelting furnace may be be divided into two kinds— that reduced by charcoal and that reduced by coke or raw coal. When charcoal iron has to be converted by charcoal, as in Sweden, it is decarburized in the charcoal refinery, with or without an intervening process. Where coal can be obtained, however, it is now usually converted by the process of puddling. Pig-iron produced by coke or coal is converted into malleable iron either by decarburation in the refinery or oxidizing hearth, and subsequent puddling, or it is converted at once in the puddling furnace by the process of boiling, which is equally effective, and is now more generally practiced.

This last process, as the one most generally adopted in Great Britain, deserves a special notice, and we are fortunate in having before us the particulars of the manner in which it is conducted by Messrs. Rushton and Eckersley of Bolton. This establishment is probably one of the most modern and complete of the kind in the kingdom; it is one that has spared no expense in the application of useful inventions, and has kept pace with every improvement that has taken place in the manufacture of bar and plate iron for the last fifteen years.

The machinery and appliances at these works consist of—

       6 Steam engines, of 180 total nominal HP.
       2 Five-ton and 2 fifty-cwt. steam hammers.
       3 Helve hammers.
       1 Set of puddled iron rolls.
       1 Set of boiler plate rolls.
       1 Merchant train and balling mill.
      16 Puddling furnaces.
      14 Balling and scrap furnaces.
And other machinery, such as plate and bar shears, lathes, etc.

At Messrs. Rushton and Eckersley's works, a small proportion of the Cumberland hæmatite ore is mixed with the crude pig-iron to be converted, as it is found to assist in the process of boiling in the puddling furnace, and in other respects to facilitate the process and improve the quality of the iron.

The crude pig-iron is assorted according to the degree and uniformity of its carburization, and classed as Nos. 1, 2, 3, etc.; No. 1 being most highly carburized, No. 2 less so, and so on to No. 4, which contains much more oxygen than the others. The carbon combined with iron gives it fusibility and fluidity, but deprives it of ductility. To render it malleable and capable of being welded, it must be deprived as far as possible of all the extraneous substances which have been mixed with it in the blast-furnace, more especially of the carbon. *Prima facie*, therefore, it would appear that the highly carburized pig-iron is the most suitable for casting, whilst that containing least carbon is best adapted for conversion into malleable iron; hence, in the trade, the crude iron is divided into foundry and forge pigs.

The pigs, however, in which carbon most predominates, and which, as a rule, have been most effectually separated from all *other* impurities during the process of smelting, are in many respects preferable for the manufacture of wrought iron; up to this time, however, great practical difficulties have attended the decarburization of iron containing so much carbon, and the white or forge iron is almost always preferred, measures having been taken for depriving it of the metals and earthy impurities not separated in the blast-furnace.

With regard to the process of refining, we may observe that the crude iron is melted in a hollow fire, and partially decarburized by the action of a blast of air forced over its surface by a fan or blowing engine. The carbon having a greater affinity for the oxygen than for the iron, combines with it, and passes off as gaseous carbonic oxide or carbonic acid. During this process, a portion of the silicum, etc., is fused out and separated from the iron. It is obvious from the above that the iron to be refined, being placed in contact with fuel at a high temperature, is liable to be deteriorated by the admixture of sulphur and other impurities of the fuel; and as the iron is only partially exposed to the action of the blast, the operation is necessarily, under these circumstances, imperfect. From the refinery the metal is run out into large moulds, and is then broken up into what is technically distinguished as "*plate-metal.*"

The process of puddling succeeds that of refining; and in this operation the reverberatory furnace is employed, with the fire separated by a partition or bridge from the hearth, on which is placed the metal to be puddled. By this arrangement the flame is conducted over the surface of the metal, creating an intense heat, though the deleterious portions of the fuel cannot mix with the iron. In this furnace the iron is kept in a state of fusion, whilst the workman, called the " puddler," by means of a rake or *rabble*, agitates the metal so as to expose, as far as he is able, the whole of the charge to the action of the oxygen passing over it from the fire. By this means the carbon is oxidized, and the metal is gradually reduced to a tough, pasty condition, and subsequently to a granular form, somewhat resembling heaps of boiled rice with the grains greatly enlarged. In this condition of the furnace the cinder or earthy impurities yield to the intense heat, and flow off from the mass over the bottom in a highly fluid state.

The iron at this stage is comparatively pure, and quickly becomes capable of agglutination; the puddler then collects the metallic granules or particles with his *rabble*, and rolls them together, backward and forward, over the furnace bottom, into balls of convenient dimensions (about the size of thirteen-inch shells), when he removes them from the furnace to be subjected to the action of the hammer or mechanical pressure necessary to give to the iron homogeneity and fiber. These processes of refining and puddling have universally been employed till recently; but improvements have rendered it simpler, and the refining process is now very generally abolished.

Shortly after the employment of the puddling process, it was found advantageous to mix a portion of crude iron with the refined plate metal, the expense of the process of refining being saved upon the iron used in the crude state ; and trusting to the decarburizing effects of the puddling furnace, it was found that the refining process might be altogether dispensed with, if the crude iron containing a proportion of oxygen and very little carbon was employed. In this single process it is to be observed, that as all the carbon has to be got rid of in the puddling furnace, the evolution of gas is much more violent, the fluid iron boiling and bubbling energetically during the period of its disengagement, and hence the operation has acquired the popular name of the " boiling" process.

In this operation the pig-iron when melted is more fluid, on account of containing a greater proportion of carbon than the metal from the refinery, and requires more labor in stirring it about and submitting it to the action of the current of air ; the process, moreover, is attended by a greater waste of iron than puddling either plate, or crude iron and plate mixed, but not so great a loss as in the two operations of refining and puddling. It must, however, be admitted that the superior fluidity of the iron in the boiling process has a more injurious action on the furnace. Notwithstanding these objections, the system of boiling without the intermediate process of refining has been gaining ground for the last ten years, and in many places has entirely superseded the use of the refinery ; recent events have therefore led to the conclusion, that in a short time the refining process will have become a thing of the past.

Numerous attempts have been made to secure a more scientific and perfect decarburization of the crude iron, but without success. One improvement, however, recently patented by Mr. James Nasmyth, gives promise of making the boiling process as nearly perfect as we may hope to see it. It has been in use for two years at the Bolton Iron Works, and from its constant employment in the puddling furnaces of that establishment, it has given direct proof of its utility, and is gradually extending itself among the large manufacturers as its advantages become known.

The invention consists of the introduction of a small quantity of steam at about 5 lbs. pressure per square inch, into the molten metal as soon as it is fused, as the oxygen of the steam has at that high temperature a greater affinity for carbon than for the hydrogen with which it is combined or for the iron, the carbon is rapidly oxidized off. The liberated hydrogen has no affinity for the iron, but unites with sulphur, phosphorus, arsenic, etc.—substances very injurious to the quality of the iron, if present even in very minute quantities, and yet frequently found in the ores and fuel.

The mode of operating is as follows: The steam is conveyed from the boiler to a vertical pipe fixed near the furnace door, having at its lower end a small tap or syphon, to let off the condensed steam, and prevent its being blown into the furnace. A cock with several jointed pieces of pipe are fastened to the flange of the vertical pipe, so as to form, as it were, jointed bracket pipes, somewhat similar to those of gas-pipes, which allow free motion in every direction. This apparatus is introduced into the furnace, immediately the iron is melted, the puddler moving it slowly about in the molten iron, while the steam pours upon it through the bent end of the tube. In the course of from five to eight minutes the mass begins to thicken, the steam pipe is withdrawn, and the operation finished in the ordinary way with the common iron *rabble*. The time saved by this process in every operation or *heat*, as it is technically called, averages from ten to fifteen minutes, and that during the hottest and most laborious part of the operation.

By means of this apparatus, the highly carburized pig-iron, which is the most free from impurities, is rendered malleable in one furnace operation, without the deteriorating adjuncts of the refining and puddling process as ordinarily practiced; in this operation no deleterious substance can combine with the iron, whilst in the refinery process the mixture of the fuel and metal is liable to deteriorate the latter with sulphur, silicum, etc. This new process, it is affirmed, has a beneficial effect in purifying the iron with greater economy and rapidity than any other process with which we are acquainted.

Irrespective of the improvements just described, there is another which is extensively used on the Continent, denominated the Silesian gas furnace. The new Silesian furnaces which are used in the manufacture of iron in that country, in place of our reverberatory air furnaces, and are said, on good authority, to be a very great improvement, not only in regard to the entire prevention of smoke and the economy of fuel, but also in simplifying the wrought-iron manufacture, and enabling a less skilled class of workmen to manage the furnaces.

The chief feature is the gas generator, which may be described as a close brick chamber with an opening at the bottom for the admission of air from a fan, by means of which the gases are driven out of the chamber into the furnace amongst the iron to be heated. At the point where the gases enter the furnace, a series of tuyeres are provided for the admission of air from the same fan. The pipes that convey the air and the gas from the retort to the tuyeres are both provided with valves in order that the attendant may modify the quantity from either source, so as to produce any intensity of flame the work may require, and also to produce perfect combustion, thus placing the entire action of the furnace under complete control. It is about eleven years since these furnaces were first introduced, and notwithstanding the prejudices that were naturally raised against them, they are said to be now extensively adopted in the Silesian district, and in great favor with both the master and the workmen.

In this description of furnace there appear to be three great advantages over the air furnace—

1st. The entire absence of smoke in consequence of complete combustion.

2d. The saving of upward of 33 per cent. in fuel, from the whole of the gaseous products being made available, and there being no necessity for the flame to pass up the chimney to produce draught, as in the case of the reverberatory furnace, which requires an inordinate supply of fuel as compared with what is wanted to work the fan.

3d. The absolute control the attendant has over the furnace, as regards the temperature and the simplicity with which it can be worked. Its operations in this respect are, according to those who have seen it at work, so perfect as to be as precise in its action as a machine.

An apparently new light has been thrown on the conversion of iron, by a paper read by Mr. H. Bessemer at the meeting of the British Association for the advancement of science, held at Cheltenham in August, 1856. In this paper the author announces to the world the discovery of an entirely new system of operations for the manufacture of malleable iron and steel. The crude metal is converted, by one simple process, directly as it comes from the blast-furnace. We should detract from its clearness did we attempt to curtail the lucid description in which Mr. Bessemer has recommended his invention to the manufacturers and the public; we therefore give the account in his own words:

Mr. Bessemer states that " for the last two years his attention has been almost exclusively directed to the manufacture of malleable iron and steel, in which, however, he had made but little progress until within the last eight or nine months. The constant pulling down and rebuilding of furnaces, and the toil of daily experiments with large charges of iron, had begun to exhaust his patience, but the numerous observations he had made during this very unpromising period, all tended to confirm an entirely new view of the subject, which at that time forced itself upon his attention, viz.—that he could produce a much more intense heat, without any furnace or fuel, than could be obtained by either of the modifications he had used, and consequently, that he should not only avoid the injurious action of mineral fuel on the iron under operation, but that he would, at the same time, avoid also the expense of the fuel. Some preliminary trials were made on from 10 lbs. to 20 lbs. of iron, and although the process was fraught with considerable difficulty, it exhibited such unmistakable signs of success, as to induce him at once to put up an apparatus capable of converting about 7 cwt. of crude pig-iron into malleable iron in 30 minutes. With such masses of metal to operate on, the difficulties which beset the small laboratory experiments of 10 lbs. entirely disappeared. On this new field of inquiry, he set out with the assumption that crude iron contains about 5 per cent. of carbon; that carbon cannot exist at a white heat in the presence of oxygen without uniting therewith, and producing combustion; that such combustion would proceed with a rapidity dependent on the amount of surface of carbon exposed; and, lastly, that the temperature which the metal would acquire would also be dependent on the rapidity with which the oxygen and carbon were made to combine, and consequently, that it was only necessary to bring the oxygen and carbon together in such a manner that a vast surface should be exposed to their mutual action, in order to produce a temperature hitherto unattainable in our largest furnaces. With a view of testing practically this theory, he constructed a cylindrical vessel of three feet in diameter, and five feet in height, somewhat like an ordinary cupola furnace, the interior of which was lined with fire-bricks, and at about two inches from the bottom of it he inserted five tuyere pipes, the nozzles of which were formed of well-burnt fire clay, the orifice of each tuyere being about three-eighths of an inch in diameter; they were put into the brick lining from the outside, so as to admit of their removal and renewal in a few minutes, when they were worn out. At one side of the vessel, about half way up from the bottom, there was a hole made for running in the crude metal, and in the opposite side was a tap-hole, stopped with loam, by means of which the iron was run out at the end of the process. In practice, this converting vessel may be made of any convenient size, but he prefers that it should not hold less than one or more than five tons of fluid iron at each charge. The vessel should be placed so near to the blast-furnace as to allow the iron to flow along a gutter into it; a small blast-cylinder is required, capable of compressing air to about 8 lbs. or 10 lbs. per square inch. A communication having been made between it and the tuyeres before mentioned, the converting vessel will be in a condition to commence work; it will, however, on the occasion of its first being used, after relining with fire-bricks, be necessary to make a fire in the interior with a few baskets of coke, so as to dry the brickwork and heat up the vessel for the first operation, after which the fire is to be carefully raked out at the tapping hole, which is again to be made good with loam. The vessel will then be in readiness to commence work, and may be so continued until the brick lining, in the course of time, is worn away, and a new lining is required. The tuyeres, as before stated, were situated nearly close to the bottom of the vessel, the fluid metal therefore rose some eighteen inches or two feet above them. It was therefore necessary, in order to prevent the metal from entering the tuyere holes, to turn on the blast before allowing the fluid crude iron to run into the vessel from the blast-furnace. This having been done, and the fluid iron run in, a rapid boiling up of the metal was heard going on within the vessel, the iron being tossed violently about, and dashed from side to side, shaking the vessel by the force with which it moved. Flame, accompanied by a few bright sparks, immediately issued from the throat of the converting vessel. This state of things lasted for about fifteen or twenty minutes, during which time the oxygen in the atmospheric air combined with the carbon contained in the iron, producing carbonic acid gas, and at the same time evolving a powerful heat. Now as this heat is generated in the interior of, and is diffusive in innumerable fiery bubbles throughout the entire mass, the vessel absorbs the greater part of it, and its temperature becomes immensely increased, and by the expiration of the fifteen or twenty minutes before named, that part of the carbon which appears mechanically mixed and diffused through the crude iron has been entirely consumed. The temperature, however, is so high that the chemically combined carbon now begins to separate from the metal, as is at once indicated by an immense increase in the volume of flame rushing out of the throat of the vessel. The metal in the vessel now rises several inches above its natural level, and a light frothy slag makes its appearance, and is thrown out in large foam-like masses. This violent eruption of cinder generally lasts about five or six minutes, when all further appearance of it ceases, a steady and powerful flame replacing the shower of sparks and cinders which always accompanies the boil. The rapid union of carbon and oxygen which thus takes place, adds still further to the temperature of the metal, while the diminished quantity of carbon present allows a part of the oxygen to combine with the iron which undergoes a combustion, and is converted into an oxide. At the excessive temperature that the metal has now acquired, the oxide, as soon as formed, undergoes fusion, and forms a powerful solvent of those earthy bases that are associated with the iron. The violent ebullition which is going on mixes most intimately the scoriæ and metal, every part of which is thus brought into contact with the fluid oxide, which will thus wash and cleanse the metal most thoroughly from the silica and other earthy bases, which are combined with the crude iron, while the sulphur and other volatile matters, which cling so tenaciously to iron at ordinary temperatures, are driven off, the sulphur combining with the oxygen, and forms sulphuric acid gas. The loss in weight of crude iron during its conversion into an ingot of malleable iron, was found, on a mean of four experiments, to be $12\frac{1}{2}$ per cent., to which will have to be added the loss of metal in the finishing rolls. This will make the entire loss probably not less than 18 per cent. instead of 28 per cent., which is the loss on the present system. A large portion of this metal is, however, recoverable by treating with carbonaceous gases the rich oxides thrown out of the furnace during the boil. These slags are found to contain innumerable small grains of metallic iron, which are mechanically held in suspension in the slags, and may be easily recovered. It has already been stated that after the boil has taken place, a steady and powerful flame succeeds, which continues without any change for about ten minutes, when it rapidly falls off. As soon as this diminution is apparent, the workman knows that the process is completed, and that the crude iron has been converted into pure malleable iron, which he will form into ingots of any suitable size and shape by simply opening the tap-hole of the converting vessel, and allowing the fluid malleable iron to flow into the iron ingot moulds placed there to receive it. The masses of iron thus formed will be perfectly free from any admixture of cinder oxide, or other extraneous matters, and will be far more pure, and in a forwarder state of manufacture, than a pile formed of ordinary puddle bars. And thus, by a single process, requiring no manipulation or particular skill, and with only one workman, from three to five tons of crude iron passes into the condition of several piles of malleable iron, in from thirty to thirty-five minutes, with the expenditure of about one-third part the blast now used in a finery furnace with an equal charge of iron, and with the consumption of no other fuel than is contained in the crude iron. To those who are best acquainted with the nature of fluid iron it may be a matter of surprise that a blast of cold air forced into melted crude iron is capable of raising its temperature to such a degree as to retain it in a perfect state of fluidity, after it has lost all its carbon, and is in the condition of malleable iron, which, in the highest heat of our forges, only becomes a pasty mass. But such is the excessive temperature that may be arrived at, with a properly shaped converting vessel, and a judicious distribution of the blast, that not only may the fluidity of the metal be retained, but so much surplus heat can be created as to remelt the crop ends, ingot, runners, and other scrap, that is made throughout the process, and thus bring them, without labor or fuel, into ingots of a quality equal to the rest of the charge of new metal. For this purpose a small arched chamber is formed immediately over the throat of the converting vessel, somewhat like the tunnel-head of the blast-furnace. This chamber has two or more openings in the side of it, and its floor is made to slope downward to the throat. As soon as a charge of fluid malleable iron has been drawn off from the converting vessel, the workman will take the scrap intended to be worked into the next charge, and proceed to introduce the several small pieces into the small chamber, piling them up round the opening of the throat. When this is done, he will run in his charge of crude metal, and again commence the process. By the time the boil commences, the bar ends or other scrap will have acquired a white heat, and by the time it is over, most of them will have melted and run down into the charge. Any pieces, however, that remain, may then be pushed in by the workman, and by the time the process is completed, they will all be melted and intimately combined with the rest of the charge; so that all scrap iron, whether cast or malleable, may thus be used up without any loss or expense. As an example of the power that iron has of generating heat in this process, Mr. Bessemer mentions that when trying how small a set of tuyeres could be used, the size he had chosen proved too small, and after blowing into the metal for one hour and three-quarters, he could not get up heat enough with them to bring on the boil. The experiment was therefore discontinued, during which time two-thirds of the metal solidified, and the rest was run off. A larger set of tuyere pipes were then put in, and a fresh charge of fluid iron run into the vessel, which had the effect of entirely remelting the former charge; and when the whole was tapped out it exhibited, as usual, that intense and dazzling brightness peculiar to the electric light.

"To persons conversant with the manufacture of iron, it will be at once apparent that the ingots of malleable metal which are produced by this process, will have no hard or steely parts, such as are found in puddled iron, requiring a great amount of rolling to blend them with the general mass, nor will such ingots require an excess of rolling to expel the cinder from the interior of the mass, since none can exist in the ingot, which is pure and perfectly homogeneous throughout, and hence requires only as much rolling as is necessary for the development of fiber; it therefore follows that instead of forming a merchant bar or rail by the union of a number of separate pieces welded together, it will be far more simple and less expensive, to make several bars or rails from a single ingot; doubtless this would have been done long ago had not the whole process been limited by the size of the ball which the puddler could make.

"The facility which the new process affords, of making large masses, will enable the manufacturer to produce bars that, on the old mode of working, it was impossible to obtain; while, at the same time, it admits of the use of some powerful machinery, whereby a great deal of labor will be saved, and the process be greatly expedited. Mr. Bessemer merely mentions this in passing, without entering into details, as the patents he has obtained for improvements in this branch of the manufacture are not yet specified. He next points out the perfectly homogeneous character of cast-steel—its freedom from sand cracks and flaws—and its greater cohesive force and elasticity, compared with the blister steel from which it is made, qualities which it derives solely from its fusion and formation into ingots—all of which properties malleable iron acquires in like manner, by its fusion and formation into ingots in the new process. Nor must it be forgotten that no amount of rolling will give to blistered steel (although formed of rolled bars) the same homogeneous character that cast-steel acquires, by a mere extension of the ingot to some ten or twelve times its original length.

"One of the most important facts connected with the new system of manufacturing malleable iron is, that all the iron so produced will be of the quality known as charcoal iron, not that any charcoal is used in its manufacture, but because the whole of the processes following the smelting of it, are conducted entirely without contact with, or the use of any mineral fuel; the iron resulting therefrom will, in consequence, be perfectly free from those injurious properties which that description of fuel never fails to impart to iron that is brought

IRON.] [PLATE 3.

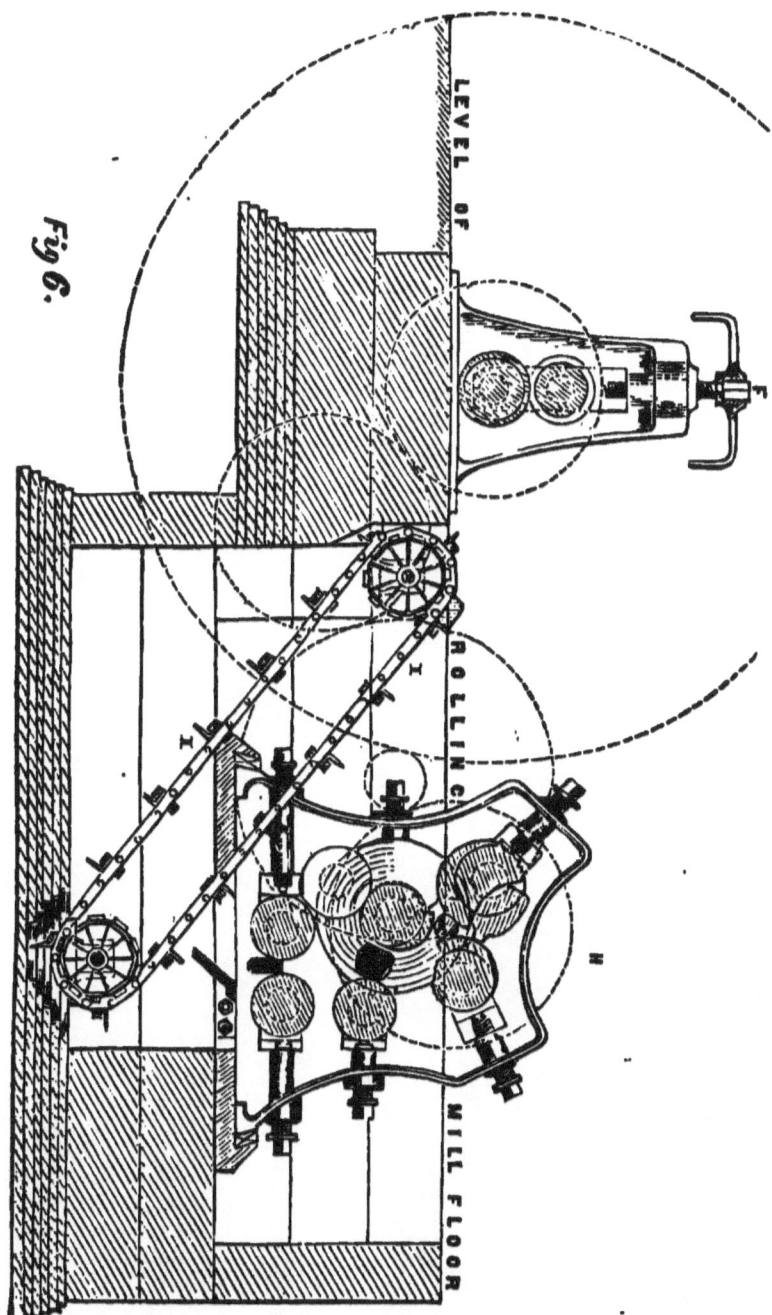

Fig 6.

IRON.] [PLATE 4.

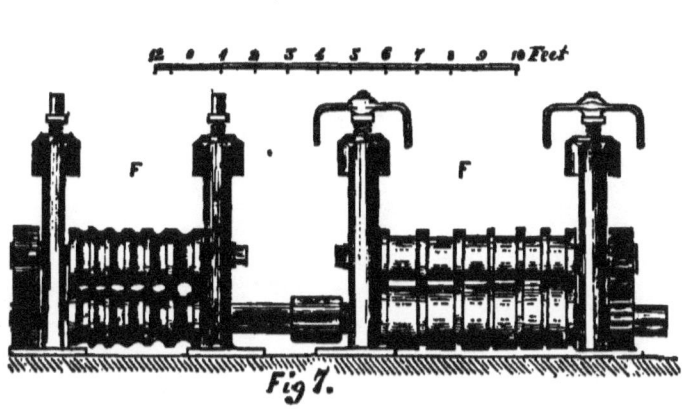

Fig 7.

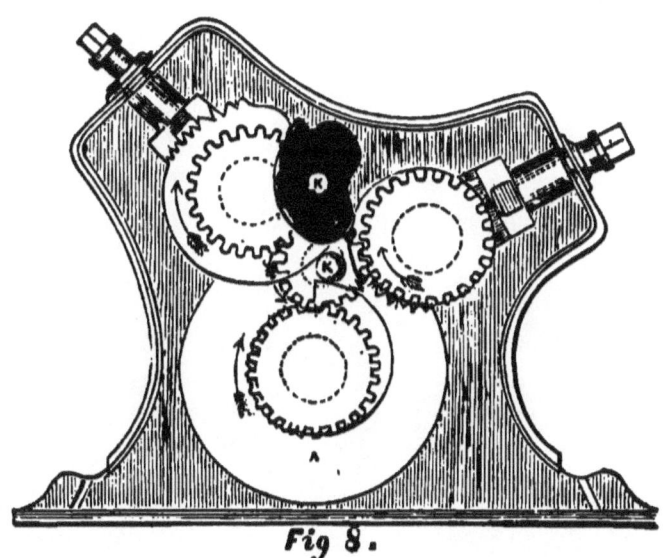

Fig 8.

under its influence. At the same time, this system of manufacturing malleable iron offers extraordinary facility for making large shafts, cranks, and other heavy masses; it will be obvious that any weight of metal that can be founded in ordinary cast-iron, by the means at present at our disposal, may also be founded in molten malleable iron, and be wrought into the forms and shapes required, provided that we increase the size and power of our machinery to the extent necessary to deal with such large masses of metal. A few minutes' reflection will show the great anomaly presented by the scale on which the processes of iron-making are at present carried on. The little furnaces originally used for smelting ore have, from time to time, increased in size, until they have assumed colossal proportions, and are made to operate on 200 or 300 tons of material at a time, giving out 10 tons of fluid metal at a single run. The manufacturer has thus gone on increasing the size of his smelting furnaces, adapting to their use the blast apparatus of the requisite proportions, and has by this means lessened the cost of production, in every way insuring a cheapness and uniformity of production, that could never have been secured by a multiplicity of small furnaces. While the manufacturer has shown himself fully alive to these advantages, he has still been under the necessity of leaving the succeeding operations to be carried out on a scale wholly at variance with the principles he has found so advantageous in the smelting department. It is true that, hitherto, no better method was known than the puddling process, in which from 4 cwt. to 5 cwt. of iron is all that can be operated upon at a time, and even this small quantity is divided into homeopathic doses of 70 lbs. or 80 lbs., each of which is moulded and fashioned by human labor, carefully watched and tended in the furnace, and removed therefrom, one at a time, to be carefully manipulated and squeezed into form. The vast extent of the manufacture, and the gigantic scale on which the early stages of its progress is conducted, it is astonishing that no effort should have been made to raise the after processes somewhat nearer to a level commensurate with the preceding ones, and thus rescue the trade from the trammels which have so long surrounded it. Mr. Bessemer then adverts to another important feature of the new process, the production of what he calls semi-steel. At the stage of the process immediately following the boil, the whole of the crude iron has passed into the condition of cast-steel of ordinary quality ; by the continuation of the process the steel so produced gradually loses its small remaining portion of carbon, and passes successively from hard to soft steel, and from softened steel to steely iron, and eventually to very soft iron ; hence, at a certain period of the process, any quality may be obtained ; there is one in particular, which, by way of distinction, he calls semi-steel, being in hardness about midway between ordinary cast-steel and soft malleable iron. This metal possesses the advantage of much greater tensile strength than soft iron ; it is also more elastic, and does not readily take a permanent set, while it is much harder, and is not worn or indented so easily as soft iron. At the same time it is not so brittle or hard to work as ordinary cast-steel. These qualities render it eminently well adapted to purposes where lightness and strength are especially required, or where there is much wear, as in the case of railway bars, which, from their softness and lamellar texture, soon become destroyed. The cost of semi-steel will be a fraction less than iron, because the loss of metal that takes place by oxidation in the converting vessel is about $2\frac{1}{2}$ per cent. less than it is with iron, but as it is a little more difficult to roll, its cost per ton may fairly be considered to be the same as iron, but as its tensile strength is some thirty or forty per cent. greater than bar iron, it follows that for most purposes a much less weight of metal may be used, so that taken in that way the semi-steel will form a much cheaper metal than any we are at present acquainted with.

"In conclusion, Mr. Bessemer observes that the facts he has discovered have not been elicited by mere laboratory experiments, but have been the result of operations on a scale nearly twice as great as is pursued in the largest iron-works, the experimental apparatus converting 7 cwt. in thirty minutes, while the ordinary puddling furnace makes only $4\frac{1}{2}$ cwt. in two hours, which is made into six separate balls ; while the ingots or blooms are smooth, even prisms ten inches square by thirty inches in length, weighing about as much as ten ordinary puddle balls."

MACHINERY OF THE MANUFACTURE.

The mechanical operations connected with the manufacture of wrought iron consist of shingling, hammering, rolling, etc., to which we may add the forging of "*uses*," that is, the forging of those peculiar forms so extensively in demand for steam-engines, railway carriages, and other works, which has lately become a large and important branch of trade.

In tracing the processes in the manufacture of wrought iron bars and plates, it will not be necessary to enlarge on those practices which have been superseded by more modern and improved machinery. Suffice it then to observe, that formerly the puddled balls were *shingled* or fashioned into oblong slabs or *blooms* by the blows of a heavy forge hammer; during this operation, the scoriæ and impurities which adhered to the balls were separated from the blooms by the force of impact, and then by a series of blows the iron was rendered malleable, dense, and compact. The blooms were then passed through a series of grooved iron rollers, which reduced them to the form of long, slender iron bars. These were cut up and piled regularly together or *fagotted*, and brought to a welding heat in the heating or *balling* furnace, when they were again passed several times through grooved rollers, and by this latter process were made into bars or plates ready for the shears.

In order to arrive at a clear conception of the mechanical operations employed in the manufacture of iron, it will be necessary to describe more at length the processes as at present practiced, with the improved and powerful machinery now employed; and as much depends upon the application of the motive power, the steam-engine claims the first notice. Until of late years, the vertical steam-engine was invariably used for giving motion to the forge hammer and rolling mill, which were placed on one side of the fly-wheel and the crank on the other; but the high-pressure, non-condensing engine is found to be decidedly preferable, as the waste heat passing off with the products of combustion from the puddling and heating furnaces, is quite sufficient to raise the steam for working the rolls and one of Brown's bloom queezers, as shown by fig. 5.

In this arrangement the cylinder A (figs. 5, 6, 7) is placed horizontally, and is supplied with steam from boilers near the puddling furnaces. The piston and slides B, and connecting rod C, give motion to the crank shaft D, on which is fixed a heavy fly-wheel E. The puddling rolls F are driven direct from the end of the fly-wheel shaft, and the bloom squeezers H, by a train of spur wheels GG. Under the lower rolls of the squeezers a Jacob's ladder or elevator I is fixed, for raising the block which has been deprived of its impurities, and reduced to an oblong shape by passing between the rollers of the squeezer. The block, on leaving the rollers, is carried in front of one of the projecting divisions of the ladder and thrown on to the platform in front of the rolls; the workman then seizes it with a pair of tongs and forces it into the largest groove in the rolls; it is then passed in succession through the other grooves till it attains the required form of the bar.

The drawings of Brown's bloom squeezers, figs. 8, 9, and 10, will sufficiently explain how the heated ball of puddled iron, K, thrown on the top, is gradually compressed between the revolving rollers as it descends and at last emerges at the bottom, where it is thrown on to the movable "Jacob's ladder," by which it is elevated to the rolls, as already described. This machine effects a considerable saving of time; will do the work of 12 or 14 furnaces, and may be kept constantly going as a feeder to one or two pairs of rolls. There are two distinct forms of this machine, one as shown in fig. 8, where the bloom receives only two compressions; and the other, which is much more effective, where it is squeezed four times before it leaves the rolls and falls upon the Jacob's ladder, as exhibited in figs. 9 and 10.

There are two other machines for preparing the blooms by compression. One is a table firmly imbedded in masonry, as shown at AA, in fig. 11, with a ledge rising up from it to a height of about two feet, so as to form an open box. Within this is a revolving box C, of a similar character, much smaller than the last and placed eccentrically in regard to it. The ball or bloom D is placed between the innermost revolving box C and the outer case AA, where the space between them is greatest, and is carried round till it emerges at E, compressed and fit for the rolls.

Another instrument, fig. 12, used for the same purpose, acts

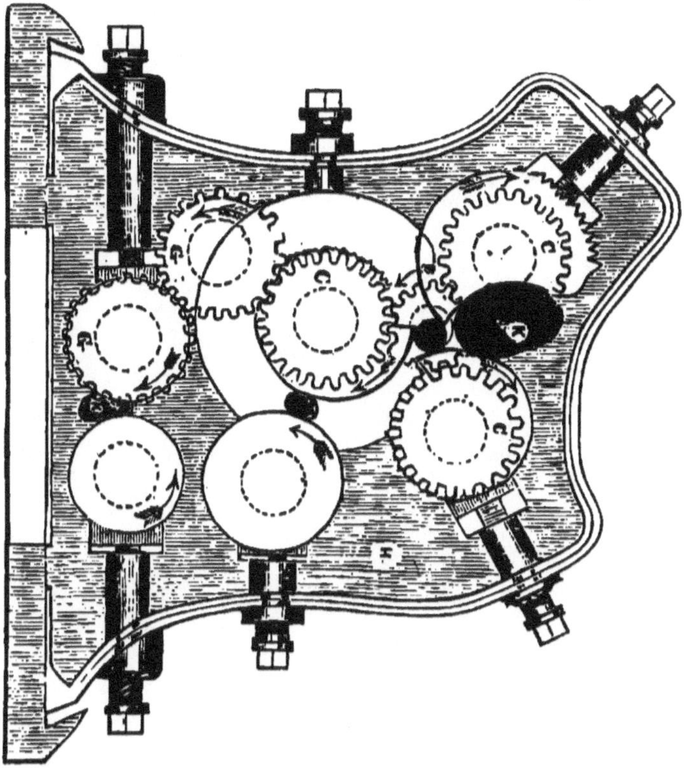

[IRON.] [PLATE 5.

Fig 9.

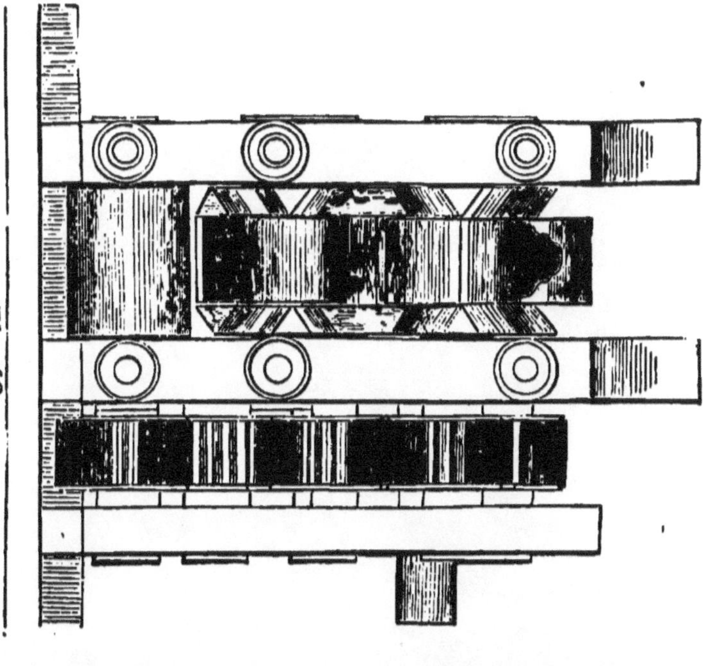

Fig 10.

[IRON.] [PLATE 6.

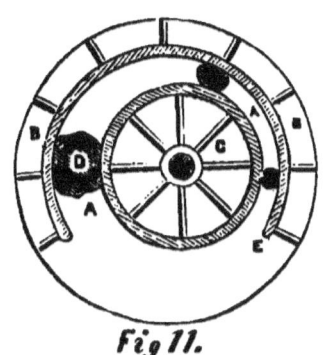

Fig 11.

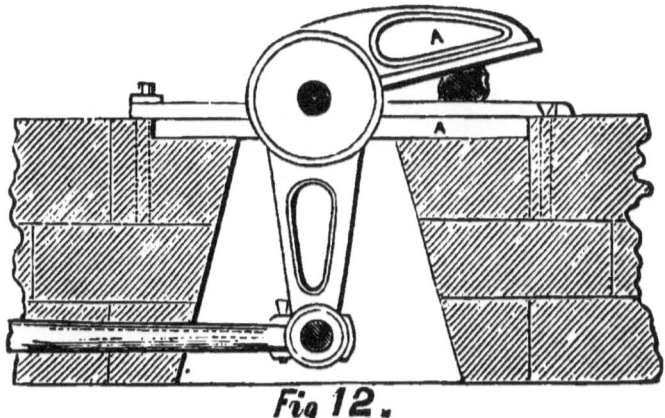

Fig 12.

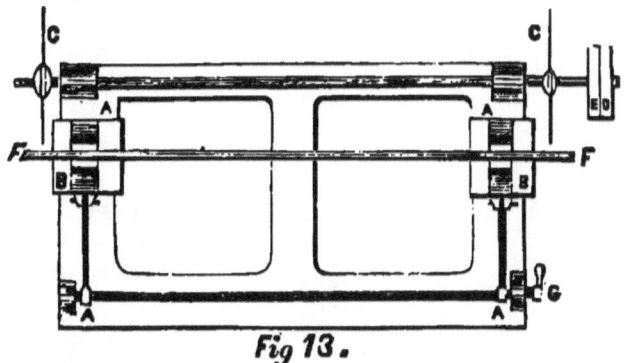

Fig 13.

as a pair of pliers, and squeezes the iron between two flat blades AA. This machine is called the Alligator, and is probably more effective than the horizontal machine, but it requires an attendant to keep the bloom rolling about under the jaws AA, and is, in other respects, inferior to Brown's patent squeezer.

We have stated that the horizontal, non-condensing steam-engine, from its compact form and convenience of handling, is admirably adapted for giving motion to the machinery of iron-works. For this object, it is superior to the beam-engine, as its speed can be regulated with the greatest nicety, by opening or shutting the valve, so as to suit all the requirements of the manufacture, under the varied conditions of the pressure of the steam, and the power required for rolling heavy plates and bars, or those of a lighter description. It is also much cheaper in its original cost, and all its parts being fixed upon a large bed-plate, require a comparatively small amount of masonry to render it solid and secure.

In regard to the manufacture of the rollers for the puddling, boiler-plate, and merchant train, the greatest care must be observed in the selection of the iron and the mode of casting. In Staffordshire there are roller-makers, but in general the manufacturer casts his own, and as much depends upon the metal, the strongest qualities are carefully selected and mixed with Welsh No. 1 or No. 2, and Staffordshire No. 2. This latter description of iron, when duly prepared, exhibits great tenacity, and is well adapted, either in the first or second melting, for such a purpose. In casting, the moulds are prepared in loam, and when dry are sunk vertically into the pit to a depth of about five feet below the floor. The moulding box is surrounded by sand firmly consolidated by beaters, and a second mould or head is placed above it, which receives an additional quantity of iron to supply the space left by shrinking, and keep the roll under pressure until it solidifies, and thus secures a great uniformity and density in the roller. The metal is run into the mould direct from the air furnace by channels cut in the sand, and immediately the mould is filled, the workman agitates the metal with a rod, in order to consolidate the mass and get rid of any air or gas which may be confined in the metal. This stirring with iron rods is continued till the metal cools to a semi-fluid state, when it is covered up and allowed slowly to cool and crystallize. This slow rate of cooling is necessary to favor a uniform degree of contraction, as the exterior closes up like a series of hoops round the core of the casting, which is always the most porous and the last to cool. In every casting of this kind it is essential to avoid unequal contraction, and this cannot be accomplished unless time is given for the arrangement of the particles by a slow process of crystallization. Rollers for boiler-plates and thin sheet iron are difficult to cast sound on account of their large size. They are subjected to very great strain, and require to be cast from the most tenacious metals. The bearings or neck should be enlarged, or turned to the shape shown at AA, and the cylindrical part B should be slightly concave, because, when the slab is first passed through the rollers, it comes in contact with a small portion only of the revolving surface. The central parts of the roller thus become highly heated, whilst their extremities are perfectly cool; the consequence is, that the expansion of the roller is greatest in the middle, and unless this be provided for by a concavity in the barrel, the plates become buckled, that is, both warped and uneven in thickness, and consequently, imperfect and unfit for the purposes of boiler making. Bar rolls are generally cast in chill, and great care is required to prevent the chill penetrating too deep so as to injure the tenacity of the metal and render it brittle.

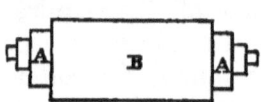

There are different kinds of rolling mills used in the iron manufacture, and they vary considerably in their dimensions according to the work they have to perform. The first, through which the puddled iron is passed, we have already described as puddling rolls. There are others for roughing down which vary from 4 to 5 feet long, and are about 18 inches diameter; those for merchant bars, about 2 feet 6 inches to 3 feet long and 18 inches in diameter, are in constant use. The boiler-plate and black sheet-iron rolls are generally of large dimensions; some of them for large plates are upward of 6 feet long and 18 to 21 inches in diameter; these require a powerful engine and the momentum of a large fly-wheel to carry the plate through the rollers, and not unfrequently when thin wide plates have to be rolled, the two combined prove unequal to the task, and the result is, the plates cool and stick fast in the middle. The greatest care is necessary in rolling plates of this kind, as any neglect of the speed of the engine or the setting of the rolls results in the breakage of the latter, or bringing the former to a complete stand.

The speed of the different kinds of rolling mills varies according to the work they have to perform. Those for merchant bars make from 60 to 70 revolutions a minute, whilst those of large size for boiler-plates are reduced to 28 or 30. Others, such as the finishing and guide rollers, run at from 120 to 400 revolutions a minute. In Stafford shire, where some of the finer kinds of iron are prepared for the manufacture of wire, the rollers are generally made of cast-steel, and run at a high velocity; such is the ductility of this description of iron, that in passing through a succession of rollers, it will have elongated to 10 or 15 times its original length, and when completely finished, will have assumed the form of a strong wire $\frac{3}{8}$ to $\frac{1}{4}$ of an inch in diameter, and 40 to 50 feet in length.

A high temperature is an indispensable condition of success in rolling. The experience of the workman enables him to judge, from the appearance of the furnace, when the pile is at a welding heat, so that when compressed in the rolls the particles will unite. Sometimes it is necessary to give a fine polish or skin to the iron as it leaves the rolls, but this can only be done when the iron cools down to a dark-red color, and by the practiced eye of an intelligent workman.

The above operations would still be incomplete unless the ironmaster had means of cutting the bars and plates to any required size or shape. The machinery for this purpose has of late been brought to a high degree of perfection, both in regard to power and precision.

The circular saw has been successfully applied for squaring and cutting the larger descriptions of bars, and does its work, particularly in railway bars, with almost mathematical precision. This machine consists of a cast-iron frame or bed AA, fig. 13, bolted down to a solid foundation, on the ends of which slide two frames BB to support the bar to be cut. The two circular saws or cutters CC are driven by straps passing over the pulleys DE, and rotate at the rate of 800 to 1000 revolutions per minute. The machine is set in motion by transferring the straps from the loose pulley D to the fast pulley E, and as soon as the required speed is attained, the frame BB is carried forward, and the bar FF along with it, by a lever G or eccentric motion, till the bar is cut through. The rate of cutting or pressure upon the saws may be regulated either by hand or weight; care must, however, be taken not to allow the saws to become too hot, and this is provided against by running them in a trough of water. By this process it is evident that the bar must always be cut square at the ends and correctly to the same length.

A great variety of shears are used for cutting iron, some driven by cams or eccentrics, and some by connecting-rods and a crank on the revolving shaft. In large iron-works it is necessary to have two or three kinds, some for cutting up scrap iron and bars for piling, and others for boiler-plates. Of the first we may notice two, one shown in fig. 14, cuts on both sides at AA, and is driven by a crank and connecting-rod B. This machine is chiefly used for cutting puddled bars from the puddling rolls, or any work required for shingling. The next machine, fig. 15, receives motion in the same manner, and also cuts on both sides, the cutters being fixed on the lever and moving with it. This is used for the same purpose as the last, and likewise for cutting scrap iron. These machines are extensively used in the manufacture of iron, and before the introduction of the plate shears, they were used, with some modifications, to cut boiler-plates, but the work was very imperfectly executed.

The demand for plates of large dimensions and greatly increased weight, such as those for the front and tube plates of locomotive and marine boilers, and those for tubular and plate bridges, created great difficulties, not only in piling, heating, and rolling, but also in cutting the plates accurately to the required size. To meet these demands, and more particularly for the manufacture of the large plates employed in the cellular top of the Britannia and Conway tubular bridges, Messrs. G. B. Thorneycroft and Co. constructed a large shearing machine which cut upward of 10 feet at one stroke. These shears have now come into general use, and are of great importance, on account of the accuracy with which they cut plates of large dimensions, square and even. Figs. 16 and 17 represent this machine; $a\,a\,a$ is the standard and table on which the plate is fixed. This table slides

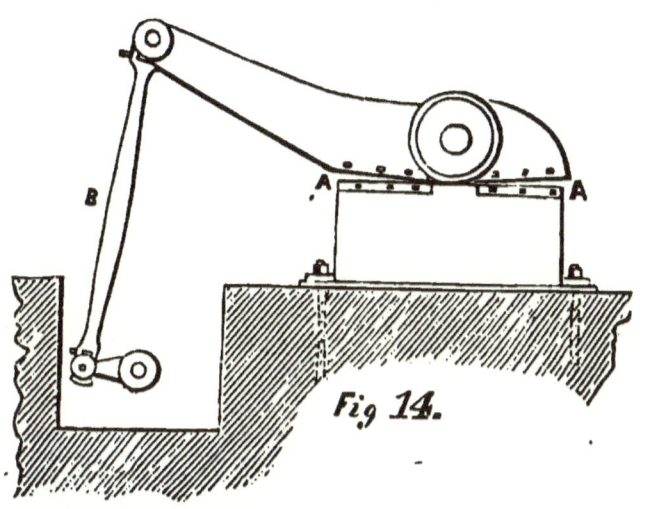

Fig 14.

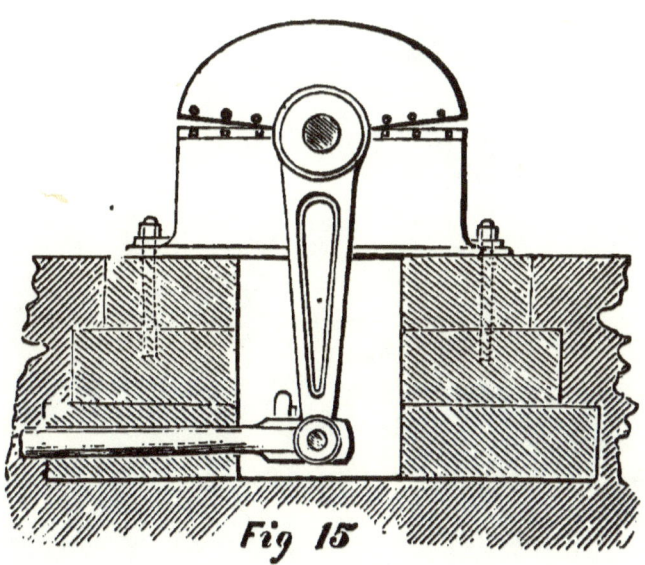

Fig 15.

[IRON.] [PLATE 8.

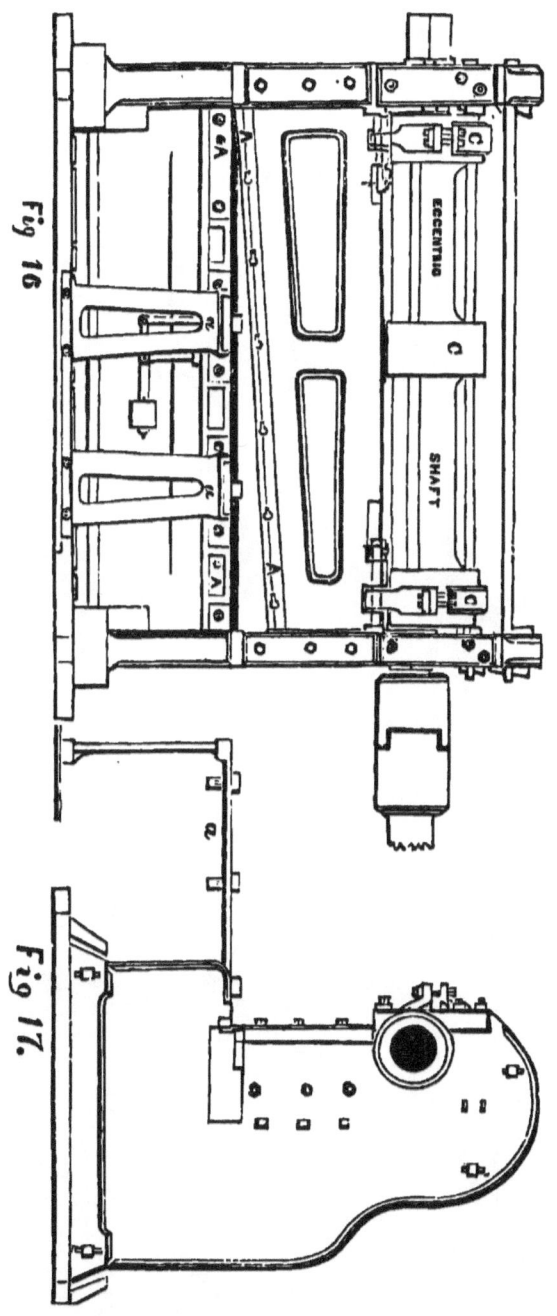

Fig 16

Fig 17.

forward at right angles to the shears or cutters AAA*A*. The top cutter descending by the action of three eccentrics ccc, which press upon the top of the frame B as it revolves, and force it down, and by one stroke, the knife AA cuts through the whole length of the plate, perfectly clean. and straight. The plate is then reversed, the newly cut edge being held against the slopes, and the sliding frame again moved forward to the required width of the plate, when another stroke cuts the other side as before. The rapidity with which the plates are cut is another advantage of this machine, as great as the precision of its cut, and when the immense quantity of plates daily produced at Messrs. Thorneycroft and Co.'s works are considered, its importance becomes evident.

At the Paris Universal Exhibition (1855), a plate-cutting machine was exhibited, from the United States, which appears to effect the same operation as Messrs. Thorneycroft and Co.'s. It consists of a strong cast-iron frame, nine or ten feet wide, having inserted along its face a steel plate, on which the iron to be cut rests and is held firmly by a faller, which descends on the upper side of the plate. On the same side of the frame a revolving steel cutter, about nine inches in diameter, traverses the whole length of the frame, and in its passage cuts the plate, by compression, in a perfectly straight line, corresponding with the steel edge below. Cutting and shaving plates by a revolving disc has been long in use, but the traversing motion in this machine is certainly new, and its application very creditable to the ingenuity of the inventor.

Having thus traced the processes for the conversion of crude into malleable iron, and the machinery employed, it only remains to give a general summary of the whole. As regards the arrangement of large iron-works, the general principle should be for the machinery to be classed and fixed in the order of the different processes, so that the products of one machine should pass at once to the next, and, in fact, the crude iron should be received at one end, and having passed through all the processes, delivered at the other in the manufactured state.

The crude iron from the smelting furnace is either refined and puddled, or subjected to the boiling process, to get rid of the combined carbon and render the iron malleable; it is then shingled by the forge hammer, by the "alligator," by Brown's squeezer, or by the other machines which have been invented for this purpose. It is then at once passed through the puddling rolls, where it is reduced to the form of a flat bar, and is then cut into convenient lengths by the shears. These pieces are again piled or faggoted together into convenient heaps and re-heated in the furnace. As soon as a faggot thus prepared has been heated to the welding temperature, it is passed through the roughing rolls to reduce it to the form of a bar, and then through the finishing rolls, where the required form and size is given to it, either round or square bars, etc. These are straightened and cut to the required sizes, and are then ready for delivery. In most large works all these operations are carried on simultaneously with the smelting process, and in some with extensive mining operations for procuring the coal, ore, and limestone required to supply a production of several thousand tons of manufactured iron per month.

THE FORGE.

The forging of iron has entered, of late years, so largely into the constructive arts, that the manufactures, however perfect in the rolling-mill, would be imperfect indeed without the forge. To the discussion of this part of the subject there are many inducements, and we cannot but wonder at the many devices, and the numerous contrivances which present themselves for the attainment of the operations of the forge. In effecting these objects, Mr. Nasmyth's steam-hammer is evidently the most effective, and to that instrument we are indebted for the formation and welding of iron upon a scale previously unknown to the workers in that metal.

Mr. Nasmyth took out his patent for this invention in 1833; and from that time up to the present, it has maintained its ground against every innovation, and has performed an important duty in almost every well-regulated work in Europe. It consists of an inverted cylinder D, figs. 18. and 19, through which the piston-rod E passes, attached to the hammer-blade F by means of bars and cross-key k, which press upon an elastic packing, to soften the blow of the hammer, which in heavy forgings and heavy blows, operates severely upon the piston-rod. The hammer-block FF is guided in its vertical descent by two planed guides or projections, extending the length of the side-standards AA, between which the hammer-block slides. The attendant gives motion to the hammer by admitting steam from the boiler to act upon the under side of the piston, by moving the regulator I by the handle d. The length of stroke is regulated by increasing or diminishing the distance between the cam N and the valve lever O o, by turning the screws P and U by the bevil wheels $q\,q$. The lever O o operates by the cam N coming in contact with the roller o. As soon as this contact takes place, the further admission of steam is not only arrested, but its escape is at the same time effected, and the hammer, left unsupported, descends by its gravity upon the work on the anvil with an energy due to the height of the fall. From this description, it will be seen that the movement of the roller o causes the shoulder of the rod P to get under the point of the trigger catch U; the valve is by these means kept closed till the whole force of the blow is struck. The instant the operation is effected, the concussion of the hammer causes the latch X to knock off the point of the trigger from the shoulder on the valve-rod P, by means of the bent lever $s\,v$, and the instant this is accomplished, the valve is re-opened to admit the steam below the piston, by the pressure of steam on the upper side of the small piston in the cylinder M, forcing down the valve rod, which, in this respect, is the active agent for opening the valve.

To arrest the motion of the hammer, it is only necessary to shut the steam-valve; during the process of forging, it is, however, desirable to give time between the blows, to enable the workman to turn and shift his work on the anvil; and to effect this reduced motion, the trigger U is held back from the shoulder of the valve-rod P, by the handle y, which at the same instant opens the valve in the case J, and thus the action of the steam in the cylinder D retards the downward motion of the hammer. The result of these changes is an easy descent of the hammer, which vibrates up and down without touching the anvil, but ready for blows of any severity the instant the trigger is elevated above the shoulder of the valve lever P. From this description, it will appear evident that Mr. Nasmyth's invention is one of the most important that has occurred in the art of forging iron. It has given an impetus to the manufacture, and affords facilities for the welding of large blocks of malleable iron that could not be accomplished by the tilt and helve hammers formerly in use; and we have only to instance the forging of the stern-posts and cutwaters of iron ships; the paddle-wheel and screw-shafts of our ocean steamers, some of them weighing upward of 20 tons, to appreciate the value as well as the intensity of action of the steam-hammer.

In addition to the machinery of the forge, the V anvil, the natural offspring of the steam-hammer, came into existence from this same fertile source. It is chiefly employed for forging round bars and shafts, and may be thus described, A being a section of the round bar or shaft to be forged, B the anvil-block, and C the hammer. From this, it is obvious that, in place of the old plan, where the work is forged upon flat surfaces, as shown in the annexed figure, and where the blows are diverging, the effect of the V anvil is a converging action, thus consolidating the mass, and enabling the forger to retain his work directly under the center of the hammer. This is the more strikingly apparent, as the blows of the hammer upon a round shaft have the effect of causing the mass to assume the elliptical form, forcing out the sides as at AA in every successive blow, and this again, when turned, produces a spongy, porous center, as shown in D. This process is, however, more clearly exemplified in Ryder's forging machines, where all the anvils are of the V form, for the forging of spindles, round bars, and bolts.

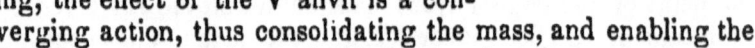

The next important discovery in the art of forging, is that of Mr. Ryder's machine, patented some years since, for forging small articles, which, on account of the rapidity and precision of its operations, demands a notice in passing. It consisted essentially of a series of small anvils about three inches square, supported from below by large screws passing through the frame of the machine. This screw was employed in order that the dis-

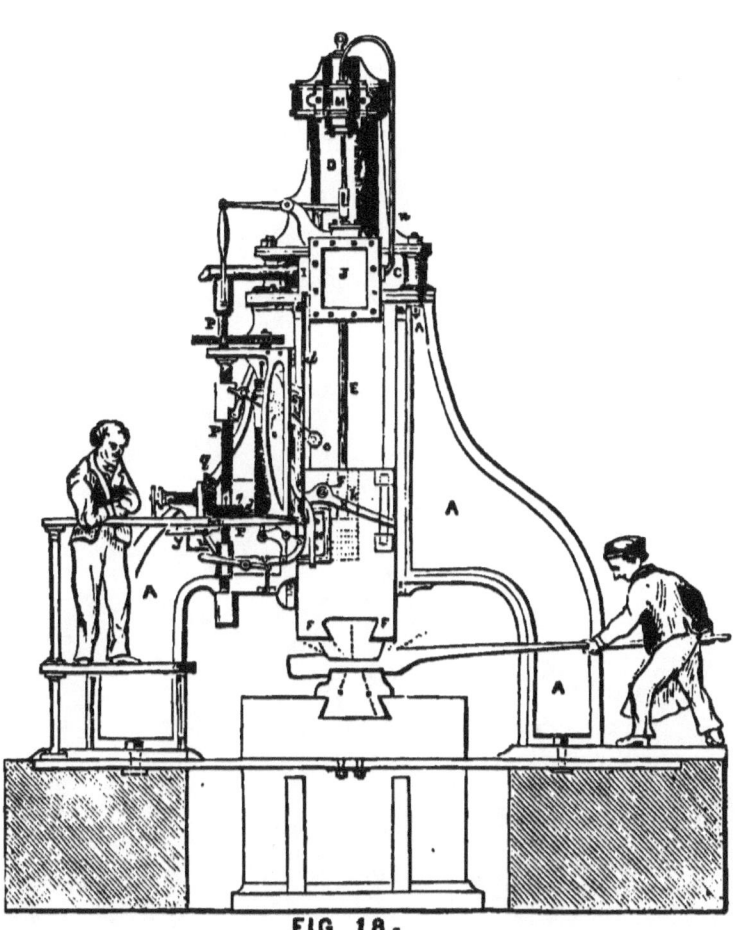

FIG 18.

[IRON.] [PLATE 10.

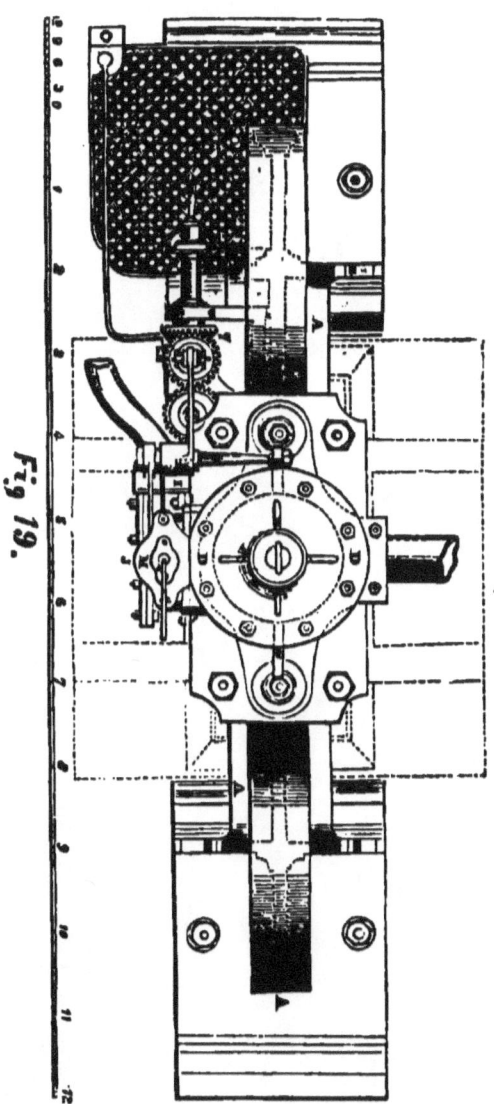

Fig. 19.

tance between the hammer and anvil might be accurately adjusted. Between the screw and the anvil, a stuffing of cork was introduced to deaden the effect of the blow. The hammers were arranged over the anvils, and slid up and down in the frame of the machine. The blow was effected by the revolution of an eccentric, acting by means of a cradle on the hammer-head, the hammer, however, being lifted again by a strong spiral spring. At the side of the machine was a cutter or shears worked by a long lever; with this the articles were cut to the required length as they were finished.

In Mr. Ryder's machine 700 strokes a minute was the maximum; but Messrs. Platt Brothers, of Oldham, by increasing the strength of the spring, run as high as 1100. A pair of knife edges, worked by the machine itself, has also been substituted for the hand-shears. These perform the work more rapidly and accurately than before, and leave the workman more at liberty. Dies are let into the surfaces of the hammers and anvils, which shape the iron as required.

The rapidity with which this machine executes all kinds of intricate work is truly remarkable; for instance, a bar about $2\frac{1}{4} \times 2\frac{1}{2}$ inches, will be reduced to $1\frac{1}{4} \times 10$ inches, and cut off in a minute. Set-screws, bolts, spindles, and all kinds of small work are produced at the same rate. Its precision is very effective; the articles are almost as true as if turned in a lathe, and very accurate as to size and weight. Other machines, called "lifts," have been, and continue to be, used for forging a variety of forms and "*uses;*" but as these partake more or less of the principle employed in Ryder's machine, it will not be necessary to furnish further examples.

In conclusion, we may observe that the facilities afforded by the present age for the forging of malleable iron, are without a parallel in the history of that material. Every known resource has been adopted, and every contrivance and device has been employed to meet the demands of a large and an intricate trade; and looking at the present resources of the country, and the admirable mechanical contrivances for the conversion of crude iron into the malleable state, it assuredly is not unreasonable to look forward to still greater improvements in the manipulations of the forge.

THE STRENGTH AND OTHER PROPERTIES OF IRON.

In this section we have to consider the tensile and transverse strengths and powers of resisting compression of cast and malleable iron as determined by direct experiment upon specimens of the material; and also to examine whether, as has been alleged, the hot-blast process injures the tenacity of the metal.

CAST IRON.—The following tables give the results of experiments undertaken by Mr. Hodgkinson and Mr. Fairbairn at the request of the British Association, to determine the tensile and transverse strengths of cast-iron derived from the hot and cold blast. The castings for ascertaining the tensile strain were made very strong at the ends, with eyes for the bolts to which the shackles were attached; the middle part, where it was intended that the specimen should break, was cast of a cruciform + transverse section. The four largest castings were broken by the chain-testing machine belonging to the corporation of Liverpool, the others by Mr. Fairbairn's lever.

TABLE I.—*Results of the Experiments on the tensile strength of Cast-Iron.*

Description of Iron.	Number of Experiments.	Mean strength per square inch of section.	
		lbs.	tons cwts.
Carron iron, No. 2, hot-blast	3	13,505	6 0½
" " cold-blast	2	16,683	7 9
" No. 3, hot-blast	2	17,755	7 18¼
" " cold-blast	2	14,200	6 7
Devon (Scotland) iron, No. 3, hot-blast	1	21,907	9 15¼
Buffery iron, No. 1, hot-blast	1	13,434	6 0
" " cold-blast	1	17,466	7 16
Coed Talon (North Wales) iron, No. 2 hot-blast	2	16,676	7 9
Do. do. cold-blast	2	18,855	8 8

From the same series of experiments we select the following tables, giving the results obtained in regard to the resistance opposed to compression by cast-iron. The specimens employed were cylinders and prisms of various dimensions, and having their faces turned accurately parallel to each other and perpendicular to the axis of the specimen. They were crushed by a lever between parallel steel discs.

TABLE II.—*Weights required to crush cylinders, etc., of Carron Iron, No. 2, Hot-Blast.*

Diameter of Cylinder in parts of an inch.	Number of Experiments.	Mean Crushing Weight.	Mean Crushing Weight per square inch	General mean per square inch.
		lbs.	lbs.	
$\frac{1}{4}$	3	6,426	130.909	
$\frac{3}{8}$	4	14,542	131,665	121,685 lbs.
$\frac{1}{2}$	5	22,110	121,605	= 54 tons
$\frac{16}{25} = \cdot 64$	1	35,888	111,560	6½ cwt.
Prism base ·50 inch square ..	3	25,104	100,416	100,738 lbs.
Prism base 1·00 × ·26	2	26,276	101,062	= 44 tons 19½ cwt.

TABLE III.—*Weights required to crush cylinders, etc., of Carron Iron, No. 2, Cold-Blast.*

Diameter of Cylinder in parts of an inch.	Number of Experiments.	Mean Crushing Weight.	Mean Crushing Weight per square inch.	General Mean per square inch.
		lbs.	lbs.	
$\frac{1}{4}$	2	6,088	124,023	
$\frac{3}{8}$	4	14,190	128,478	125,403 lbs. = 55 tons 19½ cwt.
$\frac{1}{2}$	7	24,290	123,708	
Equilateral triangle— side ·866	2	32,398	99,769	
Squares — side ½ inch	2	24,538	98,152	100,631 lbs. = 44 tons 18½ cwt.
Rectangles—base 1·00 × ·243	3	26,237	107,971	
Cylinders ·45 inch diameter, & ·75 inch high	2	15,369	96,634	

TABLE IV.—*Results of Experiments to ascertain the forces necessary to crush short cylinders, etc., of Cast Iron.*

Description of Iron.	Number of Experiments.	Mean Crushing Weight per square inch.	
		lbs.	tons. cwt.
Devon (Scotch) iron, No. 3 hot-blast .	2	145.435	64 18½
Buffery iron, No. 1, hot-bast	4	86,397	38 11¼
" " cold-blast	4	93,385	41 13½
Coed Talon, No. 2, hot-blast	4	82,734	36 18½
" " cold-blast	4	81,770	36 10
Carron iron, No. 2, hot-blast	2	108,540	48 9
" " cold-blast	3	106,375	47 9¾
Carron iron, No. 3, hot-blast	3	133,440	59 11¼
" " cold-blast	4	115,442	51 10¾

The specimens of Carron iron in table IV. were prisms, whose base was $\frac{3}{4} \times \frac{1}{3} = \frac{1}{4}$ inch, and whose height varied from ½ inch to 1 inch. The other specimens were cylinders, whose diameter was about ½ inch, and height varied from ½ inch to 2 inches.

From the above experiments, Mr. Hodgkinson concludes that "where the length is not more than about three times the diameter, the strength for a given base is pretty nearly the same." Fracture took place either by wedges sliding off, or by the top and bottom forming pyramids, and forcing out the sides; and the angle of the wedge is nearly constant, a mean of 21 cylinders being 55° 32'.

From the same series of experiments, we give the results obtained by Mr. Fairbairn, in regard to the effects of time and temperature. The bars employed were cast to be 1 inch square, and 4 feet 6 inches long, and were loaded with permanent weights as under; the deflections being taken at various intervals during a period of fifteen months. Coed-talon hot and cold-blast iron was employed.

TABLE V.—*The effects of Time on loaded bars of Hot and Cold-blast Iron in their resistance to a transverse strain.*

Permanent load in lbs.	Increase of deflection of cold-blast bars.	Increase of deflection of hot-blast bars.
280	·033	·043
336	·046	·077
392	·140	·088
449	·047	....
Mean.	.066	·069

It has been assumed by most writers on the strength of materials, that the elasticity of cast-iron remained perfect to the extent of one-third the weight that would break it. This is, however, a mere assumption, as it has been found that the elasticity of cast-iron is injured with less than one-half that weight, and the question to be solved in the above experiments was, to what extent the material could be loaded without endangering its security; or how long it would continue to support weights, varying from one-half to one-tenth of the load that would produce fracture. These experiments were continued from six to seven years, and the results obtained were, that the bars which were loaded to within $\frac{1}{18}$ of their breaking weight, would have continued to have borne the load, in the absence of any disturbing cause, *ad infinitum;* but the effect of change, either of the same or a lighter load led ultimately to fracture.

From these facts it is deduced, that so long as the molecules of the material are under strain (however severe that strain may be), they will arrange and accomodate themselves to the pressure, but with the slightest disturbance, whether produced from vibration or the increase or diminution of load, it becomes, under these influences, only a question of time when rupture ensues.

In the following experiments on the relative strengths of coed-talon hot and cold-blast iron to resist transverse strain at different temperatures, the results are reduced to those of bars 2 feet 3 inches between supports, and one inch square, as follows:

TABLE VI.

	Temperature, Fahr.	Specific Gravity.	Modulus of Elasticity.	Breaking weight.	Ultimate deflection.	Power of resisting impact.
Cold Blast, No. 2	27°	6 955	12799050	874	.4538	397.7
"	32°	6 955	14827450	919.6	.402	382.4
"	113°	6 955	14168000	812.9	.336	273.1
Hot Blast, No. 2	20°	6 968	14902900	811.69	.4002	325.0
"	82°	6 968	14003350	919.7	.429	395.0
"	84°	6 968	14500000	877.5	.421	366.4
Cold Blast, No. 2	192°	....	14398600	743.1	.301	223.7
No. 3	212°	....	....	924.5		
"	600°	....	....	1083.0		
No. 2	Red by daylight.	....	....	663.3		
"	Red in dark.	....	....	723 1		
Hot Blast, No. 2	136°	....	13046200	875.7	.389	310 6
"	187°	....	11012500	638.8	.359	229 3
"	188°	....	13369500	823 6	.363	298.9
No. 3	212°	....	....	818 4		
"	600°	....	....	875 8		
No. 2	Red in dark.	....	....	820 7		

From the above it will be seen " that a considerable failure of the strength took place after heating the No. 2 iron from 26° to 190°. At 212°, we have in the No. 3 a much greater weight sustained than by No. 2 at 190°; and at 600° there appears, in both hot and cold-bast, the anomaly of increased strength as the temperature is increased."* The

---

* This probably arises from the greater ductility of the bars at an increased temperature.

above results are, with one exception, in favor of the cold-blast, as far as strength is concerned ; and in favor of the hot-blast, with one exception, as regards power of resisting impact.

With regard to the comparative strengths of hot and cold-blast iron, the following extracts from Mr. Hodgkinson's report, read before the British Association, give the general results of his experiments:

TABLE VII.—*Carron Iron*, No. 2.

	Cold-blast.	Hot-blast.	Ratio representing Cold-blast by 1000.
Tensile strength in lbs. per inch square	16683 (2)	13505 (3)	1000 : 809
Compressive do. in lbs. per inch, from castings torn asunder	106375 (2)	108540 (2)	1000 : 1020
Do. from prisms of various forms	100631 (2)	100738 (2)	1000 : 1001
Do. from cylinders	125403 (13)	121685 (13)	1000 : 970
Transverse strength from all the experiments	(11)	(13)	1000 : 991
Power to resist impact	(9)	(9)	1000 : 1105
Transverse strength of bars 1 inch square in lbs.	476 (3)	463 (3)	1000 : 973
Ultimate deflection of do. in inches	1·313 (3)	1 337 (3)	1000 : 1018
Modulus of elasticity in lbs. per square inch	17270500 (2)	16085000 (2)	1000 : 931
Specific gravity	7·066	7·046	1000 : 997

Mean 997

TABLE VIII.—*Devon Iron*, No. 3.

	Cold-blast.	Hot-blast.	Ratio representing Cold-blast by 1000.
Tensile strength	..	21907 (1)	
Compressive do.	..	145435 (4)	
Transverse do. from the experiments generally	(5)	(5)	1000 : 1417
Power to resist impact	(4)	(4)	1000 : 2786
Transverse strength of bars 1 inch square	448 (2)	537 (2)	1000 : 1199
Ultimate deflection do.	79 (2)	1 09 (2)	1000 : 1380
Modulus of elasticity do.	22907700 (2)	22473650 (2)	1000 : 981
Specific gravity	7 295 (4)	7·229 (2)	1000 : 991

TABLE IX.—*Buffery Iron*, No. 1.

	Cold-blast.	Hot-blast.	Ratio representing Cold-blast by 1000.
Tensile strength	17466 (1)	13434 (1)	1000 : 769
Compressive do.	93366 (4)	86397 (4)	1000 : 925
Transverse do.	(5)	(5)	1000 : 931
Power to resist impact	(2)	(2)	1000 : 963
Transverse strength of bars one inch square	463 (3)	436 (3)	1000 : 942
Ultimate deflection do.	1·55 (3)	1.64 (3)	1000 : 1058
Modulus of elasticity do.	15381200 (2)	13730500 (2)	1000 : 893
Specific gravity	7·079	6·998	1000 : 989

TABLE X.—*Coed Talon Iron, No. 2.*

Tensile strength	18855 (2)	16676 (2)	1000 : 884
Compressive do.	81770 (4)	82739 (4)	1000 : 1012
Specific gravity	6·955 (4)	6·968 (3)	1000 : 1002

TABLE XI.—*Carron Iron, No. 3.*

Tensile strength	14200 (2)	17755 (2)	1000 : 1250
Compressive do.	115442 (4)	133440 (3)	1000 : 1156
Specific gravity	7·135	7·056 (1)	1000 : 989

" Beginning with No. 1 iron, of which we have a specimen from the Buffery Iron-Works, a few miles from Birmingham, we find the cold-blast iron somewhat surpassing the hot-blast in all the following particulars: direct tensile strength, compressive strength, transverse strength, power to resist impact, modulus of elasticity or stiffness, specific gravity; whilst the only numerical advantage possessed by the hot-blast iron is, that it bends a little more than the cold-blast before it breaks.

" In the irons of the quality No. 2, the case seems in some degree different; in these the advantages of the rival kinds seem to be more nearly balanced. They are still, however, rather in favor of the cold-blast.

" So far as my experiments have proceeded, the irons of No. 1 have been deteriorated by the hot-blast; those of No. 2 appear also to have been slightly injured by it, while the irons of No. 3 seem to have benefited by its mollifying powers. The Carron iron No. 3 hot-blast, resists both tension and compression with considerably more energy than that made with the cold-blast; and the No. 3 hot-blast iron from the Devon works, in Scotland, is one of the strongest cast-irons I have seen, whilst that made by the cold-blast is comparatively weak, though its specific gravity is very high, and higher than in the hot. The extreme hardness of the cold-blast Devon iron above prevented many experiments that would otherwise have been made upon it, no tools being hard enough to form the specimens. The difference of strength in the Devon irons is peculiarly striking.

" From the evidence here brought forward, it is rendered

exceedingly probable that the introduction of the heated blast in the manufacture of cast-iron has injured the softer irons, whilst it has frequently mollified and improved those of a harder nature, and, considering the small deterioration that the irons of quality No. 2 have sustained, and the apparent benefit to those of No. 3, together with the great saving effected by the heated blast, there seems good reason for the process becoming as general as it has done."

The following table gives a general summary of the results of Mr. Fairbairn's experiments on the strength of iron after successive meltings. The iron used was Eglinton No. 3 hot-blast, and was melted eighteen times, three bars being cast at each melting. These bars, which were about 1 inch square and 5 feet long, were placed upon supports 4 feet 6 apart, and broken by a transverse strain. Cubes, from the same irons, exactly 1 inch square, were then crushed between parallel steel bars, by a large wrought-iron lever.

In the following Table XII., the results on transverse strain are reduced to those on bars exactly 1 inch square and 4 feet 6 inches between supports:

No. of melting.	Specific gravity.	Mean breaking weight in lbs.	Mean ultimate deflection in inches.	Power to resist impact.	Mean crushing weight of inch cubes in tons.
1	6.966	490.0	1.440	705.6	
2	6.970	441.9	1.446	630.9	
3	6.886	401.6	1.486	596.7	
4	6.938	413.4	1.260	520.8	41.9
5	6.842	431.6	1.503	648.6	
6	6.771	438.7	1.320	579.0	
7	6.879	449.1	1.440	646.7	
8	7.025	491.3	1.753	861.2	
9	7.102	546.5	1.620	885.3	
10	7.108	566.9	1.626	921.7	
11	7.113	651.9	1.636	1066.5	64.3
12	7.160	692.1	1.666	1153.0	
13	7.134	634.8	1.646	1044.9	
14	7.530	603.4	1.513	912.9	
15	7.248	371.1	0.643	238.6	
16	7.330	351.3	0.566	198.5	82.8
17	lost.				
18	7.385	312.7	0.476	148.8	

In the above results it will be observed that the maximum of strength, elasticity, etc., is only arrived at after the metal has undergone twelve successive meltings. It is probable that other metals and their alloys may follow the same law, but that is a question that has yet to be solved, probably by a series of experiments requiring a considerable amount of time and labor to accomplish, but which we may venture to hope will be shortly forthcoming from the same author.

In the resistance of the different meltings from the same iron, to a force tending to crush them, we have the following results:

TABLE XIII.

Number of meltings.	Resistance to compression per square inch, in tons.	Remarks.
1	44.0	
2	43.6	
3	41.1	
4	40.7	
5	41.1	
6	41.1	
7	40.9	
8	41.1	
9	55.1	
10	57.7	
11 } 11 }	Mean 69.8	In this experiment the cube did not bed properly on the steel plates, otherwise it would have resisted a much greater force.
12	73.1	
13	66.0	
14	95.9	
15	76.7	
16	70.5	
18	88.0	

Nearly the whole of the specimens were fractured by wedges which split or slid off diagonally at an angle of 52° to 58°.

MALLEABLE IRON.—The greatly extended application of wrought-iron to every variety of construction renders an investigation of its properties peculiarly interesting. It is now employed more extensively than cast-iron; and, on account of its ductility and strength, nearly two-thirds of the weight of material may in many cases be saved by its employment, while great lightness and durability are secured. Its superiority is especially evident in constructions where great stiffness is not required, but on the other hand any degree of rigidity may be obtained by the employment of a tubular or cellular structure, and this may be seen in the construction of wrought-iron tubular bridges, beams, and iron ships. The material of malleable iron which is making such vast changes in the forms of construction, cannot but be interesting and important, and considering that the present is far from the limit of its application, we shall endeavor to give it that degree of attention which the importance of the subject demands.

From the forge and the rolling mill we derive two distinct qualities of iron, known as "*red short*" and "*cold short*." The former is the most ductile, and is a tough fibrous material which exhibits considerable strength when cold; the latter is more brittle, and has a highly crystalline fracture almost like cast-iron; but the fact is probably not generally known, that the brittle works as well and is as ductile under the hammer as the other when at a high temperature.

Mr. Charles Hood, in a paper read some time ago before the Institute of Civil Engineers, went into the subject of the change in the internal structure of iron, independently of and subsequently to the processes of its manufacture. After adducing several instances of tough fibrous malleable iron becoming crystalline and brittle during their employment, he attributes these changes to the influence of percussion, heat, and magnetism, but questions whether either will produce the effect *per se*. Mr. Hood continues, "The most common exemplification of the effect of heat in crystallizing fibrous iron is by breaking a wrought-iron furnace bar, which, whatever quality it was of in the first instance, will in a short time invariably be converted into crystallized iron, and by heating and rapidly cooling, by quenching with water a few times any piece of wrought-iron, the same effect may be far more speedily produced. In these cases we have at least two of the above causes in operation—heat and magnetism. In every instance of heating iron to a very high temperature, it undergoes a change in its electric or magnetic condition; for at very high temperatures iron loses its magnetic powers, which return as it gradually cools to a lower temperature. In the case of quenching the iron with water, we have a still more decisive assistance from the electric and magnetic forces; for Sir Humphrey Davy long since pointed out that all cases of vaporization produced negative electricity in the bodies in contact with the vapor; a fact which has lately excited a good deal of attention in consequence of the discovery of large quantities of negative electricity in effluent steam." Mr. Hood then proceeds to the subject of percussion. " In the manufacture of some descriptions of hammered iron, the bar is first rolled into shape, and then one-half the length of the bar is heated in the furnace, and immediately taken to the tilt-hammer and hammered, and the other end of the bar is then heated and hammered in the same manner. In order to avoid any unevenness in the bar, or any difference in its color where the two distinct operations have terminated, the workman frequently gives the bar a few blows with the hammer upon that part which he first operated upon. That part of the bar immediately becomes crystallized, and so extremely brittle that it will break to pieces by merely throwing it on the ground, though all the rest of the bar will exhibit the best and toughest quality imaginable. This change, therefore, has been produced by percussion (as the primary agent) when the bar is at a lower temperature than the welding heat. Here it must be observed that it is not the excess of hammering which produces the effect, but the absence of a sufficient degree of heat, at the time that the hammering takes place; and the evil may probably all be produced by four or five blows of the hammer if the bar happens to be of a small size. In this case we witness the combined effects of percussion, heat, and magnetism. When the bar is hammered at the proper temperature, no such crystallization takes place, because the bar is insensible to magnetism; but as soon as the bar becomes of that lower degree of temperature at which it can be affected by magnetism, the effect of the blows it receives is to produce magnetic induction, and that magnetic induction and consequent polarity of its particles, when assisted by further vibrations from additional percussion, produces a crystallized texture."

The crystallization of perfectly fibrous and ductile wrought-iron has long been a subject of dispute, and although we agree with most of Mr. Hood's views, we are not altogether prepared to admit that the causes assignd are the only ones concerned in producing the change, or that more than one is *necessary*. On the occasion of the accident on the Versailles Railway some years since, the whole array of science and practice were brought to bear upon the elucidation of the cause. Undoubtedly the broken axle presented a crystalline fracture, but it has never been ascertained how far heat and magnetism were in operation as in the case of an axle, and more especialy a crank-axle, the constant vibration caused by irregularities in the way and the weight of the engine appears to be quite sufficient to occasion the breakage without aid from the other forces. Undoubtedly in almost all cases of the sudden fracture of axles or wrought-iron bars, during employment, the fracture presents a crystalline structure, but we believe that any molecular disturbance, such as impact, can effect this, the only question being, how long will the material sustain the action before it breaks. This question has been attempted to be decided by direct experiment under the direction of the Commission on Railway Structures. It was found that with cast-iron bars subjected to long continued impacts, " when the blow was powerful enough to bend the bars through one-half of their ultimate deflection (that is to say, the deflection which corresponds to their fracture by dead pressure), no bar was able to stand 4000 of such blows in succession. But all the bars (when sound) resisted the effects of 4000 blows, each bending them through one-third of their ultimate deflection. These results were confirmed by experiments with a revolving cam which deflected the bars.

"In wrought-iron bars no very perceptible effect was produced by 10,000 successive deflections by means of a revolving cam, each deflection being due to half the weight which, when applied statically, produced a large permanent flexure." These results agree with those obtained by Mr. Fairbairn in regard to the effects of time on loaded bars of cast-iron, already given.

Arago and Wollaston have paid considerable attention to this subject, the latter having been the first to point out that native iron is disposed to break in octohedra and tetrahedra, or combinations of these forms. The law which leads to fracture in wrought-iron from changes in the molecular structure operates with more or less intensity in other bodies; repeated disturbances, in turn, destroying the cohesive force of the material by which they are held together. A French writer of eminence, Arago, appears to consider the crystallization of wrought-iron to be due to the joint action of time and vibration, but we think with Mr. Hood that time and its duration depends entirely upon the intensity of the disturbing forces, and, moreover, that the time of fracture is retarded or accelerated in a given ratio to the intensity with which these forces are applied.

From the above statements we may safely deduce the fact, that it is essential to the use of this material to consider the purposes to which it is applied, the forms to which it may be subjected, and the conditions under which it may be placed, in order to arrive at just conclusions as to the proportions, in order to afford to the structure (whatever that may be) ample security in its powers of resistance to strain.

On the subject of the strength of wrought-iron, we have before us the researches of Mr. Fairbairn, in a paper entitled, " An Inquiry into the Strength of wrought-iron plates and their riveted joints, as applied to Ship-building and Vessels exposed to severe strains."* In that communication it is shown, from direct experiments, that in plates of rolled iron there is no material difference between those torn asunder in the direction of the fiber, and those torn asunder across the fiber. This uniformity of resistance arises probably from the way in which the plates are manufactured, which is generally out of 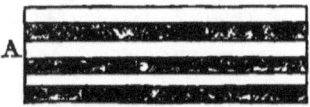 flat bars, cut and piled upon each other, as at A, one-half transversely and the other half longitudinal in the line of the pile. From this it will be seen that in preparing the bloom or shingle for the rollers, the fiber is equally divided, and the only superiority that can possibly be attained is in the rolling, which draws the shingle rather more in the direction of the length of the plate than its breadth.

* Philosophical Transactions, part ii., 1850, p. 677.

In the following table we have the results of the experiments:

Quality of Plates.	Mean breaking weight in the direction of the fiber, in tons per square inch.	Mean breaking weight across the fiber, in tons per square inch.
Yorkshire plates	25.770	27.490
Yorkshire plates	22.760	26.037
Derbyshire plates	21.680	18.650
Shropshire plates	22.826	22.000
Staffordshire plates	19.563	21.010
Mean................	22.519	23.037

Or as 22.5, 23.0, equal to about $\frac{1}{45}$ in favor of those torn across the fiber.

From the above it is satisfactory to know, so far as regards uniformity in the strength of plates, that the liability to rupture is as great when drawn in one direction as in the other; and it is not improbable, that the same properties would be exhibited, and the same resistance maintained, if the plates were drawn in any particular direction obliquely across the fibrous or laminated structure.

From the same author we select the results of a series of experiments on the tensile strength of S C ♕ bars of different lengths, and about $1\frac{3}{8}$ in diameter. The following table gives the strains required for each of four succesive breakages of the same pieces of iron. These experiments are highly interesting, as they not only confirm those made upon plates, but they indicate a progressive increase of strength, notwithstanding the elongation and the reduced sectional area of the bars. These facts are of considerable value, as they distinctly show that a severe tensile strain is not seriously injurious to the bearing powers of wrought-iron, even when carried to the extent of or increased four times repeated, as was done in these experiments. In practice it may not be prudent to test bars and chains to their utmost limit of resistance; it is nevertheless satisfactory to know that in cases of emergency those limits may be approached without incurring serious risk of injury to the ultimate strength of the material.

The following abstract gives the results of the experiments:

Length between the nippers.	Breaking Strain in tons.	Mean Elongation in inches.
Inches.		
120	32·21	26·0
42	32·125	9·8
36	32·35	8·8
24	32·00	6·2
10	32·29	4·2

"As all these experiments were made upon the same description of iron, it may be fairly inferred that the length of a bar does not in any way affect its strength."

Reduction of the above Table.

Length of bar.	Elongation.	Elongation per unit of length.
Inches.		
120	26·0	·216
42	9·8	·233
36	8·8	·244
24	6·2	·258
10	4·2	·420

"Here it appears that the rate of elongation of bars of wrought-iron increases with the decrease of their length; thus while a bar of 120 inches had an elongation of ·216 inch per unit of its length, a bar of ten inches has an elongation of ·42 inch per unit of its length, or nearly double what it is in the former case. The relation between the length of and its maximum elongation per unit, may be approximately expressed by the following formula, viz.:

$$l = \cdot 18 + \frac{2 \cdot 5}{L}$$

where L represents the length of the bar, and l the elongation per unit of the length of the bar."

The above results are not without value, as they exhibit the ductility of wrought-iron at a low temperature, as also the greatly increased strength it exhibits with a reduced sectional area under severe strain.

On the transverse strength of wrought-iron it will not be necessary to enlarge, as we have numerous examples before us in the experiments undertaken to determine the strength and form of the Britannia and Conway Tubular Bridges.* In these experiments will be found an entirely new description and form of construction, which have emanated from them, and which have led to a new era in the history of bridges, and the application of wrought-iron to other purposes besides those in connection with buildings, and its greatly extended application to the useful arts. For further information on this subject we refer the reader to Mr. Fairbairn's † and Professor Hodgkinson's works, in both of which will be found data sufficient to establish the great superiority of malleable over cast-iron, or any other material, either as regards strength or economy in its application.

On the resistance of wrought-iron plates to a force tending to burst them, Rondelet has shown that it requires a force of 70,000 lbs. per square inch to produce fracture, and Mr. Fairbairn's experiments proved that a wrought-iron plate of one-quarter of an inch thick resisted a pressure from a ball 3 inches in diameter, equal to that required to rupture a 3 inch oak plank.

At the request of the British Association, Dr. Thomson of Glasgow examined the chemical constitution of hot-blast iron, and he gives the following as the result of his inquiry:

"(1.) The specific gravity of hot-blast iron is greater than that of cold-blast.

"The following is the specific gravities of eight specimens of cold-blast iron.

1st.	Muirkirk	6·410	5th.	Muirkirk	6·775
2d.	Do.	6·435	6th.	From Pyrites	6·9444
3d.	Do.	6·493	7th.	From Carron	6·9888
4th.	Do.	6·579	8th.	Clyde Iron-Works	7·0028

"The specific gravity of the Muirkirk iron is considerably less than of that smelted at Carron and the Clyde Iron-Works; the mean of the eight specimens is 6.7034.

"It has been hitherto supposed that the difference between cast-iron and malleable iron consists in the presence of carbon in the former, and its absence from the latter; in other words, that cast-iron is a carburet of iron. But in all the specimens of cast-iron which we analyzed we constantly found several other ingredients besides iron and carbon. Manganese is pretty generally present in minute quantity, though in one specimen it amounted to no less a quantity than 7 per cent.; its average amount is 2 per cent. *Silicon* is never wanting, though its amount is exceedingly

---

* See Mr. Fairbairn's and Mr. Edwin Clark's work on the Conway and Britannia Tubular Bridges.

† "On the application of cast and wrought-iron to building purposes," and "Useful Information for Engineers."

variable, the average quantity is 1½ per cent.; some specimens contained 3½ per cent. of it, while others contained less than a half per cent. Aluminum is very rarely altogether absent, though its amount is more variable than that of silicon. Its average amount is 2 per cent.; sometimes it exceeds 4½ per cent., and sometimes it is not quite 1-5000th part of the weight of the iron.

"Calcium and magnesium are sometimes present, but very rarely, and the quantity does not exceed 1-5th per cent. In a specimen of cast-iron which I got from Mr. Neilson, and which he had smelted from pyrites, there was a trace of copper, showing that the pyrites employed was not quite free from copper; and in a specimen from the Clyde Iron-Works there was a trace of sulphur. The following table exhibits the composition of six different specimens of cast-iron, No. 1, analyzed in my laboratory, either by myself or by Mr. John Tennent.

	Muirkirk	Muirkirk.	Muirkirk	Pyrites.	Carron	Clyde.	Mean.
Iron	90·98	90·29	91·38	89·442	94·010	90·824	91·154
Copper	..	..	..	0·288	..	..	..
Manganese	..	7·14	2·00	..	0·626	2·458	2·037
Sulphur	..	..	..	..	..	0·045	..
Carbon	7·40	1·706	4·88	3·600	3·086	2·458	3·855
Silica	0·46	0·830	1·10	3·220	1·006	0·450	1·177
Aluminum	0·48	0·016	..	3·776	1·032	4·602	1·651
Calcium	..	0·018	0·20	..	..	..	..
Magnesium,	..	..	..	..	..	0·340	..

"The constant constituents of cold-blast cast-iron, No. 1, are iron, manganese, carbon, silicon, and aluminum. The occasional constituents are copper, sulphur, calcium, and magnesium. These occur so rarely, and in such minute quantity, that we may overlook them altogether.

"The constant constituents occur in the following mean atomic proportions:

22 atoms iron .... = 77·00
½ atom manganese .... = 1·75
4·36 atoms carbon .... = 3·27
1 atom silicon .... = 1·00
1⅓ aluminum .... = 1·40—84·42

"(2.) I examined only one specimen of cast-iron, No. 2. It was an old specimen, said to have come from Sweden, but I have no evidence of the correctness of this statement. Its specific gravity was 7·1633 higher than any specimens of cold-blast iron, No. 1. Its constituents were:

Iron .... 93·594
Manganese .... 0·708
Carbon .... 3·080
Silicon .... 1·262
Aluminum .... 0·732
Sulphur .... 0·038—99·414

"The presence of sulphur in this specimen leads to the suspicion that it is not a Swedish specimen; for as the Swedish ore is magnetic iron, and the fuel charcoal, the presence of sulphur in the iron is very unlikely."

* I have been told by Mr. Mushet that the Swedes add sulphur to their iron No. 2

"In this specimen, the atoms of iron and manganese are to those of carbon, silicon, and aluminum, in the proportion of 4½ to one, instead of 3½ to one, as in cast-iron No. 1.

"The atoms of carbon, silicon, and aluminum, approach the proportions of 7, 2, and 1, so that in cast-iron, No. 2, judging from one specimen, there is a greater proportion of carbon, compared with the silicon and aluminum, than in cast-iron No. 1.

"Mr. Tennent analyzed a specimen of hot-blast iron, No. 2, from Gartsherry. Its specific gravity was 6·9156, and its constituents,

		Atoms.	
Iron	90·542	25·86	3·72
Manganese	2·764	0·78	
Carbon	3·094	4·05	
Silicon	0·680	0·68	1·
Aluminum	2·894	2·31	
Sulphur	0·023	0·011	
	99.997		

So that it resembles cast-iron, No. 1, in the proportion of its constituents. The carbon is almost the same as in cold-blast iron, No. 2, but the proportion of aluminum is four times as great, while the silicon is little more than half as much. The atomic ratios are, carbon, 4·ʼ; silicon, 0·67; aluminum, 2·28.

"(3.) Five specimens of hot-blast cast-iron, No. 1, were analyzed. Two of these were from Carron, and three from the Clyde Iron-Works, where the hot-blast originally began; and where, of course, it has been longest in use. The specific gravity of these specimens was found to be as follows:

1st. From Clyde Works .... 7·0028
2d. From Carron .... 7·0721
3d. From Carron .... 7·0721
4th. From Clyde Works .... 7·1022

Mean .... 7·0623

"It appears from this, that the hot-blast increases the specific gravity of cast-iron by about 1-22d part. It approaches nearer the specific gravity of cast-iron, No. 2, smelted by cold air, than to that of No. 1.

The following table exhibits the constituents of these four specimens:

	Clyde.	Carron.	Carron.	Clyde.	Clyde.
Iron	97·096	95·422	96·09	94·966	94·345
Manganese	0·332	0·336	0·41	0·160	3·120
Carbon	2·460	2·400	2·48	1·560	1·416
Silicon	0·280	1·820	1·49	1·322	0·520
Aluminum	0·385	0·488	0·26	1·374	0·599
Magnesium	......	....	....	0·792	....
	100·55	100·466	100·73	100·174	100·

The mean of these analyses gives us,

		Atoms.	
Iron	95·584 or	27·31	⎫
Manganese	0·871 or	0·249	⎬ 6·5
Carbon	2·099 or	2·79	⎫
Silicon..................	1·086 or	1·086	⎬ 1·
Aluminum	0·422 or	0·337	⎭
	101.285		

Or, in the proportion of $6\frac{1}{2}$ atoms of iron and manganese to 1 atom of carbon, silicon, and aluminum. In the cold-blast cast-iron we have,

	Iron.	Carbon, etc.
In No. 1	$3\frac{1}{2}$ atoms	1 atom.
In No. 2	$4\frac{1}{2}$	1 "
In hot-blast	$6\frac{1}{2}$	1 "

"Thus it appears, that when iron is smelted by the hot-blast its specific gravity is increased, and it contains a greater proportion of iron, and a smaller proportion of carbon, silicon, and aluminum, than when smelted by the cold-blast."

THE STATISTICS OF THE IRON TRADE.

This article has already extended so much beyond the limits of our inquiry, that we must confine ourselves to an exceedingly brief notice of the statistics of this important manufacture. In 1740 the iron trade suffered a sudden check from a falling off in the supply of charcoal, coal or coke not having been employed at that time for smelting. The annual production seems to have decreased from 180,000 to about 17,350 tons per annum.

Furnaces			59
Tons			17,350
	Tons.	cwt.	qrs.
Annual average for each furnace	294	1	1
Weekly do. do.	5	13	0

Soon afterward the difficulties in the way of using coal were overcome, and the manufacture extended rapidly. The number of charcoal furnaces decreased, but the quantity produced by each was considerably increased. The following table shows the state of the trade in 1788 compared with 1740:

Total quantity of charcoal iron, in Britain, in 1788 14,500
Do. coke do. do. 53,800

Total quantity of iron, in Britain, in 1788 68,300
Do. do. 1740 17,350

Increased produce of pig iron 50,950

About the year 1796 it was contemplated by Mr. Pitt to add to the revenue by a tax on coal. This met with a powerful opposition on the part of the manufacturers and consumers, especially those in the iron trade. A committee was appointed, witnesses were examined, and the measure abandoned as unwise and impracticable.

The following table shows the comparative make of pig iron in 1820 and 1827 :

1820.
Tons 400,000
Furnaces 284
1827.
Tons 690,500

From that time to the present the manufacture has steadily increased. The following table gives the state of the trade in Great Britain in 1854 :

No. of Works 228
No. of Furnaces erected 724
No. of Furnaces in blast 555
Total produce in tons 3,069,874

In connection with the above, we insert the following table from Mr. Kenyon Blackwell's paper on the Iron Industry of Great Britain, read before the Society of Arts. It gives the estimated production of crude iron in the various countries.

	Tons.		Tons.
Great Britain	3,000,000	Sweden	150,000
France	750,000	Various German States	100,000
United States	750,000		
Prussia	300,000	Other Countries	300,000
Austria	250,000		
Belgium	200,000		6,000,000
Russia	200,000		

www.ingramcontent.com/pod-product-compliance
Lightning Source LLC
Chambersburg PA
CBHW031953300426
44117CB00008B/744